The Little Book of Bigger Primes

Second Edition

Springer

New York
Berlin
Heidelberg
Hong Kong
London
Milan
Paris
Tokyo

Paulo Ribenboim

The Little Book of Bigger Primes

Second Edition

 Springer

Paulo Ribenboim
Department of Mathematics and Statistics
Queen's University
Kingston, ON K7L 3N6
Canada

Mathematics Subject Classification (2000): 11A41, 11B39, 11A51

Library of Congress Cataloging-in-Publication Data
Ribenboim, Paulo.
    The little book of bigger primes / Paulo Ribenboim.
      p. cm.
    Includes bibliographical references and index.
    ISBN 0-387-20169-6 (alk. paper)
     1. Numbers, Prime. I. Title.

    QA246.R473 2004
    512.7′23—dc22   2003066220

ISBN 0-387-20169-6   Printed on acid-free paper.

See first edition © 1991 Paulo Ribenboim.

Printed in the United States of America.   (EB)

9 8 7 6 5 4 3 2 1   SPIN 10940969

Springer-Verlag is a part of *Springer Science+Business Media*

springeronline.com

Nel mezzo del cammin di nostra vita
mi ritrovai per una selva oscura
che la diritta via era smarrita

Dante Alighieri, *L'Inferno*

Preface

This book could have been called "Selections from the New Book of Prime Number Records." However, I preferred the title which propelled you on the first place to open it, and perhaps (so I hope) to buy it!

But the book is not very different from its parent. Like a bonsai, which has all the main characteristics of the full-sized tree, this paperback should exert the same fatal attraction. I wish it to be as dangerous as the other one. I wish you, young student, teacher or retired mathematician, engineer, computer buff, all of you who are friends of numbers, to be driven into thinking about the beautiful theory of prime numbers, with its inherent mystery. I wish you to exercise your brain and fingers—not vice-versa.

This second edition is still a little book, but the primes have "grown bigger". An irrepressible activity of computation specialists has pushed records to levels previously unthinkable. These endeavours generated—or were possible by—new algorithms and great advances in programming techniques and hardware developments. A fruitful interplay for the intended aim, to produce large, awesome numbers.

These updated records are reported; they are like a snapshot taken May 2003. However, only limited progress was made in the theoretical results. They are explained in the appropriate place. The old classical problems remain open and continue defying our great minds. With an inner smile: "If you solve me, you'll become idle". Not knowing that we, mathematicians, invent more problems than we can solve. Idle, we shall not be.

Paulo Ribenboim

Acknowledgements

First and foremost, I wish to express my gratitude to Wilfrid Keller. He spent uncountable hours working on this book, informing me of the newest records, discussing my text to great depths, with judicious comments. He also took up the arduous task of preparing the camera-ready copy. Like the proud Buenos Aires tailor who was not happy until the jacket fitted to perfection.

I have also obtained great support from many colleagues who explained patiently their results. As a consequence, their names are included in the text.

Finally, Chris Caldwell maintains a rich, well selected, informative website on prime numbers, which I consulted often with great profit.

Contents

Guiding the Reader

If a notation, which is not self-explanatory, appears without explanation on, say, page 107, look at the Index of Notations, which is organized by page number; the definition of the notation should appear before or at page 107.

If you wish to see where and how often your name is quoted in this book, turn to the Index of Names, at the end of the book. Should I say that there is no direct relation between achievement and number of quotes earned?

If, finally, you do not want to read the book but you just want to have some information about Cullen numbers—which is perfectly legitimate, if not laudable—go quickly to the Subject Index. Do not look under the heading *Numbers*, but rather *Cullen*. And for a subject like *Strong Lucas pseudoprimes*, you have exactly three possibilities ... .

Index of Notations

The following traditional notations are used in the text without explanation:

Notation	Explanation
$m \mid n$	the integer m divides the integer n
$m \nmid n$	the integer m does not divide the integer n
$p^e \parallel n$	p is a prime, $p^e \mid n$ but $p^{e+1} \nmid n$
$\gcd(m, n)$	greatest common divisor of the integers m, n
$\mathrm{lcm}(m, n)$	least common multiple of the integers m, n
$\log x$	natural logarithm of the real number $x > 0$
$\mathbb{Z}$	ring of integers
$\mathbb{Q}$	field of rational numbers
$\mathbb{R}$	field of real numbers
$\mathbb{C}$	field of complex numbers

The following notations are listed as they appear in the book:

Page	Notation	Explanation
134	$\mu(x)$	Möbius function
140	Δ	fundamental discriminant associated to $d \neq 0, 1$
140	$\mathbb{Q}(\sqrt{d})$	$= \mathbb{Q}(\sqrt{\Delta})$, quadratic field
141	Cl_d or Cl_Δ	class group of $\mathbb{Q}(\sqrt{d})$
141	h_d or h_Δ	class number of $\mathbb{Q}(\sqrt{d})$
141	e_d	exponent of the class group Cl_d
148	$\pi^*_{f(X)}(N)$	$\#\{n \mid 0 \leq n \leq N, |f(n)| \text{ is a prime}\}$
150	$P_0[m]$	smallest prime factor of $m > 1$
150	$P_0[f(X)]$	$= \min\{P_0[f(k)] \mid k = 0, 1, 2, \dots\}$
158	$f(x) \sim h(x)$	f, h are asymptotically equal
159	$f(x) = g(x)$ $+O(h(x))$	the difference $f(x) - g(x)$ is ultimately bounded by a constant multiple of $h(x)$
159	$f(x) = g(x)$ $+o(h(x))$	the difference $f(x) - g(x)$ is negligible in comparison to $h(x)$
159	$\zeta(s)$	Riemann's zeta function
161	B_k	Bernoulli number
162	$S_k(n)$	$= \sum_{j=1}^n j^k$
162	$B_k(X)$	Bernoulli polynomial
163	$\mathrm{Li}(x)$	logarithmic integral
164	$\theta(x)$	$= \sum_{p \leq x} \log p$, Tschebycheff's function
165	$\mathrm{Re}(s)$	real part of s
165	$\Gamma(s)$	gamma function
166	γ	Euler's constant
167	$J(x)$	weighted prime-power counting function
168	$R(x)$	Riemann's function
169	$\Lambda(x)$	von Mangoldt's function
170	$\psi(x)$	summatory function of the von Mangoldt function
172	$M(x)$	Mertens' function
174	$\varphi(x, m)$	$\#\{a \mid 1 \leq a \leq x, a \text{ is not a multiple of } 2, 3, \dots, p_m\}$
177	ρ_n	nth zero of $\zeta(s)$ in the upper half of the critical strip
177	$N(T)$	$\#\{\rho = \sigma + it \mid 0 \leq \sigma \leq 1, \zeta(\rho) = 0, 0 < t \leq T\}$
186	d_n	$= p_{n+1} - p_n$

Page	Notation	Explanation
235	Rn	$= \dfrac{10^n - 1}{9}$, repunit
240	Cn	$= n \times 2^n + 1$, Cullen number
241	$C\pi(x)$	$\#\{n \mid Cn \leq x \text{ and } Cn \text{ is prime}\}$
241	Wn	$= n \times 2^n - 1$, Woodall number or Cullen number of the second kind
244	$\mathcal{P}(T)$	set of primes p dividing some term of the sequence $T = (T_n)_{n \geq 0}$
244	$\pi_H(x)$	$\#\{p \in \mathcal{P}(H) \mid p \leq x\}$
247	S_{2n+1}	NSW number
254	$\pi_{f(X)}(x)$	$\#\{n \geq 1 \mid |f(n)| \leq x \text{ and } |f(n)| \text{ is prime}\}$
261	$p(f)$	smallest integer $m \geq 1$ such that $|f(m)|$ is a prime
263	$\pi_{X,X+2k}(x)$	$\#\{p \text{ prime} \mid p + 2k \text{ prime and } p + 2k \leq x\}$
264	$\pi_{X^2+1}(x)$	$\#\{p \text{ prime} \mid p \text{ is of the form } p = m^2 + 1 \text{ and } p \leq x\}$
265	$\pi_{aX^2+bX+c}(x)$	$\#\{p \text{ prime} \mid p \text{ is of the form } p = am^2 + bm + c \text{ and } p \leq x\}$

Introduction

The *Guinness Book of Records* became famous as an authoritative source of information to settle amiable disputes between drinkers of, it was hoped, the Guinness brand of stout. Its immense success in recording all sorts of exploits, anomalies, endurance performances, and so on has in turn influenced and sparked these very same performances. So one sees couples dancing for countless hours or persons buried in coffins with snakes, for days and days—just for the purpose of having their name in this bible of trivia. There are also records of athletic performances, extreme facts about human size, longevity, procreation, etc.

Little is found in the scientific domain. Yet scientists—mathematicians in particular—also like to chat while sipping wine or drinking a beer in a bar. And when the spirits mount, bets may be exchanged about the latest advances, for example, about recent discoveries concerning numbers.

Frankly, if I were to read in the *Whig-Standard* that a brawl in one of our pubs began with a heated dispute concerning which is the largest known pair of twin prime numbers, I would find this highly civilized.

However, not everybody agrees that fights between people are desirable, even for such all-important reasons. So, maybe I should reveal some of these records. Anyone who knows better should not hesitate to pass me additional information.

I will restrict my discussion to *prime* numbers: these are natural numbers, like $2, 3, 5, 7, 11, \ldots$, which are not multiples of any smaller natural number (except 1). If a natural number is neither 1 nor a prime, it is called a *composite* number.

Prime numbers are important, since the *fundamental theorem in arithmetic* states that every natural number greater than 1 is a product of prime numbers, and moreover, in an essentially unique way.

Without further ado it is easy to answer the following question: "Which is the oddest prime number?" It is 2, because it is the only even prime number!

There will be plenty of opportunities to encounter other prime numbers, like 1093 and 608 981 813 029, possessing interesting distinctive properties. Prime numbers are like cousins, members of the same family, resembling one another, but not quite alike.

Facing the task of presenting the records on prime numbers, I was led to think how to organize this volume. In other words, to classify the main lines of investigation and development of the theory of prime numbers.

It is quite natural, when studying a set of numbers—in this case the set of prime numbers—to ask the following questions, which I phrase informally as follows:

How many? How to decide whether an arbitrary given number is in the set? How to describe them? What is the distribution of these numbers, both at large and in short intervals? Then, to focus attention on distinguished types of such numbers, as well as to experiment with these numbers and make predictions—just as in any science.

Thus, I have divided the presentation into the following topics:

(1) How many prime numbers are there?

(2) How to recognize whether a natural number is a prime?

(3) Are there functions defining prime numbers?

(4) How are the prime numbers distributed?

(5) Which special kinds of primes have been considered?

(6) Heuristic and probabilistic results about prime numbers.

The discussion of these topics will lead me to indicate the relevant records.

1
How Many Prime Numbers Are There?

The answer to the question of how many prime numbers exist is given by the fundamental theorem:

There exist infinitely many prime numbers.

I shall give several proofs of this theorem (plus four variants), by famous, but also by forgotten, mathematicians. Some proofs suggest interesting developments; others are just clever or curious. There are of course more (but not quite infinitely many) proofs of the existence of infinitely many primes.

I Euclid's Proof

Suppose that $p_1 = 2 < p_2 = 3 < \cdots < p_r$ are all the primes. Let $P = p_1 p_2 \cdots p_r + 1$ and let p be a prime dividing P; then p cannot be any of $p_1, p_2, \ldots, p_r$, otherwise p would divide the difference $P - p_1 p_2 \cdots p_r = 1$, which is impossible. So this prime p is still another prime, and $p_1, p_2, \ldots, p_r$ would not be all the primes. □

I shall write the infinite increasing sequence of primes as

$$p_1 = 2, \ p_2 = 3, \ p_3 = 5, \ p_4 = 7, \ \ldots, \ p_n, \ \ldots.$$

An elegant variant of Euclid's proof was given by Kummer in 1878.

Kummer's proof. Suppose that there exist only finitely many primes $p_1 < p_2 < \cdots < p_r$. Let $N = p_1 p_2 \cdots p_r > 2$. The integer $N - 1$, being a product of primes, has a prime divisor p_i in common with N; so, p_i divides $N - (N - 1) = 1$, which is absurd!   □

This proof, by an eminent mathematician, is like a pearl, round, bright, and beautiful in its simplicity.

A proof similar to Kummer's was given in 1890 by Stieltjes, another great mathematician.

Did you like Kummer's proof? Then compare with the one which follows, even more beautiful and simpler. My attention to this proof was called by W. Narkiewicz. It was published by H. Brocard in 1915, in the *Intermédiaire des Mathématiciens* 22, page 253, and attributed to Hermite. It is of course another variant of the proof of Euclid:

It suffices to show that for every natural number n there exists a prime number p bigger than n. For this purpose one considers any prime p dividing $n! + 1!$ (The second ! applies to the proof, but if you did not like it, you may apply it to 1.)

Euclid's proof is pretty simple; however, it does not give any information about the new prime found in each stage, only that it is at most equal to the number $P = p_1 p_2 \cdots p_n + 1$. Thus, it may be that P is itself a prime (for some n), or that it is composite (for other indices n).

For every prime p, let $p\#$ denote the product of all primes q such that $q \leq p$. Following a suggestion of Dubner (1987), $p\#$ may be called the *primorial* of p.

The answers to the following questions are unknown:

Are there infinitely many primes p for which $p\# + 1$ is prime?

Are there infinitely many primes p for which $p\# + 1$ is composite?

RECORD

The largest known prime numbers of the form $p\# + 1$ are:

Prime	Digits	Year	Discoverer
$392113\# + 1$	169966	2001	D. Heuer et al.
$366439\# + 1$	158936	2001	D. Heuer et al.
$145823\# + 1$	63142	2000	A.E. Anderson, D.E. Robinson et al.

The numbers $p\#+1$ had been tested for all $p < 120000$ by Caldwell & Gallot (2002). They were found to be prime for $p = 2, 3, 5, 7,$ 11, 31, 379, 1019, 1021, 2657, 3229, 4547, 4787, 11549, 13649, 18523, 23801, 24029 and 42209, and for no other prime p in the tested range. Previous work was done by Borning (1972), Templer (1980), Buhler, Crandall & Penk (1982), Caldwell & Dubner (1993), and Caldwell (1995).

A similar search for prime numbers of the form $p\# - 1$ has also been undertaken. In the article by Caldwell & Gallot it is reported that the only primes of the form $p\# - 1$, with $p < 120000$, occur for $p = 3, 5, 11, 13, 41, 89, 317, 337, 991, 1873, 2053, 2377, 4093, 4297,$ 4583, 6569, 13033, 15877.

Euclid's proof suggests other problems. Here is one: Consider the sequence $q_1 = 2$, $q_2 = 3$, $q_3 = 7$, $q_4 = 43$, $q_5 = 139$, $q_6 = 50\,207$, $q_7 = 340\,999$, $q_8 = 2\,365\,347\,734\,339, \ldots$, where q_{n+1} is the highest prime factor of $q_1 q_2 \cdots q_n + 1$ (so $q_{n+1} \neq q_1, q_2, \ldots, q_n$). In 1963, Mullin asked: Does the sequence $(q_n)_{n \geq 1}$ contain all the prime numbers? Does it exclude at most finitely many primes? Is the sequence monotonic?

Concerning the first question, it is easy to see that 5 does not appear in Mullin's sequence. In 1968, Cox & van der Poorten found congruence criteria sufficient to decide if a given prime is excluded from the sequence. In this manner, they could establish that 2, 3, 7, and 43 are the only primes not exceeding 47 which belong to Mullin's sequence. The proof is given, in all details, in the recent book of Narkiewicz (2000).

For the second question the prevailing thoughts are that there exist infinitely many primes which do not belong to Mullin's sequence. Finally, by extending previous computations, Naur showed in 1984 that $q_{10} < q_9$, so the sequence $(q_n)_{n \geq 1}$ is not monotonic.

In 1991, Shanks considered the similar sequence $l_1 = 2$, $l_2 = 3$, $l_3 = 7$, $l_4 = 43$, $l_5 = 13$, $l_6 = 53$, $l_7 = 5$, $l_8 = 6\,221\,271, \ldots$. More generally, l_{n+1} is the smallest prime factor of $l_1 l_2 \cdots l_n + 1$. Shanks conjectured that every prime belongs to the sequence, but the truth of this assertion is still undecided. Wagstaff (1993) computed all terms l_n for $n \leq 43$, continuing previous calculations by Guy & Nowakowski (1975).

The calculation of the terms of these sequences requires the determination of the smallest prime factor or just the complete factorization of numbers of substantial size. This becomes increasingly difficult to perform, as the numbers grow. I shall discuss the matter of factorization in Chapter 2, Section XI, D.

In 1985, Odoni considered a similar sequence:

$$w_1 = 2, \ w_2 = 3, \ \ldots, \ w_{n+1} = w_1 w_2 \cdots w_n + 1,$$

and he showed that there exist infinitely many primes which do not divide any number of the sequence, and of course, there exist infinitely many primes which divide some number of the sequence.

II   Goldbach Did It Too!

The idea behind the proof is very simple and fruitful. It is enough to find an infinite sequence of natural numbers $1 < a_1 < a_2 < a_3 < \cdots$ that are pairwise relatively prime (i.e., without a common prime factor). So, if p_1 is a prime dividing a_1, if p_2 is a prime dividing a_2, etc., then $p_1, p_2, \ldots$, are all different.

The point is that the greatest common divisor is calculated by successive euclidean divisions and this does not require knowledge of the prime factors of the numbers.

Nobody seems to be the first to have a good idea—especially if it is simple. I thought it was due to Pólya & Szegö (see their book, 1924). E. Specker called my attention to the fact that Pólya used an exercise by Hurwitz (1891). But W. Narkiewicz told me that in a letter to Euler (July 20/31, 1730), Goldbach wrote the proof given below using Fermat numbers—this may well be the only written proof of Goldbach.

The Fermat numbers $F_n = 2^{2^n} + 1$ (for $n \geq 0$) are pairwise relatively prime.

Proof. It is easy to see, by induction on m, that $F_m - 2 = F_0 F_1 \cdots F_{m-1}$; hence, if $n < m$, then F_n divides $F_m - 2$.

If a prime p would divide both F_n and F_m, then it would divide $F_m - 2$ and F_m, hence also 2, so $p = 2$. But F_n is odd, hence not divisible by 2. This shows that the Fermat numbers are pairwise relatively prime.   □

Explicitly, the first Fermat numbers are $F_0 = 3$, $F_1 = 5$, $F_2 = 17$, $F_3 = 257$, $F_4 = 65537$, and it is easy to show that they are prime numbers. F_5 already has 10 digits, and each subsequent Fermat number is about the square of the preceding one, so the sequence grows very quickly. An important problem has been to find out whether F_n is a prime, or at least to find a prime factor of it. I shall return to this point in Chapter 2.

It would be desirable to find other infinite sequences of pairwise relatively prime integers, without already assuming the existence of infinitely many primes. In a paper of 1964, Edwards examined this question and indicated various sequences, defined recursively, having this property. For example, if S_0, a are relatively prime integers, with $S_0 > a \geq 1$, the sequence defined by the recursive relation

$$S_n - a = S_{n-1}(S_{n-1} - a) \quad \text{(for } n \geq 1)$$

consists of pairwise relatively prime natural numbers. In the best situation, that is, when $S_0 = 3$, $a = 2$, the sequence is in fact the sequence of Fermat numbers: $S_n = F_n = 2^{2^n} + 1$.

Similarly, if S_0 is odd and

$$S_n = S_{n-1}^2 - 2 \quad \text{(for } n \geq 1),$$

then, again, the integers S_n are pairwise relatively prime.

This sequence, which grows essentially just as quickly, has been considered by Lucas, and I shall return to it in Chapter 2.

In 1947, Bellman gave the following method to produce infinite sequences of pairwise relatively prime integers, without using the fact that there exist infinitely many primes. One begins with a nonconstant polynomial $f(X)$, with integer coefficients, such that $f(0) \neq 0$, and such that if n, $f(0)$ are relatively prime integers, then $f(n)$, $f(0)$ are also relatively prime integers. Then, let $f_1(X) = f(X)$ and for $m \geq 1$, let $f_{m+1}(X) = f(f_m(X))$.

If it happens that $f_m(0) = f(0)$ for every $m \geq 1$ and if n, $f(0)$ are relatively prime, then the integers n, $f_1(n)$, $f_2(n), \ldots, f_m(n), \ldots$ are pairwise relatively prime. For example, $f(X) = (X - 1)^2 + 1$ satisfies the required conditions and, in fact, $f_n(-1) = 2^{2^n} + 1$—so, back to Fermat numbers!

The following variant based on Hurwitz's idea was kindly communicated to me by P. Schorn.

Schorn's Proof. First, one notes that if $1 \leq i < j \leq n$, then

$$\gcd\big((n!)i + 1, (n!)j + 1\big) = 1.$$

Indeed, writing $j = i + d$ then $1 \leq d < n$, so

$$\gcd\big((n!)i + 1), (n!)j + 1\big) = \gcd\big((n!)i + 1, (n!)d\big) = 1,$$

because every prime p dividing $(n!)d$ is at most equal to n.

Now, if the number of primes would be m, taking $n = m + 1$, the above remark implies that the $m + 1$ integers $(m + 1)!i + 1$ $(1 \leq i \leq m + 1)$ are pairwise relatively prime, so there exist at least $m + 1$ distinct primes, contrary to the hypothesis.   □

III   Euler's Proof

This is a rather indirect proof, which, in some sense, is unnatural; but, on the other hand, as I shall indicate, it leads to the most important developments.

Euler showed that there must exist infinitely many primes because a certain expression formed with all the primes is infinite.

If p is any prime, then $1/p < 1$; hence, the sum of the geometric series is

$$\sum_{k=0}^{\infty} \frac{1}{p^k} = \frac{1}{1 - (1/p)}.$$

Similarly, if q is another prime, then

$$\sum_{k=0}^{\infty} \frac{1}{q^k} = \frac{1}{1 - (1/q)}.$$

Multiplying these equalities:

$$1 + \frac{1}{p} + \frac{1}{q} + \frac{1}{p^2} + \frac{1}{pq} + \frac{1}{q^2} + \cdots = \frac{1}{1 - (1/p)} \times \frac{1}{1 - (1/q)}.$$

Explicitly, the left-hand side is the sum of the inverses of all natural numbers of the form $p^h q^k$ ($h \geq 0$, $k \geq 0$), each counted only once, because every natural number has a unique factorization as a product of primes. This simple idea is the basis of the proof.

Euler's Proof. Suppose that $p_1, p_2, \ldots, p_n$ are all the primes. For each $i = 1, \ldots, n$

$$\sum_{k=0}^{\infty} \frac{1}{p_i^k} = \frac{1}{1 - (1/p_i)}.$$

Multiplying these n equalities, one obtains

$$\prod_{i=1}^{n} \left(\sum_{k=0}^{\infty} \frac{1}{p_i^k} \right) = \prod_{i=1}^{n} \frac{1}{1 - (1/p_i)},$$

and the left-hand side is the sum of the inverses of all natural numbers, each counted once—this follows from the fundamental theorem that every natural number is equal, in a unique way, to the product of primes.

But the series $\sum_{n=1}^{\infty}(1/n)$ is divergent; being a series of positive terms, the order of summation is irrelevant, so the left-hand side is infinite, while the right-hand side is clearly finite. This is absurd.

$\square$

In Chapter 4, I will return to developments along this line.

IV   Thue's Proof

Thue's proof uses only the fundamental theorem of unique factorization of natural numbers as products of prime numbers.

Thue's Proof. First, let $n, k \geq 1$ be integers such that $(1+n)^k < 2^n$. Let $p_1 = 2$, $p_2 = 3$, $\ldots$, p_r be all the primes satisfying $p_i \leq 2^n$. Suppose that $r \leq k$.

By the fundamental theorem, every integer m, $1 \leq m \leq 2^n$, may be written in a unique way in the form

$$m = 2^{e_1} \cdot 3^{e_2} \cdots p_r^{e_r},$$

where $0 \leq e_1 \leq n, 0 \leq e_2 \leq n, \ldots, 0 \leq e_r \leq n$.

Counting all the possibilities, it follows that $2^n \leq (n+1)n^{r-1} < (n+1)^r \leq (n+1)^k < 2^n$, and this is absurd. So $r \geq k+1$.

Choose $n = 2k^2$. From $1 + 2k^2 < 2^{2k}$ for every $k \geq 1$, it follows that

$$(1 + 2k^2)^k \leq 2^{2k^2} = 4^{k^2}.$$

Thus, there exist at least $k + 1$ primes p such that $p < 4^{k^2}$. Since k may be taken arbitrarily large, this shows that there are infinitely many primes. □

Actually, the proof also shows that $k + 1$ is a lower bound for the number of primes less than 4^{k^2}. This is a quantitative result, which is, of course, very poor. In Chapter 4, I shall further investigate this kind of questions.

V   Three Forgotten Proofs

The next proofs are by Perott, Auric, and Métrod. Who remembers these names? If it were not for Dickson's *History of the Theory of Numbers*, they would be totally forgotten. As I shall show, these proofs are very pleasant and ingenious; yet, they do not add new insights.

A   PEROTT'S PROOF

Perott's proof dates from 1881.

It is required to know that $\sum_{n=1}^{\infty}(1/n^2)$ is convergent with sum smaller than 2. (As a matter of fact, it is a famous result of Euler that the sum is exactly $\pi^2/6$, and I shall return to this point in Chapter 4.) Indeed,

$$\sum_{n=1}^{\infty} \frac{1}{n^2} < 1 + \sum_{n=1}^{\infty} \frac{1}{n(n+1)} = 1 + \sum_{n=1}^{\infty} \left(\frac{1}{n} - \frac{1}{n+1} \right) = 1 + 1 = 2.$$

Suppose that there exist only r prime numbers $p_1 < p_2 < \cdots < p_r$. Let N be any integer such that $p_1 p_2 \cdots p_r < N$. The number of integers $m \leq N$ that are not divisible by a square is therefore 2^r (which is the number of all possible sets of distinct primes), because every integer is, in a unique way, the product of primes. The number of integers $m \leq N$ divisible by p_i^2 is at most N/p_i^2, so the number of integers $m \leq N$ divisible by some square is at most $\sum_{i=1}^{r}(N/p_i^2)$. Hence,

$$N \leq 2^r + \sum_{i=1}^{r} \frac{N}{p_i^2} < 2^r + N \left(\sum_{n=1}^{\infty} \frac{1}{n^2} - 1 \right) = 2^r + N(1 - \delta),$$

where $\delta > 0$.

By choosing N such that $N\delta \geq 2^r$, it follows a contradiction.   □

B   Auric's Proof

Auric's proof, which appeared in 1915, is very simple.

Suppose that there exist only r primes $p_1 < p_2 < \cdots < p_r$. Let $t \geq 1$ be any integer and let $N = p_r^t$. By the unique factorization theorem, each integer m, $1 \leq m \leq N$, is written $m = p_1^{f_1} p_2^{f_2} \cdots p_r^{f_r}$ and the sequence $(f_1, f_2, \ldots, f_r)$, with each $f_i \geq 0$, is uniquely defined. Note also that $p_i^{f_i} \leq p_i^{f_i} \leq m \leq N = p_r^t$. Then, for $i = 1, 2, \ldots, r$, we have $f_i \leq tE$, where $E = (\log p_r)/(\log p_1)$. Thus, the number N (of integers m, $1 \leq m \leq N$) is at most the number of sequences $(f_1, f_2, \ldots, f_r)$; hence $p_r^t = N < (tE + 1)^r < t^r (E + 1)^r$. If t is sufficiently large, this inequality cannot hold, which shows that the number of primes must be infinite.   □

C   Métrod's Proof

Métrod's proof of 1917 is also very simple.

Assume that there exist only r primes $p_1 < p_2 < \cdots < p_r$. Let $N = p_1 p_2 \cdots p_r$, and for each $i = 1, 2, \ldots, r$, let $Q_i = N/p_i$. Note that p_i does not divide Q_i for each i, while p_i divides Q_j for all $j \neq i$. Let $S = \sum_{i=1}^{r} Q_i$. If q is any prime dividing S, then $q \neq p_i$ because p_i divides Q_j (for $i \neq j$) but p_i does not divide Q_i. Thus there exists yet another prime!   □

VI   Washington's Proof

Washington's proof (1980) is via commutative algebra. The ingredients are elementary facts of the theory of principal ideal domains, unique factorization domains, Dedekind domains, and algebraic numbers, and may be found in any textbook on the subject, such as Samuel's (1967) book: there is no mystery involved. First, I recall the needed facts:

1. In every number field (of finite degree) the ring of algebraic integers is a Dedekind domain: every nonzero ideal is, in a unique way, the product of prime ideals.

2. In every number field (of finite degree) there are only finitely many prime ideals that divide any given prime number p.

3. A Dedekind domain with only finitely many prime ideals is a principal ideal domain, and as such, every nonzero element is, up to units, the product of prime elements in a unique way.

Washington's Proof. Consider the field of all numbers of the form $a + b\sqrt{-5}$, where a, b are rational numbers. The ring of algebraic integers in this field consists of the numbers of the above form, with a, b ordinary integers. It is easy to see that 2, 3, $1 + \sqrt{-5}$, $1 - \sqrt{5}$ are prime elements of this ring, since they cannot be decomposed into factors that are algebraic integers, unless one of the factors is a "unit" 1 or -1. Note also that

$$(1 + \sqrt{-5})(1 - \sqrt{-5}) = 2 \times 3,$$

the decomposition of 6 into a product of primes is not unique up to units, so this ring is not a unique factorization domain; hence, it is not a principal ideal domain. So, it must have infinitely many prime ideals (by fact 3 above) and (by fact 2 above) there exist infinitely many prime numbers. $\qquad\square$

VII   Furstenberg's Proof

This is an ingenious proof based on topological ideas. Since it is so short, I cannot do any better than transcribe it verbatim; it appeared in 1955:

> In this note we would like to offer an elementary "topological" proof of the infinitude of the prime numbers. We introduce a topology into the space of integers S, by using the arithmetic progressions (from $-\infty$ to $+\infty$) as a basis. It is not difficult to verify that this actually yields a topological space. In fact, under this topology, S may be shown to be normal and hence metrizable. Each arithmetic progression is closed as well as open, since its complement is the union of other arithmetic progressions (having the same difference). As a result, the union of any finite number of arithmetic progressions is closed.

Consider now the set $A = \bigcup A_p$, where A_p consists of all multiples of p, and p runs through the set of primes ≥ 2. The only numbers not belonging to A are -1 and 1, and since the set $\{-1, 1\}$ is clearly not an open set, A cannot be closed. Hence A is not a finite union of closed sets which proves that there are an infinity of primes. $\square$

Golomb developed further the idea of Furstenberg, and wrote an interesting short paper in 1959.

2
How to Recognize Whether a Natural Number is a Prime

In the article 329 of *Disquisitiones Arithmeticae*, Gauss (1801) wrote:

> The problem of distinguishing prime numbers from composite numbers and of resolving the latter into their prime factors is known to be one of the most important and useful in arithmetic.... The dignity of the science itself seems to require that every possible means be explored for the solution of a problem so elegant and so celebrated.

The first observation concerning the problem of primality and factorization is clear: there is an algorithm for both problems. By this, I mean a procedure involving finitely many steps, which is applicable to every number N and which will indicate whether N is a prime, or, if N is composite, which are its prime factors. Namely, given the natural number N, try in succession every number $n = 2, 3, \ldots$ up to $[\sqrt{N}]$ (the largest integer not greater than $\sqrt{N}$) to see whether it divides N. If none does, then N is a prime. If, say, N_0 divides N, write $N = N_0 N_1$, so $N_1 < N$, and then repeat the same procedure with N_0 and with N_1. Eventually this gives the complete factorization into prime factors.

What I have just said is so evident as to be irrelevant. It should, however, be noted that for large numbers N, it may take a long time with this algorithm to decide whether N is prime or composite.

This touches the most important practical aspect, the need to find an efficient algorithm—one which involves as few operations as possible, and therefore requires less time and costs less money to be performed.

It is my intention to divide this chapter into several sections in which I will examine various approaches, as well as explain the required theoretical results.

I   The Sieve of Eratosthenes

As I have already said, it is possible to find if N is a prime using trial division by every number n such that $n^2 \leq N$.

Since multiplication is an easier operation than division, Eratosthenes (in the 3rd century BC) had the idea of organizing the computations in the form of the well-known sieve. It serves to determine all the prime numbers, as well as the factorizations of composite numbers, up to any given number N. This is illustrated now for $N = 101$.

Do as follows: write all the numbers up to 101; cross out all the multiples of 2, bigger than 2; in each subsequent step, cross out all the multiples of the smallest remaining number p, which are bigger than p. It suffices to do it for $p^2 < 101$.

$$
\begin{array}{rrrrrrrrrr}
 & 2 & 3 & \cancel{4} & 5 & \cancel{6} & 7 & \cancel{8} & \cancel{9} & \cancel{10} \\
11 & \cancel{12} & 13 & \cancel{14} & \cancel{15} & \cancel{16} & 17 & \cancel{18} & 19 & \cancel{20} \\
\cancel{21} & \cancel{22} & 23 & \cancel{24} & \cancel{25} & \cancel{26} & \cancel{27} & \cancel{28} & 29 & \cancel{30} \\
31 & \cancel{32} & \cancel{33} & \cancel{34} & \cancel{35} & \cancel{36} & 37 & \cancel{38} & \cancel{39} & \cancel{40} \\
41 & \cancel{42} & 43 & \cancel{44} & \cancel{45} & \cancel{46} & 47 & \cancel{48} & \cancel{49} & \cancel{50} \\
\cancel{51} & \cancel{52} & 53 & \cancel{54} & \cancel{55} & \cancel{56} & \cancel{57} & \cancel{58} & 59 & \cancel{60} \\
61 & \cancel{62} & \cancel{63} & \cancel{64} & \cancel{65} & \cancel{66} & 67 & \cancel{68} & \cancel{69} & \cancel{70} \\
71 & \cancel{72} & 73 & \cancel{74} & \cancel{75} & \cancel{76} & \cancel{77} & \cancel{78} & 79 & \cancel{80} \\
\cancel{81} & \cancel{82} & 83 & \cancel{84} & \cancel{85} & \cancel{86} & \cancel{87} & \cancel{88} & 89 & \cancel{90} \\
\cancel{91} & \cancel{92} & \cancel{93} & \cancel{94} & \cancel{95} & \cancel{96} & 97 & \cancel{98} & \cancel{99} & \cancel{100} \\
101 & & & & & & & & &
\end{array}
$$

Thus, all the multiples of $2, 3, 5, 7 < \sqrt{101}$ are sifted away. The number 53 is prime because it remained. Thus the primes up to 101 are 2, 3, 5, 7, 11, 13, 17, 19, 23, 29, 31, 37, 41, 43, 47, 53, 59, 61, 67, 71, 73, 79, 83, 89, 97, 101.

This procedure is the basis of sieve theory, which has been developed to provide estimates for the number of primes satisfying given conditions.

II   Some Fundamental Theorems on Congruences

In this section, I intend to describe some classical methods to test primality and to find factors. They rely on theorems on congruences, especially Fermat's little theorem, the old theorem of Wilson, as well as Euler's generalization of Fermat's theorem. I shall also include a subsection on quadratic residues, a topic of central importance, which is also related with primality testing, as I shall have occasion to indicate.

A   FERMAT'S LITTLE THEOREM AND PRIMITIVE ROOTS MODULO A PRIME

Fermat's Little Theorem. *If p is a prime number and if a is an integer, than $a^p \equiv a \pmod{p}$. In particular, if p does not divide a then $a^{p-1} \equiv 1 \pmod{p}$.*

Euler published the first proof of Fermat's little theorem.

Proof. It is true for $a = 1$. Assuming that it is true for a, then, by induction, $(a+1)^p \equiv a^p + 1 \equiv a + 1 \pmod{p}$. So the theorem is true for every natural number a.   □

The above proof required only the fact that if p is a prime number and if $1 \leq k \leq p - 1$, then the binomial coefficient $\binom{p}{k}$ is a multiple of p.

Note the following immediate consequence: if $p \nmid a$ and p^n is the highest power of p dividing $a^{p-1} - 1$, then p^{n+e} is the highest power of p dividing $a^{p^e(p-1)} - 1$ (where $e \geq 1$); in this statement, if $p = 2$, then n must be at least 2.

It follows from the theorem that for any integer a, which is not a multiple of the prime p, there exists the smallest exponent $h \geq 1$, such that $a^h \equiv 1 \pmod{p}$. Moreover, $a^k \equiv 1 \pmod{p}$ if and only if h divides k; in particular, h divides $p - 1$. This exponent h is called the *order of a modulo p*. Note that $a \bmod p$, $a^2 \bmod p$, ... , $a^{h-1} \bmod p$, and $1 \bmod p$ are all distinct.

It is a basic fact that for every prime p there exists at least one integer g, not a multiple of p, such that the order of g modulo p is equal to $p - 1$. Then, the set $\{1 \bmod p, 2 \bmod p, \ldots, g^{p-2} \bmod p\}$ is equal to the set $\{1 \bmod p, 2 \bmod p, \ldots, (p-1) \bmod p\}$.

Every integer g, $1 \leq g \leq p - 1$, such that $g \bmod p$ has order $p - 1$, is called a *primitive root modulo p*. I note this proposition:

Let p be any odd prime, $k \geq 1$, and $S = \sum_{j=1}^{p-1} j^k$. Then

$$S \equiv \begin{cases} -1 \bmod p, & \text{when } p - 1 \mid k, \\ 0 \bmod p, & \text{when } p - 1 \nmid k. \end{cases}$$

Proof. Indeed, if $p - 1$ divides k, then $j^k \equiv 1 \pmod{p}$ for $j = 1, 2, \ldots, p - 1$; so $S \equiv p - 1 \equiv -1 \pmod{p}$. If $p - 1$ does not divide k, let g be a primitive root modulo p. Then $g^k \not\equiv 1 \pmod{p}$. Since the sets of residue classes $\{1 \bmod p, 2 \bmod p, \ldots, (p-1) \bmod p\}$ and $\{g \bmod p, 2g \bmod p, \ldots, (p-1)g \bmod p\}$ are the same, then

$$g^k S \equiv \sum_{j=1}^{p-1} (gj)^k \equiv \sum_{j=1}^{p-1} j^k \equiv S \pmod{p}.$$

Hence $(g^k - 1)S \equiv 0 \pmod{p}$ and, since p does not divide $g^k - 1$, then $S \equiv 0 \pmod{p}$. $\qquad\square$

The determination of a primitive root modulo p may be effected by a simple method indicated by Gauss in articles 73, 74 of *Disquisitiones Arithmeticae*.

Proceed as follows:

Step 1. Choose any integer a, $1 < a < p$, for example, $a = 2$, and write the residues modulo p of a, a^2, a^3, ... Let t be the smallest exponent such that $a^t \equiv 1 \pmod{p}$. If $t = p - 1$, then a is a primitive root modulo p. Otherwise, proceed to the next step.

Step 2. Choose any number b, $1 < b < p$, such that $b \not\equiv a^i \pmod{p}$ for $i = 1, \ldots, t$; let u be the smallest exponent such that $b^u \equiv 1 \pmod{p}$. It is simple to see that u cannot be a factor of t, otherwise $b^t \equiv 1 \pmod{p}$; but $1, a, a^2, \ldots, a^{t-1}$ are t pairwise incongruent solutions of the congruence $X^t \equiv 1 \pmod{p}$; so they are all the possible solutions, and therefore $b \equiv a^m \pmod{p}$, for some m, $0 \leq m \leq t - 1$, which is contrary to the hypothesis. If $u = p - 1$, then b is a primitive root modulo p. If $u \neq p - 1$, let v be the least common multiple of t, u; so $v = mn$ with m dividing t, n dividing u, and $\gcd(m, n) = 1$. Let $a' \equiv a^{t/m} \pmod{p}$, $b' \equiv b^{u/n} \pmod{p}$ so $c = a'b'$ has order $mn = v$ modulo p. If $v = p - 1$, then c is a primitive root modulo p. Otherwise, proceed to the next step, which is similar to step 2.

Note that $v > t$, so in each step either one reaches a primitive root modulo p, or one constructs an integer with a bigger order modulo p. The process must stop; one eventually reaches an integer with order $p - 1$ modulo p, that is, a primitive root modulo p.

Gauss also illustrated the procedure with the example $p = 73$, and found that $g = 5$ is a primitive root modulo 73.

The above construction leads to a primitive root modulo p, but not necessarily to the smallest integer g_p, $1 < g_p < p$, which is a primitive root modulo p.

The determination of g_p is done by trying successively the various integers $a = 2, 3, \ldots$ and computing their orders modulo p. There is no uniform way of predicting, for all primes p, which is the smallest primitive root modulo p. However, several results were known about the size of g_p. In 1944, Pillai proved that there exist infinitely many primes p, such that $g_p > C \log \log p$ (where C is a positive constant). In particular, $\limsup_{p \to \infty} g_p = \infty$. A few years later, using a very deep theorem of Linnik (see Chapter 4) on primes in arithmetic progressions, Fridlender (1949), and independently Salié (1950), proved that $g_p > C \log p$, for some constant C and infinitely many primes p. On the other hand, g_p does not grow too fast, as proved by Burgess in 1962:

$$g_p \leq C p^{1/4 + \varepsilon}$$

(for $\varepsilon > 0$, a constant $C > 0$, and p sufficiently large).

Grosswald made Burgess' result explicit in 1981: if $p > e^{e^{24}}$ then $g_p < p^{0.499}$.

The proof of the weaker result (with $1/2$ in place of $1/4$), attributed to Vinogradov, is in Landau's *Vorlesungen über Zahlentheorie*, Part VII, Chapter 14 (see General References).

The proof of the following result is elementary (problem proposed by Powell in 1983, solution by Kearnes in 1984):

For any positive integer M, there exist infinitely many primes p such that $M < g_p < p - M$.

As an illustration, the following table gives the smallest primitive root modulo p, for each prime $p < 1000$.

Table 1.   The smallest primitive root modulo p

p	g_p	p	g_p	p	g_p	p	g_p	p	g_p	p	g_p
2	1	127	3	283	3	467	2	661	2	877	2
3	2	131	2	293	2	479	13	673	5	881	3
5	2	137	3	307	5	487	3	677	2	883	2
7	3	139	2	311	17	491	2	683	5	887	5
11	2	149	2	313	10	499	7	691	3	907	2
13	2	151	6	317	2	503	5	701	2	911	17
17	3	157	5	331	3	509	2	709	2	919	7
19	2	163	2	337	10	521	3	719	11	929	3
23	5	167	5	347	2	523	2	727	5	937	5
29	2	173	2	349	2	541	2	733	6	941	2
31	3	179	2	353	3	547	2	739	3	947	2
37	2	181	2	359	7	557	2	743	5	953	3
41	6	191	19	367	6	563	2	751	3	967	5
43	3	193	5	373	2	569	3	757	2	971	6
47	5	197	2	379	2	571	3	761	6	977	3
53	2	199	3	383	5	577	5	769	11	983	5
59	2	211	2	389	2	587	2	773	2	991	6
61	2	223	3	397	5	593	3	787	2	997	7
67	2	227	2	401	3	599	7	797	2		
71	7	229	6	409	21	601	7	809	3		
73	5	233	3	419	2	607	3	811	3		
79	3	239	7	421	2	613	2	821	3		
83	2	241	7	431	7	617	3	823	3		
89	3	251	6	433	5	619	2	827	2		
97	5	257	3	439	15	631	3	829	2		
101	2	263	5	443	2	641	3	839	11		
103	5	269	2	449	3	643	11	853	2		
107	2	271	6	457	13	647	5	857	3		
109	6	277	5	461	2	653	2	859	2		
113	3	281	3	463	3	659	2	863	5		

A simple glance at the table suggests the following question: Is 2 a primitive root for infinitely many primes? More generally, if the integer $a \neq \pm 1$ is not a square, is it a primitive root modulo infinitely many primes? This is a difficult problem and I shall return to it in Chapter 4.

B   THE THEOREM OF WILSON

Wilson's Theorem. *If p is a prime number, then*

$$(p - 1)! \equiv -1 \pmod{p}.$$

Proof. This is just a corollary of Fermat's little theorem. Indeed, $1, 2, \ldots, p - 1$ are roots of the congruence $X^{p-1} - 1 \equiv 0 \pmod{p}$. But a congruence modulo p cannot have more roots than its degree. Hence,

$$X^{p-1} - 1 \equiv (X - 1)(X - 2) \cdots (X - (p - 1)) \pmod{p}.$$

Comparing the constant terms, $-1 \equiv (-1)^{p-1}(p - 1)! = (p - 1)!$ $\pmod{p}$. (This is also true if $p = 2$.)   □

Wilson's theorem gives a characterization of prime numbers. Indeed, if $N > 1$ is a natural number that is not a prime, then $N = mn$, with $1 < m, n < N - 1$, so m divides N and $(N - 1)!$, and therefore $(N - 1)! \not\equiv -1 \pmod{N}$.

However, Wilson's characterization of the prime numbers is not of practical value to test the primality of N, since there is no known algorithm to rapidly compute $N!$, say, in $\log N$ steps.

C   THE PROPERTIES OF GIUGA AND OF WOLSTENHOLME

Now, I shall consider other properties that are satisfied by prime numbers.

The property of Giuga

First, I note that if p is a prime, then by Fermat's little theorem (as already indicated)

$$1^{p-1} + 2^{p-1} + \cdots + (p - 1)^{p-1} \equiv -1 \pmod{p}.$$

In 1950, Giuga asked whether the converse is true: If $n > 1$ and n divides $1^{n-1} + 2^{n-1} + \cdots + (n - 1)^{n-1} + 1$, then is n a prime number?

It is easy to show that n satisfies Giuga's condition if and only if, for every prime p dividing n, $p^2(p-1)$ divides $n - p$. Indeed, writing $n = pt$, Giuga's condition becomes

$$A = 1 + \sum_{j=1}^{pt-1} j^{pt-1} \equiv 0 \pmod{p},$$

while the condition that $p^2(p-1)$ divides $pt - p$ is equivalent to the conjunction of both conditions: $p \mid t - 1$ and $p - 1 \mid t - 1$. But $pt - 1 = (p-1)t + (t-1)$; hence, by Fermat's little theorem,

$$A \equiv 1 + \sum_{j=1}^{pt-1} j^{t-1} \equiv 1 + tS \pmod{p},$$

where $S = \sum_{j=1}^{p-1} j^{t-1}$. Hence,

$$A \equiv \begin{cases} 1 - t \pmod{p}, & \text{when} \quad p - 1 \mid t - 1 \\ 1 \pmod{p}, & \text{when} \quad p - 1 \nmid t - 1. \end{cases}$$

Thus, if $A \equiv 0 \pmod{p}$, then $p-1 \mid t-1$ and $p \mid t-1$. But, conversely, these latter conditions imply that $A \equiv 0 \pmod{p}$ and $p \nmid t$, so n is squarefree and therefore $A \equiv 0 \pmod{n}$.   □

It follows at once that $n \equiv p \equiv 1 \pmod{p - 1}$; so, if $p \mid n$, then $p - 1 \mid n - 1$. A composite number n having this property is called a *Carmichael number*.

In Section IX, I shall indicate that this condition is severely restrictive. At any rate, it is now known that if there exists a composite integer n satisfying Giuga's condition, then n must have at least 12000 digits; see Bedocchi (1985) and Borwein, Borwein, Borwein & Girgensohn (1996).

The property of Wolstenholme

In 1862, Wolstenholme proved the following interesting result: If p is a prime, $p \geq 5$, then the numerator of

$$1 + \frac{1}{2} + \frac{1}{3} + \cdots + \frac{1}{p - 1}$$

is divisible by p^2, and the numerator of

$$1 + \frac{1}{2^2} + \frac{1}{3^2} + \cdots + \frac{1}{(p-1)^2}$$

is divisible by p.

For a proof, see Hardy & Wright (1938, p. 88, General References). Based on this property, it is not difficult to deduce that if $n \geq 5$ is a prime number, then

$$\binom{2n-1}{n-1} \equiv 1 \pmod{n^3}.$$

Is the converse true? This question, still unanswered today, has been asked by J.P. Jones for many years. An affirmative reply would provide an interesting and formally simple characterization of prime numbers.

The problem leads naturally to the following concepts and questions. Let $n \geq 5$ be odd, and let

$$A(n) = \binom{2n-1}{n-1}.$$

For each $k \geq 1$ we may consider the set

$$W_k = \{n \text{ odd}, n \geq 5 \mid A(n) \equiv 1 \pmod{n^k}\}.$$

Thus $W_1 \supset W_2 \supset W_3 \supset W_4 \supset \ldots$. From Wolstenholme's theorem, every prime number greater than 3 belongs to W_3. Jones' question is whether W_3 is just the set of these prime numbers.

A prime number belonging to W_4 is called a *Wolstenholme prime*. Only two Wolstenholme primes are known today: 16843, indicated by Selfridge & Pollack in 1964, and 2124679, discovered by Crandall, Ernvall and Metsänkylä in 1993. In 1995, McIntosh determined by calculation that there is no other Wolstenholme prime $p < 5 \times 10^8$.

The set of composite integers in W_2 contains the squares of Wolstenholme's primes. McIntosh conjectured that these sets coincide and verified that this is true up to 10^9: the only composite $n \in W_2$, $n < 10^9$, is $n = 283686649 = 16843^2$.

It is believed, and was suggested by McIntosh, that there exist infinitely many Wolstenholme primes. The proof of this assertion would be very difficult.

D   THE POWER OF A PRIME DIVIDING A FACTORIAL

In 1808, Legendre determined the exact power p^m of the prime p that divides a factorial $a!$ (so p^{m+1} does not divide $a!$).

There is a very nice expression of m in terms of the p-adic development of a:

$$a = a_k p^k + a_{k-1} p^{k-1} + \cdots + a_1 p + a_0,$$

where $p^k \leq a < p^{k+1}$ and $0 \leq a_i \leq p - 1$ (for $i = 0, 1, \ldots, k$). The integers $a_0, a_1, \ldots, a_k$ are the digits of a in base p.

For example, in base $5, 328 = 2 \times 5^3 + 3 \times 5^2 + 3$, so the digits of 328 in base 5 are $2, 3, 0, 3$. Using the above notation:

Legendre's Theorem.

$$m = \sum_{i=1}^{\infty} \left[\frac{a}{p^i} \right] = \frac{a - (a_0 + a_1 + \cdots + a_k)}{p - 1}.$$

Proof. By definition $a! = p^m b$, where $p \nmid b$. Let $a = q_1 p + r_1$ with $0 \leq q_1$, $0 \leq r_1 < p$; so $q_1 = [a/p]$. The multiples of p, not bigger than a, are $p, 2p, \ldots, q_1 p \leq a$. So $p^{q_1}(q_1!) = p^m b'$, where $p \nmid b'$. Thus $q_1 + m_1 = m$, where p^{m_1} is the exact power of p which divides $q_1!$. Since $q_1 < a$, by induction,

$$m_1 = \left[\frac{q_1}{p} \right] + \left[\frac{q_1}{p^2} \right] + \left[\frac{q_1}{p^3} \right] + \cdots .$$

But

$$\left[\frac{q_1}{p^i} \right] = \left[\frac{[a/p]}{p^i} \right] = \left[\frac{a}{p^{i+1}} \right],$$

as may be easily verified. So

$$m = \left[\frac{a}{p} \right] + \left[\frac{a}{p^2} \right] + \left[\frac{a}{p^3} \right] + \cdots .$$

Now, I derive the second expression, involving the p-adic digits of $a = a_k p^k + \cdots + a_1 p + a_0$. Then

$$\left[\frac{a}{p}\right] = a_k p^{k-1} + \cdots + a_1,$$

$$\left[\frac{a}{p^2}\right] = a_k p^{k-2} + \cdots + a_2,$$

$$\vdots$$

$$\left[\frac{a}{p^k}\right] = a_k.$$

So

$$\sum_{i=0}^{\infty} \left[\frac{a}{p^i}\right] = a_1 + a_2(p+1) + a_3(p^2 + p + 1) + \cdots$$

$$+ a_k(p^{k-1} + p^{k-2} + \cdots + p + 1)$$

$$= \frac{1}{p-1}\{a_1(p-1) + a_2(p^2-1) + \cdots + a_k(p^k-1)\}$$

$$= \frac{1}{p-1}\{a - (a_0 + a_1 + \cdots + a_k)\}. \qquad \square$$

In 1852, Kummer used Legendre's result to determine the exact power p^m of p dividing a binomial coefficient

$$\binom{a+b}{a} = \frac{(a+b)!}{a!b!},$$

where $a \geq 1$, $b \geq 1$.
    Let

$$a = a_0 + a_1 p + \cdots + a_t p^t,$$
$$b = b_0 + b_1 p + \cdots + b_t p^t,$$

where $0 \leq a_i \leq p-1$, $0 \leq b_i \leq p-1$, and either $a_t \neq 0$ or $b_t \neq 0$. Let $S_a = \sum_{i=0}^{t} a_i$, $S_b = \sum_{i=0}^{t} b_i$ be the sums of p-adic digits of a, b. Let c_i, $0 \leq c_i \leq p-1$, and $\varepsilon_i = 0$ or 1, be defined successively as follows:

$$a_0 + b_0 = \varepsilon_0 p + c_0,$$
$$\varepsilon_0 + a_1 + b_1 = \varepsilon_1 p + c_1,$$

$$\vdots$$

$$\varepsilon_{t-1} + a_t + b_t = \varepsilon_t p + c_t.$$

Multiplying these equations successively by 1, p, $p^2, \ldots$ and adding them:

$$a + b + \varepsilon_0 p + \varepsilon_1 p^2 + \cdots + \varepsilon_{t-1} p^t$$
$$= \varepsilon_0 p + \varepsilon_1 p^2 + \cdots + \varepsilon_{t-1} p^t + \varepsilon_t p^{t+1} + c_0 + c_1 p + \cdots + c_t p^t.$$

So, $a + b = c_0 + c_1 p + \cdots + c_t p^t + \varepsilon_t p^{t+1}$, and this is the expression of $a + b$ in the base p. Similarly, by adding those equations:

$$S_a + S_b + (\varepsilon_0 + \varepsilon_1 + \cdots + \varepsilon_{t-1}) = (\varepsilon_0 + \varepsilon_1 + \cdots + \varepsilon_t) p + S_{a+b} - \varepsilon_t.$$

By Legendre's result

$$(p - 1)m = (a + b) - S_{a+b} - a + S_a - b + S_b$$
$$= (p - 1)(\varepsilon_0 + \varepsilon_1 + \cdots + \varepsilon_t).$$

Hence the following result:

Kummer's Theorem. *The exponent of the exact power of p dividing $\binom{a+b}{a}$ is equal to $\varepsilon_0 + \varepsilon_1 + \cdots + \varepsilon_t$, which is the number of "carry-overs" when performing the addition of a, b, written in the base p.*

This theorem of Kummer was rediscovered by Lucas in 1878. In 1991, Frasnay extended the result replacing integers by p-adic integers.

The results of Legendre and Kummer have found many applications, in p-adic analysis, and also, for example, in Chapter 3, Section III.

E   THE CHINESE REMAINDER THEOREM

Even though my paramount interest is in prime numbers, there is no way to escape dealing with arbitrary integers also—which essentially amounts, in many questions, to the simultaneous consideration of several primes, because of the decomposition, in a unique way, of integers into the product of prime powers.

One of the keys connecting results for integers n, and for their prime power factors, is very old; indeed, it was known to the ancient Chinese, and it is therefore called the Chinese remainder theorem.

However, according to A. Zachariou (private communication) it was known even before them by the Greeks, but since the Greeks discovered so many theorems, I will keep the traditional name for this one. I am sure that every one of my readers knows it already:

If $n_1, n_2, \ldots, n_k$ are pairwise relatively prime integers, greater than 1, and if $a_1, a_2, \ldots, a_k$ are any integers, then there exists an integer a such that

$$\begin{cases} a \equiv a_1 \quad (\mathrm{mod}\ n_1) \\ a \equiv a_2 \quad (\mathrm{mod}\ n_2) \\ \quad\vdots \\ a \equiv a_k \quad (\mathrm{mod}\ n_k). \end{cases}$$

Another integer a' also satisfies the same congruences as a if and only if $a \equiv a'$ (mod $n_1 n_2 \cdots n_k$). So, there exists a unique integer a, as above, with $0 \leq a < n_1 n_2 \cdots n_k$.

The proof is indeed very simple; it is in many books and also in a short note by Mozzochi (1967).

The Chinese remainder theorem has numerous applications. It is conceivable that one of these might have been the way the Chinese generals counted their troops:

Line up 7 by 7!   (Not factorial of 7, but a SCREAMED
                     military command.)

Line up 11 by 11!

Line up 13 by 13!

Line up 17 by 17!

Counting only the remainders in the incomplete rows, the intelligent generals could know the exact number of their soldiers.[1]

Here is another application of the Chinese remainder theorem. If $n = p_1 p_2 \cdots p_k$ is a product of distinct primes, if g_i is a primitive root modulo p_i (for each i), if g is such that $1 \leq g \leq n - 1$ and $g \equiv g_i$ (mod p_i) for every $i = 1, 2, \ldots, k$, then the order of g modulo p_i is $p_i - 1$ for each $i = 1, 2, \ldots, k$ and the order of g modulo n is $\prod_{i=1}^{k}(p_i - 1)$.

---

[1]In between us, this may never have been practiced. The existence of intelligent generals remains a wide open question.

F EULER'S FUNCTION

Euler generalized Fermat's little theorem by introducing the *totient* or *Euler's function*.

For every $n \geq 1$, let $\varphi(n)$ denote the number of integers a, $1 \leq a < n$, such that $\gcd(a, n) = 1$. Thus, if $n = p$ is a prime, then $\varphi(p) = p - 1$; also

$$\varphi(p^k) = p^{k-1}(p - 1) = p^k \left(1 - \frac{1}{p} \right).$$

Moreover, if $m, n \geq 1$ and $\gcd(m, n) = 1$, then $\varphi(mn) = \varphi(m)\varphi(n)$, that is, φ is a multiplicative function. Hence, for any integer $n = \prod_p p^k$ (product for all primes p dividing n, and $k \geq 1$), then

$$\varphi(n) = \prod_p p^{k-1}(p - 1) = n \prod_p \left(1 - \frac{1}{p} \right).$$

Another simple property is: $n = \sum_{d|n} \varphi(d)$.

Euler proved the following:

Euler's Theorem. *If $\gcd(a, n) = 1$, then $a^{\varphi(n)} \equiv 1 \pmod{n}$.*

Proof. Let $r = \varphi(n)$ and let $b_1, \dots, b_r$ be integers, pairwise incongruent modulo n, such that $\gcd(b_i, n) = 1$ for $i = 1, \dots, r$.

Then $ab_1, \dots, ab_r$ are again pairwise incongruent modulo n and $\gcd(ab_i, n) = 1$ for $i = 1, \dots, r$. Therefore, the sets $\{b_1 \bmod n, \dots, b_r \bmod n\}$ and $\{ab_1 \bmod n, \dots, ab_r \bmod n\}$ are equal. Now,

$$a^r \prod_{i=1}^{r} b_i \equiv \prod_{i=1}^{r} ab_i \equiv \prod_{i=1}^{r} b_i \pmod{n}.$$

Hence,

$$(a^r - 1) \prod_{i=1}^{r} b_i \equiv 0 \pmod{n} \quad \text{and so} \quad a^r \equiv 1 \pmod{n}. \qquad \square$$

Just like for Fermat's little theorem, it follows also from Euler's theorem that there exists the smallest positive exponent e such that $a^e \equiv 1 \pmod{n}$. It is called the *order of a modulo n*. If n is a prime number, this definition coincides with the previous one. Note also that $a^m \equiv 1 \pmod{n}$ if and only if m is a multiple of the order e of $a \bmod n$; thus, in particular, e divides $\varphi(n)$.

Once again, it is natural to ask: Given $n > 2$ does there always exist an integer a, relatively prime to n, such that the order of a $\bmod n$ is equal to $\varphi(n)$? Recall that when $n = p$ is a prime, such numbers exist, namely, the primitive roots modulo p. If $n = p^e$, a power of an odd prime, it is also true. More precisely, the following assertions are equivalent:

(i) g is a primitive root modulo p and $g^{p-1} \not\equiv 1 \pmod{p^2}$;

(ii) g is a primitive root modulo p^2;

(iii) for every $e \geq 2$, g is a primitive root modulo p^e.

Note that 10 is a primitive root modulo 487, but $10^{486} \equiv 1 \pmod{487^2}$, so 10 is not a primitive root modulo 487^2. This is the smallest example illustrating this possibility, when the base is 10. Another example is 14 modulo 29.

However, if n is divisible by $4p$, or pq, where p, q are distinct odd primes, then there is no number a, relatively prime to n, with order equal to $\varphi(n)$. Indeed, it is easy to see that the order of $a \bmod n$ is at most equal to $\lambda(n)$, where $\lambda(n)$ is the following function, defined by Carmichael in 1912:

$$\lambda(1) = 1, \; \lambda(2) = 1, \; \lambda(4) = 2,$$

$$\lambda(2^r) = 2^{r-2} \quad \text{(for } r \geq 3\text{)},$$

$$\lambda(p^r) = p^{r-1}(p-1) = \varphi(p^r) \quad \text{for any odd prime } p \text{ and } r \geq 1,$$

$$\lambda\bigl(2^r p_1^{r_1} p_2^{r_2} \cdots p_s^{r_s}\bigr) = \operatorname{lcm}\bigl\{\lambda(2^r), \lambda(p_1^{r_1}), \ldots, \lambda(p_s^{r_s})\bigr\}$$

(lcm denotes the least common multiple).

Note that $\lambda(n)$ divides $\varphi(n)$, but may be smaller, and that there is an integer a, relatively prime to n, with order of $a \bmod n$ equal to $\lambda(n)$.

I shall use this opportunity to study Euler's function in more detail. First I shall consider Lehmer's problem, and thereafter the values of φ, the valence, the values avoided, the average of the function, etc.

Lehmer's problem

Recall that if p is a prime, then $\varphi(p) = p - 1$. In 1932, Lehmer asked whether there exists any composite integer n such that $\varphi(n)$ divides $n - 1$. This question remains open and its solution seems as remote today as it was when Lehmer raised it seven decades ago. If the answer is negative, it will provide a characterization of prime numbers.

What can one say, anyway, when it is not possible to solve the problem? Only that the existence of composite integers n, for which $\varphi(n)$ divides $n - 1$, is unlikely, for various reasons:

(a) any such number must be very large (if it exists at all);

(b) any such number must have many prime factors
    (if it exists at all);

(c) the number of such composite numbers, smaller than any given
    real number x, is bounded by a very small function of x.

Thus, Lehmer showed in 1932 that if n is composite and $\varphi(n)$ divides $n - 1$, then n is odd and square-free, and the number of its distinct prime factors is $\omega(n) \geq 7$. Subsequent work by Schuh (1944) gave $\omega(n) \geq 11$. In 1970, Lieuwens showed that if $3 \mid n$, then $\omega(n) \geq 213$ and $n > 5.5 \times 10^{570}$; if $30 \nmid n$, then $\omega(n) \geq 13$.

RECORD

In 1980, Cohen and Hagis showed that if n is composite and $\varphi(n)$ divides $n - 1$, then $n > 10^{20}$ and $\omega(n) \geq 14$. Wall (1980) showed that if $\gcd(30, n) = 1$, then $\omega(n) \geq 26$, while if $3 \mid n$, the best result is still Lieuwens'.

In 1977, Pomerance showed that for every sufficiently large positive real number x, the number $L(x)$ of composite n such that $\varphi(n)$ divides $n - 1$ and $n \leq x$, satisfies

$$L(x) \leq x^{1/2}(\log x)^{3/4}.$$

Moreover, if $\omega(n) = k$, then $n < k^{2^k}$.

Values of Euler's function

Not every even integer $m > 1$ is a value of Euler's function—a fact which is not difficult to establish. For example, Schinzel showed in 1956 that, for every $k \geq 1$, 2×7^k is not a value of Euler's function.

In 1976, Mendelsohn showed that there exist infinitely many primes p such that, for every $k \geq 1$, $2^k p$ is not a value of the function φ. Concerning interesting values assumed by Euler's function, Erdös in 1946 proposed as a problem to show that for every $k \geq 1$ there exists n such that $\varphi(n) = k!$. A solution by Lambek was proposed in 1948; the same result was given later by Gupta (1950).

The next results tell how erratic is the behaviour of Euler's function. Thus, in 1950, Somayajulu showed that

$$\limsup_{n \to \infty} \frac{\varphi(n+1)}{\varphi(n)} = \infty \quad \text{and} \quad \liminf_{n \to \infty} \frac{\varphi(n+1)}{\varphi(n)} = 0.$$

This result was improved by Schinzel and Sierpiński, see Schinzel (1954): the set of all numbers $\varphi(n+1)/\varphi(n)$ is dense in the set of all real positive numbers.

Schinzel & Sierpiński (1954) and Schinzel (1954) also proved the following:

For every m, $k \geq 1$, there exist n, $h \geq 1$ such that

$$\frac{\varphi(n+i)}{\varphi(n+i-1)} > m \quad \text{and} \quad \frac{\varphi(h+i-1)}{\varphi(h+i)} > m$$

for $i = 1, 2, \ldots, k$. It is also true that the set of all numbers $\varphi(n)/n$ is dense in the interval $(0, 1)$.

The valence of Euler's function

Now I shall examine the "valence" of Euler's function; in other words, how often a value $\varphi(n)$ is assumed. In order to explain the results in a systematic way, it is better to introduce some notation. If $m \geq 1$, let

$$V_\varphi(m) = \#\{n \geq 1 \mid \varphi(n) = m\}.$$

What are the possible values of $V_\varphi(m)$? I have already said that there are infinitely many even integers m for which $V_\varphi(m) = 0$. It is also true that if $m = 2 \times 3^{6k+1}$ ($k \geq 1$), then $\varphi(n) = m$ exactly when $n = 3^{6k+2}$ or $n = 2 \times 3^{6k+2}$. Hence, there are infinitely many integers m such that $V_\varphi(m) = 2$.

It is not difficult to show that $V_\varphi(m) \neq \infty$ for every $m \geq 1$.

Schinzel gave a simpler proof (in 1956) of the following result of Pillai (1929):

$$\sup\{V_\varphi(m)\} = \infty.$$

In other words, for every $k \geq 1$ there exists an integer m_k such that there exist at least k integers n with $\varphi(n) = m_k$.

The above result is weaker than the long-standing conjecture of Sierpiński: For every integer $k \geq 2$ there exists $m > 1$ such that $k = V_\varphi(m)$. With very sophisticated methods, this conjecture has now been proved by Ford (1999).

Carmichael's conjecture

The conjecture that dominates the study of the valence of φ was proposed by Carmichael in 1922: V_φ does not assume the value 1. In other words, given $n \geq 1$, there exists $n' \geq 1$, $n' \neq n$, such that $\varphi(n') = \varphi(n)$.

This conjecture was studied by Klee, who showed in 1947 that it holds for every integer n such that $\varphi(n) < 10^{400}$. Masai & Valette (1982), using Klee's method, showed that $\varphi(n) < 10^{10000}$. In 1994, still basically using Klee's method, but with extensive calculations, Schlafly & Wagon have brilliantly increased the lower bound for a counterexample to Carmichael's conjecture: if $V_\varphi(n) = 1$, so $n > 10^{10^7}$. With much more powerful methods, Ford (1998) further improved the lower bound to reach $n > 10^{10^{10}}$.

An article about Carmichael's conjecture, also written by Wagon, had appeared earlier in *The Mathematical Intelligencer* (1986). Numerical evidence points to the truth of Carmichael's conjecture. However, Pomerance (1974) has shown the following: Suppose that m is a natural number such that if p is any prime and $p-1$ divides $\varphi(m)$, then p^2 divides m. Then $V_\varphi(\varphi(m)) = 1$.

Of course, if there exists a number m satisfying the above condition, then Carmichael's conjecture would be false. However, the existence of such a number m is far from established, and perhaps unlikely.

The most important recent work on Carmichael's conjecture is due to K. Ford (1998). For every $x > 0$ let $E(x) = \#\{n \mid 1 \leq n < x$ such that there exists $k > 1$ with $\varphi(k) = n\}$ and $E_1(x) = \#\{n \mid 1 \leq n < x$ such that there exists a unique k with $\varphi(k) = n\}$. Carmichael's conjecture says that $E_1(x) = 0$ for every $x > 0$. Ford showed that if

Carmichael's conjecture is false, then there exists $C > 0$ such that for every sufficiently large x we have $E(x) \leq C\,E_1(x)$. It follows that Carmichael's conjecture is equivalent to the statement

$$\liminf_{x \to \infty} \frac{E_1(x)}{E(x)} = 0.$$

Ford also showed that $E_1(10^{10^{10}}) = 0$.

Finally, in variance with Carmichael's conjecture, it is reasonable to expect that every $s > 1$ is a value of V_φ; this was conjectured by Sierpiński. As a matter of fact, I shall indicate in Chapter 6, Section II, that this statement follows from an unproved and very interesting hypothesis.

And how about the valence of the valence function V_φ? I have already said that there exist infinitely many m that are not values of φ, for which $V_\varphi(m) = 0$. So V_φ assumes the value 0 infinitely often.

This was generalized by Erdös in 1958: If $s \geq 1$ is a value of V_φ, then it is assumed infinitely often. (Try to phrase this statement directly using Euler's function, to see whether you understand my notation.)

The growth of Euler's function

I have not yet considered the growth of the function φ. Since $\varphi(p) = p - 1$ for every prime p, then $\limsup \varphi(n) = \infty$. Similarly, from $\varphi(p) = p - 1$, $\limsup \varphi(n)/n = 1$.

I shall postpone the indication of other results about the growth of φ until Chapter 4: they depend on methods that will be discussed in that chapter.

G   Sequences of Binomials

The preceding considerations referred to congruences modulo a given integer $n > 1$, and a was any positive integer relatively prime to n.

Another point of view is very illuminating. This time, let $a > 1$ be given, and consider the sequence of integers $a^n - 1$ (for $n \geq 1$), as well as the companion sequence of integers $a^n + 1$ (for $n \geq 1$). More generally, if $a > b \geq 1$ with $\gcd(a, b) = 1$, one may consider the sequences $a^n - b^n$ $(n \geq 1)$ and $a^n + b^n$ $(n \geq 1)$.

A first natural question, with an immediate answer, is the following: to determine all primes p, such that there exists $n \geq 1$ for which

p divides $a^n - b^n$. These are primes p not dividing ab because a, b are relatively prime. Conversely, if $p \nmid ab$, if $bb' \equiv 1 \pmod{p}$ and n is the order of $ab' \bmod p$, then p divides $a^n - b^n$.

It is more complicated for the binomials $a^n + b^n$. If $p \neq 2$ and there exists $n \geq 1$ such that p divides $a^n + b^n$, then $p \nmid ab(a - b)$. The converse is false; for example, 7 does not divide $2^n + 1$ for every $n \geq 1$.

Primitive prime factors

If $n \geq 1$ is the smallest integer such that p divides $a^n - b^n$ (resp. $a^n + b^n$), then p is called a *primitive prime factor* of the sequence of binomials in question. In this case, by Fermat's little theorem, n divides $p - 1$; this was explicitly observed by Legendre.

So, every prime $p \nmid ab$ appears as a primitive factor of some binomial $a^n - b^n$. Does, conversely, every binomial have a primitive factor?

In 1892, Zsigmondy proved the following theorem, which is very interesting and has many applications:

If $a > b \geq 1$ and $\gcd(a, b) = 1$, then every number $a^n - b^n$ has a primitive prime factor—the only exceptions being $a - b = 1$, $n = 1$; $2^6 - 1 = 63$; and $a^2 - b^2$, where a, b are odd and $a + b$ is a power of 2.

Equally, if $a > b \geq 1$, then every number $a^n + b^n$ has a primitive prime factor—with the exception of $2^3 + 1 = 9$.

The special case, where $b = 1$, had been proved by Bang in 1886. Later, this theorem, or Bang's special case, was proved again, sometimes unknowingly, by a long list of mathematicians: Birkhoff & Vandiver (1904), Carmichael (1913), Kanold (1950), Artin (1955), Lüneburg (1981), and probably others.

The proof is definitely not so obvious; however, it is very easy to write up such sequences and watch the successive appearance of new primitive prime factors.

It is interesting to consider the primitive part t_n^* of $a^n - b^n$; namely, write $a^n - b^n = t_n^* t_n'$ with $\gcd(t_n^*, t_n') = 1$ and a prime p divides t_n^* if and only if p is a primitive factor of $a^n - b^n$.

By experimenting numerically with sequences $a^n - b^n$, it is observed that, apart from a few initial terms, t_n^* is composite. In fact, Schinzel indicated the following theorem in 1962.

Let $k(m)$ denote the square-free kernel of m, that is, m divided by its largest square factor. Let

$$e = \begin{cases} 1, & \text{if } k(ab) \equiv 1 \pmod{4}, \\ 2, & \text{if } k(ab) \equiv 2 \text{ or } 3 \pmod{4}. \end{cases}$$

If $n/ek(ab)$ is integral and odd, and if $n > 1$, then $a^n - b^n$ has at least two distinct primitive prime factors, with only a few exceptions (of which the largest possible is $n = 20$). When $n > 1$ and $b = 1$, the exceptions are:

$$\text{if } a = 2: \quad n = 4, 12, 20;$$
$$\text{if } a = 3: \quad n = 6;$$
$$\text{if } a = 4: \quad n = 3.$$

Therefore, there are infinitely many n such that the primitive part of $a^n - b^n$ is composite.

Schinzel also proved that if $ab = c^h$ with $h \geq 3$, or $h = 2$ and $k(c)$ odd, then there are infinitely many n such that the primitive part of $a^n - b^n$ has at least three prime factors.

For the sequence of binomials $a^n + b^n$, it follows at once that if $n/ek(ab)$ is odd, and $n > 10$, then the primitive part of $a^n + b^n$ is composite. Just note that each primitive prime factor of $a^{2n} - b^{2n}$ is also a primitive prime factor of $a^n + b^n$.

Here are some questions that are very difficult to answer:

Are there infinitely many n such that the primitive part of $a^n - b^n$ is prime?

Are there infinitely many n such that the primitive part $a^n - b^n$ is square-free?

And how about the seemingly easier questions:

Are there infinitely many n such that the primitive part t_n^* of $a^n - b^n$ has a prime factor p such that p^2 does not divide $a^n - b^n$?

Are there infinitely many n such that t_n^* has a square-free kernel $k(t_n^*) \neq 1$?

These questions, for the special case when $b = 1$, are ultimately related, in a very surprising way, to Fermat's last theorem!

The largest prime factor

It is also an interesting problem to estimate the size of the largest prime factor of $a^n - b^n$, where $a > b \geq 1$ and $\gcd(a, b) = 1$. The following notation will be used: $P[m]$ designates the largest prime factor of $m \geq 1$.

It is not difficult to show, using Zsigmondy's theorem, that $P[a^n - b^n] \geq n + 1$ when $n > 2$.

In 1962, Schinzel showed that $P[a^n - b^n] \geq 2n + 1$ in the following cases, with $n > 2$: $4 \nmid n$, with exclusion $a = 2$, $b = 1$, $n = 6$; $k(ab) \mid n$ or $k(ab) = 2$, with exclusions $a = 2$, $b = 1$, $n = 4, 6$, or 12.

Erdös conjectured in 1965 that $\lim_{n \to \infty} P[2^n - 1]/n = \infty$. Despite very interesting work, this conjecture has not yet been settled completely; but there are very good partial results, which I report now.

In 1975, using Baker's inequalities for linear forms of logarithms, Stewart showed the following. Let $0 < r < 1/\log 2$, and let S_r be the set of integers n having at most $r \log \log n$ distinct prime factors (the set S_r has density 1); then

$$\lim_{\substack{n \to \infty \\ n \in S_r}} \frac{P[a^n - b^n]}{n} = \infty.$$

How fast does the expression increase? This was answered by Stewart in 1977, with sharper inequalities of Baker's type:

$$\frac{P[a^n - b^n]}{n} > C \frac{(\log n)^{\lambda}}{\log \log \log n},$$

where $\lambda = 1 - r \log 2$, $C > 0$ is a convenient constant, and $n \in S_r$.

Stewart also showed that, for every sufficiently large prime p, $P[a^p - b^p]/p > C \log p$ $(C > 0)$. The special case of Mersenne numbers $2^p - 1$ had been established in 1976 by Erdös and Shorey.

There is also a close connection between the numbers $a^n - 1$, the values of the cyclotomic polynomials, and primes in certain arithmetic progressions, but I cannot explain everything at the same time—so be patient and wait until I consider this matter again in Chapter 4, Section IV.

H   QUADRATIC RESIDUES

In the study of quadratic diophantine equations, developed by Fermat, Euler, Legendre, and Gauss, it was very important to determine when an integer a is a square modulo a prime $p > 2$.

If $p > 2$ does not divide a, and if there exists an integer b such that $a \equiv b^2 \pmod{p}$, then a is called a *quadratic residue modulo p*; otherwise, it is a *nonquadratic residue modulo p*.

Legendre introduced the following practical notation:

$$\left(\frac{a}{p}\right) = (a \mid p) = \begin{cases} +1 & \text{if } a \text{ is a quadratic residue modulo } p, \\ -1 & \text{otherwise.} \end{cases}$$

It is also convenient to define $(a \mid p) = 0$ when p divides a.

I shall now indicate the most important properties of the Legendre symbol. References are plentiful—practically every book in elementary number theory.

If $a \equiv a' \pmod{p}$, then

$$\left(\frac{a}{p}\right) = \left(\frac{a'}{p}\right).$$

For any integers a, a':

$$\left(\frac{aa'}{p}\right) = \left(\frac{a}{p}\right)\left(\frac{a'}{p}\right).$$

So, for the computation of the Legendre symbol, it suffices to calculate $(q \mid p)$, where $q = -1$, 2, or any odd prime different from p.

Euler proved the following congruence:

$$\left(\frac{a}{p}\right) \equiv a^{(p-1)/2} \pmod{p}.$$

In particular,

$$\left(\frac{-1}{p}\right) = \begin{cases} +1 & \text{when } p \equiv 1 \pmod{4}, \\ -1 & \text{when } p \equiv -1 \pmod{4}, \end{cases}$$

and

$$\left(\frac{2}{p}\right) = \begin{cases} +1 & \text{when } p \equiv \pm 1 \pmod{8}, \\ -1 & \text{when } p \equiv \pm 3 \pmod{8}. \end{cases}$$

The computation of the Legendre symbol $(q \mid p)$, for any odd prime $q \neq p$, can be performed with an easy, explicit, and fast algorithm (needing only Euclidean division), by using Gauss' *reciprocity law*:

$$\left(\frac{p}{q}\right) = \left(\frac{q}{p}\right)(-1)^{\frac{p-1}{2} \times \frac{q-1}{2}}.$$

The importance of Legendre's symbol was such that it prompted Jacobi to consider the following generalization, now called the Jacobi symbol. Again, references are abundant, for example, Grosswald's book (1966, 2nd edition 1984), or (why not?) my own book (1972, enlarged edition 2001).

Let a be a nonzero integer, and let b be an odd integer, such that $\gcd(a, b) = 1$. The Jacobi symbol $(a \mid b)$ is defined as an extension of Legendre's symbol, in the following manner. Let $b = \prod_{p \mid b} p^{e_p} > 0$ (with $e_p \geq 1$). Then

$$\left(\frac{a}{b}\right) = \prod_{p \mid b} \left(\frac{a}{p}\right)^{e_p},$$

$$\left(\frac{a}{-b}\right) = \begin{cases} \left(\dfrac{a}{b}\right), & \text{if } a > 0, \\ -\left(\dfrac{a}{b}\right), & \text{if } a < 0. \end{cases}$$

Therefore, $(a \mid b)$ is equal to $+1$ or -1. Note that

$$\left(\frac{a}{1}\right) = \left(\frac{a}{-1}\right) = +1 \quad \text{when } a > 0.$$

Here are some of the properties of the Jacobi symbol (under the assumptions of its definition):

$$\left(\frac{aa'}{b}\right) = \left(\frac{a}{b}\right)\left(\frac{a'}{b}\right),$$

$$\left(\frac{a}{bb'}\right) = \left(\frac{a}{b}\right)\left(\frac{a}{b'}\right)$$

$$\left(\frac{-1}{b}\right) = (-1)^{(b-1)/2} = \begin{cases} +1 & \text{if } b \equiv 1 \pmod{4}, \\ -1 & \text{if } b \equiv -1 \pmod{4}, \end{cases}$$

$$\left(\frac{2}{b}\right) = (-1)^{(b^2-1)/8} = \begin{cases} +1 & \text{if } b \equiv \pm 1 \pmod{8}, \\ -1 & \text{if } b \equiv \pm 3 \pmod{8}. \end{cases}$$

For the calculation of the Jacobi symbol, the key result is the reciprocity law, which follows easily from Gauss' reciprocity law for the Legendre symbol:

If a, b are relatively prime odd integers, then

$$\left(\frac{a}{b}\right) = \varepsilon \left(\frac{b}{a}\right)(-1)^{\frac{a-1}{2} \times \frac{b-1}{2}},$$

where

$$\varepsilon = \begin{cases} +1 & \text{if } a > 0 \text{ or } b > 0 \\ -1 & \text{if } a < 0 \text{ and } b < 0. \end{cases}$$

Finally, if $b \geq 3$, and if a is a square modulo b, then $(a \mid b) = +1$.

III   Classical Primality Tests Based on Congruences

After the discussion of the theorems of Fermat, Wilson, and Euler, I am ready. For me, the classical primality tests based on congruences are those indicated by Lehmer, extending or using previous tests by Lucas, Pocklington, and Proth. I reserve another section for classical tests based on recurring sequences.

Wilson's theorem, which characterizes prime numbers, might seem very promising, but it has to be discarded as a practical test, since the computation of factorials is very time consuming.

Fermat's little theorem says that if p is a prime and a is any natural number not a multiple of p, then $a^{p-1} \equiv 1 \pmod{p}$. However, I note right away that a crude converse of this theorem is not true—because there exist composite integers N, and $a \geq 2$, such that $a^{N-1} \equiv 1 \pmod{N}$. I shall devote Section VIII to the study of these numbers, which are very important in primality questions.

Nevertheless, a true converse of Fermat's little theorem was discovered by Lucas in 1876. It says:

Test 1. Let $N > 1$. Assume that there exists an integer $a > 1$ such that:

(i) $a^{N-1} \equiv 1 \pmod{N}$,

(ii) $a^m \not\equiv 1 \pmod{N}$ for $m = 1, 2, \ldots, N - 2$.

Then N is a prime.

Defect of this test: it might seem perfect, but it requires $N - 2$ successive multiplications by a, and finding residues modulo N—too many operations.

Proof. It suffices to show that every integer m, $1 \leq m < N$, is prime to N, that is, $\varphi(N) = N - 1$. For this purpose, it suffices to show that there exists a, $1 \leq a < N$, $\gcd(a, N) = 1$, such that the order of $a \bmod N$ is $N - 1$. This is exactly spelled out in the hypothesis. $\square$

In 1891, Lucas gave the following test:

Test 2. Let $N > 1$. Assume that there exists an integer $a > 1$ such that:

(i) $a^{N-1} \equiv 1 \pmod{N}$,

(ii) $a^m \not\equiv 1 \pmod{N}$ for every $m < N$, such that m divides $N - 1$.

Then N is a prime.

Defect of this test: it requires the knowledge of all factors of $N - 1$, thus it is only easily applicable when $N - 1$ can be factored, like $N = 2^n + 1$, or $N = 3 \times 2^n + 1$.

The proof of Test 2 is, of course, the same as that of Test 1.

In 1967, Brillhart & Selfridge made Lucas' test more flexible; see also the paper by Brillhart, Lehmer & Selfridge in 1975:

Test 3. Let $N > 1$. Assume that for every prime factor q of $N - 1$ there exists an integer $a = a(q) > 1$ such that

(i) $a^{N-1} \equiv 1 \pmod{N}$,

(ii) $a^{(N-1)/q} \not\equiv 1 \pmod{N}$.

Then N is a prime.

Defect of this test: once again, it is necessary to know the prime factors of $N - 1$, but fewer congruences have to be satisfied.

An observant reader should note that, after all, to verify that $a^{N-1} \equiv 1 \pmod{N}$ it is necessary in particular to calculate, as one goes, the residue of a^n modulo N (for every $n \leq N - 1$), and so the first Lucas test could have been used. The point is that there is a fast algorithm to find the power a^n, hence also $a^n \bmod N$, without computing all the preceding powers. It runs as follows.

Write the exponent n in base 2:

$$n = n_0 2^k + n_1 2^{k-1} + \cdots + n_{k-1} 2 + n_k,$$

where each n_i is equal to 0 or 1, and $n_0 = 1$.

Define the integers $r_0, r_1, r_2, \ldots$ successively, putting $r_0 = a$ and, for $j \geq 0$:

$$r_{j+1} = \begin{cases} r_j^2 & \text{if } n_{j+1} = 0, \\ ar_j^2 & \text{if } n_{j+1} = 1. \end{cases}$$

Then $a^n = r_k$.

So, it is only necessary to perform at most $2k$ operations, which are either a squaring or a multiplication by a. If the computation is of $a^n \bmod N$, then it is even easier; at each stage r_j is to be replaced by its residue modulo N. Now, k is equal to

$$\left[\frac{\log n}{\log 2}\right].$$

Therefore, if $n = N - 1$, then only about

$$2 \left[\frac{\log N}{\log 2}\right]$$

operations are needed to find $a^{N-1} \bmod N$, and there is no requirement of computing all powers $a^n \bmod N$.

Why don't you try calculating $2^{1092} \bmod 1093^2$ in this way? You should find $2^{1092} \equiv 1 \pmod{1093^2}$—if you really succeed! This has nothing to do directly with primality—but it will appear much later, in Chapter 5.

I return to Brillhart and Selfridge's Test 3 and give its proof.

Proof of Test 3. It is enough to show that $\varphi(N) = N - 1$, and since $\varphi(N) \leq N - 1$, it suffices to show that $N - 1$ divides $\varphi(N)$. If this is false, there exists a prime q and $r \geq 1$ such that q^r divides $N - 1$, but q^r does not divide $\varphi(N)$. Let $a = a(q)$ and let e be the order of $a \bmod N$. Thus e divides $N - 1$ and e does not divide $(N - 1)/q$, so q^r divides e. Since $a^{\varphi(N)} \equiv 1 \pmod{N}$, then e divides $\varphi(N)$, so $q^r \mid \varphi(N)$, which is a contradiction, and concludes the proof. $\square$

In the section on Fermat numbers, I will derive Pepin's primality test for Fermat numbers, as a consequence of Test 3.

To make the primality tests more efficient, it is desirable to avoid the need to find all prime factors of $N - 1$. So there are tests that only require a partial factorization of $N - 1$. The basic result was proved by Pocklington in 1914, and it is indeed very simple:

Let $N - 1 = q^n R$, where q is a prime, $n \geq 1$, and q does not divide R. Assume that there exists an integer $a > 1$ such that:

(i) $a^{N-1} \equiv 1 \pmod{N}$,

(ii) $\gcd(a^{(N-1)/q} - 1, N) = 1$.

Then each prime factor of N is of the form $mq^n + 1$, with $m \geq 1$.

Proof. Let p be a prime factor of N, and let e be the order of a mod p, so e divides $p - 1$; by condition (ii), e cannot divide $(N-1)/q$, because p divides N; hence, q does not divide $(N-1)/e$; so q^n divides e, and a fortiori, q^n divides $p - 1$.   □

The above statement looks more like a result on factors than a primality test. However, if it may be verified that each prime factor $p = mq^n + 1$ is greater than $\sqrt{N}$, then N is a prime. When q^n is fairly large, this verification is not too time consuming.

Pocklington gave also the following refinement of his result above:

Let $N - 1 = FR$, where $\gcd(F, R) = 1$ and the factorization of F is known. Assume that for every prime q dividing F there exists an integer $a = a(q) > 1$ such that:

(i) $a^{N-1} \equiv 1 \pmod{N}$,

(ii) $\gcd(a^{(N-1)/q} - 1, N) = 1$.

Then each prime factor of N is of the form $mF + 1$, with $m \geq 1$.

The same comments apply here. So, if $F > \sqrt{N}$, then N is a prime.

This result is very useful to prove the primality of numbers of certain special form. The old criterion of Proth (1878) is easily deduced:

Test 4. Let $N = 2^n h + 1$ with h odd and $2^n > h$. Assume that there exists an integer $a > 1$ such that $a^{(N-1)/2} \equiv -1 \pmod{N}$. Then N is prime.

Proof. $N - 1 = 2^n h$, with h odd and $a^{N-1} \equiv 1 \pmod{N}$. Since N is odd, then $\gcd(a^{(N-1)/2} - 1, N) = 1$. By the above result, each prime factor p of N is of the form $p = 2^n m + 1 > 2^n$. But $N = 2^n h + 1 < 2^{2n}$, hence $\sqrt{N} < 2^n < p$ and so N is prime. $\qquad\square$

In the following test (using the same notation) it is required to know that R (the nonfactored part of $N - 1$) has no prime factor less than a given bound B. Precisely:

Test 5. Let $N - 1 = FR$, where $\gcd(F, R) = 1$, the factorization of F is known, B is such that $FB > \sqrt{N}$, and R has no prime factors less than B. Assume:

  (i) For each prime q dividing F there exists an integer $a = a(q) > 1$ such that $a^{N-1} \equiv 1 \pmod{N}$ and $\gcd(a^{(N-1)/q} - 1, N) = 1$.

  (ii) There exists an integer $b > 1$ such that $b^{N-1} \equiv 1 \pmod{N}$ and $\gcd(b^F - 1, N) = 1$.

Then N is a prime.

Proof. Let p be any prime factor of N, let e be the order of b modulo N, so e divides $p - 1$ and also e divides $N - 1 = FR$. Since e does not divide F, then $\gcd(e, R) \neq 1$, so there exists a prime q such that $q \mid e$ and $q \mid R$; hence, $q \mid p - 1$. However, by the previous result of Pocklington, F divides $p - 1$; since $\gcd(F, R) = 1$, then qF divides $p - 1$. So $p - 1 \geq qF \geq BF > \sqrt{N}$. This implies that $p = N$, so N is a prime. $\qquad\square$

The paper of Brillhart, Lehmer & Selfridge (1975) contains other variants of these tests, which have been put to good use to determine the primality of numbers of the form $2^r + 1$, $2^{2r} \pm 2^r + 1$, $2^{2r-1} \pm 2^r + 1$.

I have already said enough and will make only one further comment: these tests require prime factors of $N - 1$. Later, using linear recurring sequences, other tests will be presented, requiring prime factors of $N + 1$.

IV   Lucas Sequences

Let P, Q be nonzero integers.

Consider the polynomial $X^2 - PX + Q$; its discriminant is $D = P^2 - 4Q$ and the roots are

$$\left.\begin{array}{c} \alpha \\ \beta \end{array}\right\} = \frac{P \pm \sqrt{D}}{2}.$$

So

$$\begin{cases} \alpha + \beta = P, \\ \quad \alpha\beta = Q, \\ \alpha - \beta = \sqrt{D}. \end{cases}$$

I shall assume that $D \neq 0$. Note that $D \equiv 0 \pmod{4}$ or $D \equiv 1 \pmod{4}$. Define the sequences of numbers

$$U_n(P, Q) = \frac{\alpha^n - \beta^n}{\alpha - \beta} \quad \text{and} \quad V_n(P, Q) = \alpha^n + \beta^n, \quad \text{for } n \geq 0.$$

In particular, $U_0(P, Q) = 0$, $U_1(P, Q) = 1$, while $V_0(P, Q) = 2$, $V_1(P, Q) = P$.

The sequences

$$U(P, Q) = (U_n(P, Q))_{n \geq 0} \quad \text{and} \quad V(P, Q) = (V_n(P, Q))_{n \geq 0}$$

are called the *Lucas sequences associated to the pair* (P, Q). Special cases had been considered by Fibonacci, Fermat, and Pell, among others. Many particular facts were known about these sequences; however, the general theory was first developed by Lucas in a seminal paper, which appeared in Volume I of the *American Journal of Mathematics*, 1878. It is a long memoir with a rich content, relating Lucas sequences to many interesting topics, like trigonometric functions, continued fractions, the number of divisions in the algorithm of the greatest common divisor, and also, primality tests. It is for this latter reason that I discuss Lucas sequences. If you are curious about the other connections that I have mentioned, look at the references at the end of the book and/or consult the paper in the library.

I should, however, warn that despite the importance of the paper, the methods employed are often indirect and cumbersome, so it is advisable to read Carmichael's long article of 1913, where he corrected errors and generalized results.

The first thing to note is that, for every $n \geq 2$,

$$\begin{cases} U_n(P,Q) = P\,U_{n-1}(P,Q) - Q\,U_{n-2}(P,Q), \\ V_n(P,Q) = P\,V_{n-1}(P,Q) - Q\,V_{n-2}(P,Q). \end{cases}$$

(just check it). So, these sequences deserve to be called *linear recurring sequences of order* 2 (each term depends linearly on the two preceding terms). Conversely, if P, Q are as indicated, and $D = P^2 - 4Q \neq 0$, if $W_0 = 0$ (resp. 2), $W_1 = 1$ (resp. P), if $W_n = PW_{n-1} - QW_{n-2}$ for $n \geq 2$, then Binet showed (in 1843) that

$$W_n = \frac{\alpha^n - \beta^n}{\alpha - \beta} \quad (\text{resp. } W_n = \alpha^n + \beta^n) \quad \text{for } n \geq 0;$$

here α, β are the roots of the polynomial $X^2 - PX + Q$. This is trivial, because the sequences of numbers

$$(W_n)_{n\geq 0} \quad \text{and} \quad \left(\frac{\alpha^n - \beta^n}{\alpha - \beta}\right)_{n\geq 0} \quad (\text{resp. } (\alpha^n + \beta^n)_{n\geq 0}),$$

have the first two terms equal and both have the same linear second-order recurrence definition.

Before I continue, here are the main special cases that had been considered before the full theory was developed.

The sequence corresponding to $P = 1$, $Q = -1$, $U_0 = U_0(1,-1) = 0$, and $U_1 = U_1(1,-1) = 1$ was first considered by Fibonacci, and it begins as follows:

$$0, \ 1, \ 1, \ 2, \ 3, \ 5, \ 8, \ 13, \ 21, \ 34, \ 55, \ 89, \ 144, \ 233,$$
$$377, \ 610, \ 987, \ 1597, \ 2584, \ 4181, \ 6765, \ \ldots$$

These numbers appeared for the first time in a problem in Fibonacci's *Liber Abaci*, published in 1202. It was also in this book that Arabic figures were first introduced in Europe. The problem, now reproduced in many elementary books, concerned rabbits having certain reproductive patterns. I do not care for such an explanation. As regards rabbits, I rather prefer to eat a good plate of "lapin chasseur" with fresh noodles.

The companion sequence of Fibonacci numbers, still with $P = 1$, $Q = -1$, is the sequence of Lucas numbers: $V_0 = V_0(1,-1) = 2$,

$V_1 = V_1(1, -1) = 1$, and it begins as follows:

$$2, \ 1, \ 3, \ 4, \ 7, \ 11, \ 18, \ 29, \ 47, \ 76, \ 123, \ 199, \ 322,$$
$$521, \ 843, \ 1364, \ 2207, \ 3571, \ 5778, \ 9349, \ 15127, \ \ldots$$

If $P = 3$, $Q = 2$, then the sequences obtained are

$$U_n(3, 2) = 2^n - 1 \quad \text{and} \quad V_n(3, 2) = 2^n + 1, \quad \text{for } n \geq 0.$$

These sequences were the cause of many sleepness nights for Fermat (see details in Sections VI and VII). The sequences associated to $P = 2$, $Q = -1$, are called the Pell sequences; they begin as follows:

$$U_n(2, -1): \quad 0, \ 1, \ 2, \ 5, \ 12, \ 29, \ 70, \ 169, \ 408, \ 985,$$
$$2378, \ 5741, \ 13860, \ \ldots$$
$$V_n(2, -1): \quad 2, \ 2, \ 6, \ 14, \ 34, \ 82, \ 198, \ 478, \ 1154,$$
$$2786, \ 6726, \ 16238, \ 39202, \ \ldots$$

Lucas noted a great similarity between the sequences of numbers $U_n(P, Q)$ (resp. $V_n(P, Q)$) and $(a^n - b^n)/(a - b)$ (resp. $a^n + b^n$), where a, b are given, $a > b \geq 1$, $\gcd(a, b) = 1$ and $n \geq 0$. No wonder, one is a special case of the other. Just observe that for the pair $(a + b, ab)$, $D = (a - b)^2 \neq 0$, $\alpha = a$, $\beta = b$, so

$$U_n(a + b, ab) = \frac{a^n - b^n}{a - b}, \quad V_n(a + b, ab) = a^n + b^n.$$

It is clearly desirable to extend the main results about the sequence of numbers $(a^n - b^n)/(a - b)$, $a^n + b^n$ (in what relates to divisibility and primality) for the wider class of Lucas sequences.

I shall therefore present the generalizations of Fermat's little theorem, Euler's theorem, etc., to Lucas sequences. There is no essential difficulty, but the development requires a surprising number of steps—true enough, all at an elementary level. In what follows, I shall record, one after the other, the facts needed to prove the main results. If you wish, work out the details. But I am also explicitly giving the beginning of several Lucas sequences, so you may be happy just to check my statements numerically (see the tables at the end of the section).

First, the algebraic facts, then the divisibility properties. To simplify the notations, I write only $U_n = U_n(P, Q)$, $V_n = V_n(P, Q)$.

We have the following algebraic properties:

(IV.1) $U_n = PU_{n-1} - QU_{n-2}$ $(n \geq 2)$, $U_0 = 0$, $U_1 = 1$,
$\qquad V_n = PV_{n-1} - QV_{n-2}$ $(n \geq 2)$, $V_0 = 2$, $V_1 = P$.

(IV.2) $U_{2n} = U_n V_n$,
$\qquad V_{2n} = V_n^2 - 2Q^n$.

(IV.3) $U_{m+n} = U_m V_n - Q^n U_{m-n}$,
$\qquad V_{m+n} = V_m V_n - Q^n V_{m-n}$ (for $m \geq n$).

(IV.4) $U_{m+n} = U_m U_{n+1} - QU_{m-1} U_n$,
$\qquad 2V_{m+n} = V_m V_n + DU_m U_n$.

(IV.5) $DU_n = 2V_{n+1} - PV_n$,
$\qquad V_n = 2U_{n+1} - PU_n$.

(IV.6) $U_n^2 = U_{n-1} U_{n+1} + Q^{n-1}$,
$\qquad V_n^2 = DU_n^2 + 4Q^n$.

(IV.7) $U_m V_n - U_n V_m = 2Q^n U_{m-n}$ (for $m \geq n$),
$\qquad U_m V_n + U_n V_m = 2U_{m+n}$.

(IV.8) $2^{n-1} U_n = \binom{n}{1} P^{n-1} + \binom{n}{3} P^{n-3} D + \binom{n}{5} P^{n-5} D^2 + \cdots$,

$\qquad 2^{n-1} V_n = P^n + \binom{n}{2} P^{n-2} D + \binom{n}{4} P^{n-4} D^2 + \cdots$.

(IV.9) If m is odd and $k \geq 1$, then

$$D^{(m-1)/2} U_k^m = U_{km} - \binom{m}{1} Q^k U_{k(m-2)} + \binom{m}{2} Q^{2k} U_{k(m-4)} - \cdots$$

$$\pm \binom{m}{(m-1)/2} Q^{\frac{m-1}{2}k} U_k,$$

$$V_k^m = V_{km} + \binom{m}{1} Q^k V_{k(m-2)} + \binom{m}{2} Q^{2k} V_{k(m-4)} + \cdots$$

$$+ \binom{m}{(m-1)/2} Q^{\frac{m-1}{2}k} V_k.$$

If m is even and $k \geq 1$, then

$$D^{m/2} U_k^m = \left[V_{km} - \binom{m}{1} Q^k V_{k(m-2)} + \binom{m}{2} Q^{2k} V_{k(m-4)} - \cdots \right.$$

$$+(-1)^{m/2}\binom{m}{m/2}Q^{(m/2)k}V_0\Big] - (-1)^{m/2}\binom{m}{m/2}Q^{(m/2)\,k},$$

$$V_k^m = \Big[V_{km} + \binom{m}{1}Q^k V_{k(m-2)} + \binom{m}{2}Q^{2k}V_{k(m-4)} + \cdots$$

$$+\binom{m}{m/2}Q^{(m/2)k}V_0\Big] - \binom{m}{m/2}Q^{(m/2)k}.$$

(IV.10) $U_m = V_{m-1} + QV_{m-3} + Q^2 V_{m-5} + \cdots + \text{(last summand)},$
where

$$\text{last summand} = \begin{cases} Q^{(m-2)/2}\,P & \text{if } m \text{ is even}, \\ Q^{(m-1)/2} & \text{if } m \text{ is odd}. \end{cases}$$

$$P^m = V_m + \binom{m}{1}QV_{m-2} + \binom{m}{2}Q^2 V_{m-4} + \cdots + \text{(last summand)},$$

where

$$\text{last summand} = \begin{cases} \binom{m}{m/2}Q^{m/2} & \text{if } m \text{ is even}, \\ \binom{m}{(m-1)/2}Q^{(m-1)/2}\,P & \text{if } m \text{ is odd}. \end{cases}$$

The following identity of Lagrange, dating from 1741, is required for the next property:

$$X^n + Y^n = (X+Y)^n - \frac{n}{1}XY(X+Y)^{n-2}$$

$$+ \frac{n}{2}\binom{n-3}{1}X^2Y^2(X+Y)^{n-4}$$

$$- \frac{n}{3}\binom{n-4}{2}X^3Y^3(X+Y)^{n-6} + \cdots$$

$$+ (-1)^r \frac{n}{r}\binom{n-r-1}{r-1}X^r Y^r(X+Y)^{n-2r} \pm \cdots,$$

where the sum is extended for $2r \le n$. Note that each coefficient is an integer.

(IV.11)   If $m \geq 1$ and q is odd,

$$U_{mq} = D^{(q-1)/2} U_m^q + \frac{q}{1} Q^m D^{(q-3)/2} U_m^{q-2}$$

$$+ \frac{q}{2} \binom{q-3}{1} Q^{2m} D^{(q-5)/2} U_m^{q-4} + \cdots$$

$$+ \frac{q}{r} \binom{q-r-1}{r-1} Q^{mr} D^{(q-2r-1)/2} U_m^{q-2r} + \cdots$$

$$+ \text{ last summand},$$

where the last summand is

$$\frac{q}{(q-1)/2} \binom{(q-1)/2}{(q-3)/2} Q^{\frac{q-1}{2}m} U_m = q\, Q^{\frac{q-1}{2}m} U_m.$$

Now, I begin to indicate, one after the other, the divisibility properties, in the order in which they may be proved.

(IV.12) $U_n \equiv V_{n-1} \pmod{Q},$

$$V_n \equiv P^n \pmod{Q}.$$

Hint: Use (IV.10) or proceed by induction.

(IV.13)   Let p be an odd prime, then

$$U_{kp} \equiv D^{\frac{p-1}{2}} U_k \pmod{p}$$

and, for $e \geq 1$,

$$U_{p^e} \equiv D^{\frac{p-1}{2}e} \pmod{p}.$$

In particular,

$$U_p \equiv \left(\frac{D}{p}\right) \pmod{p}.$$

Hint: Use (IV.9).

(IV.14) $V_p \equiv P \pmod{p}.$

Hint: Use (IV.10).

(IV.15)   If $n, k \geq 1$, then U_n divides U_{kn}.

Hint: Use (IV.3).

(IV.16)   If $n, k \geq 1$ and k is odd, then V_n divides V_{kn}.

Hint: Use (IV.9).

Notation. If $n \geq 2$ and if there exists $r \geq 1$ such that n divides U_r, denote by $\rho(n) = \rho(n, U)$ the smallest such r.

(IV.17)   Assume that $\rho(n)$ exists and $\gcd(n, 2Q) = 1$. Then $n \mid U_k$ if and only if $\rho(n) \mid k$.

Hint: Use (IV.15) and (IV.7).

It will be seen that $\rho(n)$ exists, for many—not for all—values of n, such that $\gcd(n, 2Q) = 1$.

(IV.18)   If Q is even and P is even, then U_n is even (for $n \geq 2$) and V_n is even (for $n \geq 1$).
    If Q is even and P is odd, then U_n, V_n are odd (for $n \geq 1$).
    If Q is odd and P is even, then $U_n \equiv n \pmod{2}$ and V_n is even.
    If Q is odd and P is odd, then U_n, V_n are even if 3 divides n, while U_n, V_n are odd, otherwise.
    In particular, if U_n is even, then V_n is even.

Hint: Use (IV.12), (IV.5), (IV.2), (IV.6), and (IV.1).

Here is the first main result, which is a companion of (IV.18) and generalizes Fermat's little theorem:

(IV.19)   Let p be an odd prime.
    If $p \mid P$ and $p \mid Q$, then $p \mid U_k$ for every $k > 1$.
    If $p \mid P$ and $p \nmid Q$, then $p \mid U_k$ exactly when k is even.
    If $p \nmid P$ and $p \mid Q$, then $p \nmid U_k$ for every $k \geq 1$.
    If $p \nmid P$, $p \nmid Q$, and $p \mid D$, then $p \mid U_k$ exactly when $p \mid k$.
    If $p \nmid PQD$, then $p \mid U_{\psi(p)}$, where $\psi(p) = p - (D \mid p)$, and $(D \mid p)$ denotes the Legendre symbol.

Proof. If $p \mid P$ and $p \mid Q$, by (IV.1) $p \mid U_k$ for every $k > 1$.
    If $p \mid P = U_2$, by (IV.15) $p \mid U_{2k}$ for every $k \geq 1$. Since $p \nmid Q$, and $U_{2k+1} = PU_{2k} - QU_{2k-1}$, by induction, $p \nmid U_{2k+1}$.
    If $p \nmid P$ and $p \mid Q$, by induction and (IV.1), $p \nmid U_k$ for every $k \geq 1$.
    If $p \nmid PQ$ and $p \mid D$, by (IV.8), $2^{p-1}U_p \equiv 0 \pmod{p}$ so $p \mid U_p$. On the other hand, if $p \nmid n$, then by (IV.8), $2^{n-1}U_n \equiv nP^{n-1} \not\equiv 0 \pmod{p}$, so $p \nmid U_n$.

Finally the more interesting case: assume $p \nmid PQD$.
If $(D \mid p) = -1$, then by (IV.8)

$$2^p U_{p+1} = \binom{p+1}{1} P^p + \binom{p+1}{3} P^{p-2} D + \cdots$$
$$+ \binom{p+1}{p} PD^{(p-1)/2} \equiv P + PD^{(p-1)/2} \equiv 0 \pmod{p},$$

so $p \mid U_{p+1}$.

If $(D \mid p) = 1$, there exists C such that $P^2 - 4Q = D \equiv C^2$ (mod p); hence, $P^2 \not\equiv C^2$ (mod p) and $p \nmid C$. By (IV.8), noting that

$$\binom{p-1}{1} \equiv -1 \pmod{p}, \quad \binom{p-1}{3} \equiv -1 \pmod{p}, \quad \ldots$$

we see that

$$2^{p-2} U_{p-1} = \binom{p-1}{1} P^{p-2} + \binom{p-1}{3} P^{p-4} D$$
$$+ \binom{p-1}{5} P^{p-6} D^2 + \cdots + \binom{p-1}{p-2} PD^{(p-3)/2}$$
$$\equiv -[P^{p-2} + P^{p-4} D + P^{p-6} D^2 + \cdots + PD^{(p-3)/2}]$$
$$\equiv -P \left(\frac{P^{p-1} - D^{(p-1)/2}}{P^2 - D} \right)$$
$$\equiv -P \frac{P^{p-1} - C^{p-1}}{P^2 - C^2} \equiv 0 \pmod{p}.$$

So $p \mid U_{p-1}$. $\square$

If I want to use the notation $\rho(p)$ introduced before, some of the assertions of (IV.19) may be restated as follows:

If p is an odd prime and $p \nmid Q$, then:
If $p \mid P$, then $\rho(p) = 2$.
If $p \nmid P$, $p \mid D$, then $\rho(p) = p$.
If $p \nmid PD$, then $\rho(p)$ divides $\psi(p)$.

Don't conclude hastily that, in this latter case, $\rho(p) = \psi(p)$. I shall return to this point, after I list the main properties of the Lucas sequences.

For the special Lucas sequence $U_n(a+1, a)$, the discriminant is $D = (a-1)^2$; so if $p \nmid a(a^2 - 1)$, then

$$\left(\frac{D}{p}\right) = 1 \quad \text{and} \quad p \mid U_{p-1} = \frac{a^{p-1} - 1}{a - 1},$$

so $p \mid a^{p-1} - 1$ (this is trivial if $p \mid a^2 - 1$)—which is Fermat's little theorem.

(IV.20)   Let $e \geq 1$, and let p^e be the exact power of p dividing U_m. If $p \nmid k$ and $f \geq 1$, then p^{e+f} divides U_{mkp^f}.

Moreover, if $p \mid Q$ and $p^e \neq 2$, then p^{e+f} is the exact power of p dividing U_{mkp^f}, while if $p^e = 2$ then $U_{mk}/2$ is odd.

Hint: Use (IV.19), (IV.18), (IV.11), and (IV.6).

And now the generalization of Euler's theorem:

If α, β are roots of $X^2 - PX + Q$, define the symbol:

$$\left(\frac{\alpha, \beta}{2}\right) = \begin{cases} 1 & \text{if } Q \text{ is even,} \\ 0 & \text{if } Q \text{ is odd, } P \text{ even,} \\ -1 & \text{if } Q \text{ is odd, } P \text{ odd} \end{cases}$$

and for $p \neq 2$:

$$\left(\frac{\alpha, \beta}{p}\right) = \left(\frac{D}{p}\right)$$

(so it is 0 if $p \mid D$). Put

$$\psi_{\alpha,\beta}(p) = p - \left(\frac{\alpha, \beta}{p}\right)$$

for every prime p, also

$$\psi_{\alpha,\beta}(p^e) = p^{e-1}\psi_{\alpha,\beta}(p) \quad \text{for } e \geq 1.$$

If $n = \prod_{p|n} p^e$, define the Carmichael function

$$\lambda_{\alpha,\beta}(n) = \text{lcm}\{\psi_{\alpha,\beta}(p^e)\}$$

(where lcm denotes the least common multiple), and define the generalized Euler function

$$\psi_{\alpha,\beta}(n) = \prod_{p|n} \psi_{\alpha,\beta}(p^e).$$

So $\lambda_{\alpha,\beta}(n)$ divides $\psi_{\alpha,\beta}(n)$.

It is easy to check that $\psi_{a,1}(p) = p-1 = \varphi(p)$ for every prime p not dividing a; so if $\gcd(a, n) = 1$, then $\psi_{a,1}(n) = \varphi(n)$ and also $\lambda_{a,1}(n) = \lambda(n)$, where $\lambda(n)$ is the function, also defined by Carmichael, and considered in Section II.

And here is the extension of Euler's theorem:

(IV.21)   If $\gcd(n, Q) = 1$, then n divides $U_{\lambda_{\alpha,\beta}(n)}$; hence, also n divides $U_{\psi_{\alpha,\beta}(n)}$.

Hint: Use (IV.19) and (IV.20).

It should be said that the divisibility properties of the companion sequence $(V_n)_{n\geq 1}$ are not so simple to describe. Note, for example,

(IV.22)   If $p \nmid 2QD$, then $V_{p-(D|p)} \equiv 2Q^{\frac{1}{2}[1-(D|p)]} \pmod{p}$.

Hint: Use (IV.5), (IV.13), (IV.19), and (IV.14).

This may be applied to give divisibility results for $U_{\psi(p)/2}$ and $V_{\psi(p)/2}$.

(IV.23)   Assume that $p \nmid 2QD$. Then

$$p \mid U_{\psi(p)/2} \quad \text{if and only if} \quad (Q \mid p) = 1,$$
$$p \mid V_{\psi(p)/2} \quad \text{if and only if} \quad (Q \mid p) = -1.$$

Hint: For the first assertion, use (IV.2), (IV.6), (IV.22) and the congruence $(Q \mid p) \equiv Q^{(p-1)/2} \pmod{p}$. For the second assertion, use (IV.2), (IV.19), the first assertion, and also (IV.6).

For the next results, I shall assume that $\gcd(P, Q) = 1$.

(IV.24) $\gcd(U_n, Q) = 1$ and $\gcd(V_n, Q) = 1$, for every $n \geq 1$.

Hint: Use (IV.12).

(IV.25) $\gcd(U_n, V_n) = 1$ or 2.

Hint: Use (IV.16) and (IV.24).

(IV.26)   If $d = \gcd(m, n)$, then $U_d = \gcd(U_m, U_n)$.

Hint: Use (IV.15), (IV.7), (IV.24), (IV.18), and (IV.6). This proof is actually not so easy, and requires the use of the Lucas sequence $\left(U_n(V_d, Q^d)\right)_{n \geq 0}$.

(IV.27)   If $\gcd(m, n) = 1$, then $\gcd(U_m, U_n) = 1$.

No hint for this one.

(IV.28)   If $d = \gcd(m, n)$ and m/d, n/d are odd, then
$$V_d = \gcd(V_m, V_n).$$

Hint: Use the same proof as for (IV.26).

And here is a result similar to (IV.17), but with the assumption that $\gcd(P, Q) = 1$:

(IV.29)   Assume that $\rho(n)$ exists. Then $n \mid U_k$ if and only if $\rho(n) \mid k$.

Hint: Use (IV.15), (IV.24), and (IV.3).

I pause to write explicitly what happens for the Fibonacci numbers U_n and Lucas numbers V_n; now $P = 1$, $Q = -1$, $D = 5$.

Property (IV.18) becomes the *law of appearance* of p; even though I am writing this text on Halloween's evening, it would hurt me to call it the "apparition law" (as it was badly translated from the French *loi d'apparition*; in all English dictionaries "apparition" means "ghost"). Law of apparition (oops!, appearance) of p:

$$p \mid U_{p-1} \text{ if } (5 \mid p) = 1, \quad \text{that is, } p \equiv \pm 1 \pmod{10},$$
$$p \mid U_{p+1} \text{ if } (5 \mid p) = -1, \quad \text{that is, } p \equiv \pm 3 \pmod{10}.$$

Property (IV.19) is the *law of repetition*.

For the Lucas numbers, the following properties hold:

$$p \mid V_{p-1} - 2 \text{ if } (5 \mid p) = 1, \quad \text{that is, } p \equiv \pm 1 \pmod{10},$$
$$p \mid V_{p+1} + 2 \text{ if } (5 \mid p) = -1, \quad \text{that is, } p \equiv \pm 3 \pmod{10}.$$

Jarden showed in 1958 that, for the Fibonacci sequence, the function
$$\frac{\psi(p)}{\rho(p)} = \frac{p - (5 \mid p)}{\rho(p)}$$
is unbounded (when the prime p tends to infinity).

This result was generalized by Kiss & Phong in 1978: there exists $C > 0$ (depending only on P, Q) such that $\psi(p)/\rho(p)$ is unbounded, but still $\psi(p)/\rho(p) < C[p/(\log p)]$ (when the prime p tends to infinity).

Now I shall indicate the behaviour of Lucas sequences modulo a prime p.

If $p = 2$, this is as described in (IV.18). For example, if P, Q are odd, then the sequences $(U_n \bmod 2)_{n \geq 0}$, $(V_n \bmod 2)_{n \geq 0}$ are equal to

$$0, \ 1, \ 1, \ 0, \ 1, \ 1, \ 0, \ 1, \ 1, \ 0, \ \ldots.$$

It is more interesting when p is an odd prime.

(IV.30)   If $p \nmid 2QD$ and $(D \mid p) = 1$, then

$$U_{n+p-1} \equiv U_n \pmod{p},$$
$$V_{n+p-1} \equiv V_n \pmod{p}.$$

Thus, the sequences $(U_n \bmod p)_{n \geq 0}$, $(V_n \bmod p)_{n \geq 0}$ have period $p-1$.

Proof. By (IV.4), $U_{n+p-1} = U_n U_p - Q U_{n-1} U_{p-1}$; by (IV.19), $\rho(p)$ divides $p - (D \mid p) = p - 1$; by (IV.15), $p \mid U_{p-1}$; this is also true if $p \mid P$, $p \nmid Q$, because then $p - 1$ is even, so $p \mid U_{p-1}$, by (IV.19). By (IV.13),

$$U_p \equiv (D \mid p) \equiv 1 \pmod{p}.$$

So $U_{n+p-1} \equiv U_n \pmod{p}$.

Now, by (IV.5), $V_{n+p-1} = 2U_{n+p} - PU_n \equiv 2U_{n+1} - PU_n \equiv V_n$ $\pmod{p}$. □

The companion result is the following:

(IV.31)   Let $p \nmid 2QD$, let e be the order of $Q \bmod p$. If $(D \mid p) = -1$, then

$$U_{n+e(p+1)} \equiv U_n \pmod{p},$$
$$V_{n+e(p+1)} \equiv V_n \pmod{p}.$$

Thus, the sequences $(U_n \bmod p)_{n \geq 0}$, $(V_n \bmod p)_{n \geq 0}$ have period $e(p+1)$.

Proof. If $p \nmid P$, then by (IV.19), (IV.15),

$$p \mid U_{p-(D|p)} = U_{p+1}.$$

This is also true when $p \mid P$.

By (IV.22), $V_{p+1} \equiv 2Q \pmod{p}$. Now I show, by induction on $r \geq 1$, that $V_{r(p+1)} \equiv 2Q^r \pmod{p}$.

If this is true for $r \geq 1$, then by (IV.4)

$$2V_{(r+1)(p+1)} = V_{r(p+1)}V_{p+1} + DU_{r(p+1)}U_{p+1} \equiv 4Q^{r+1} \pmod{p},$$

so $V_{(r+1)(p+1)} \equiv 2Q^{r+1} \pmod{p}$. In particular, $V_{e(p+1)} \equiv 2Q^e \equiv 2 \pmod{p}$. By (IV.7),

$$U_{n+e(p+1)}V_{e(p+1)} - U_{e(p+1)}V_{n+e(p+1)} = 2Q^{e(p+1)}U_n,$$

hence $2U_{n+e(p+1)} \equiv 2U_n \pmod{p}$ and the first congruence is established.

The second congruence follows using (IV.5).   □

It is good to summarize some of the preceding results, in terms of the sets

$$\mathcal{P}(U) = \{p \text{ prime} \mid \text{there exists } n \text{ such that } U_n \neq 0 \text{ and } p \mid U_n\},$$
$$\mathcal{P}(V) = \{p \text{ prime} \mid \text{there exists } n \text{ such that } V_n \neq 0 \text{ and } p \mid V_n\}.$$

These are the sets of prime divisors of the sequences $U = (U_n)_{n \geq 1}$ and $V = (V_n)_{n \geq 1}$, respectively.

The parameters (P, Q) are assumed to be nonzero, relatively prime integers and the discriminant is $D = P^2 - 4Q \neq 0$.

A first case arises if there exists $n > 1$ such that $U_n = 0$; equivalently, $\alpha^n = \beta^n$, that is α/β is a root of unity. If n is the smallest such index, then $U_r \neq 0$ for $r = 1, \ldots, n-1$ and $U_{nk+r} = \alpha^{nk}U_r$ (for every $k \geq 1$), so $\mathcal{P}(U)$ consists of the prime divisors of $U_2 \cdots U_{n-1}$. Similarly, $\mathcal{P}(V)$ consists of the prime numbers dividing $V_1 V_2 \cdots V_{n-1}V_n$.

The more interesting case is when α/β is not a root of unity, so $U_n \neq 0$, $V_n \neq 0$, for every $n \geq 1$. Then $\mathcal{P}(U) = \{p \text{ prime} \mid p \text{ does not divide } Q\}$.

This follows from (IV.18) and (IV.19). In particular, for the sequence of Fibonacci numbers, $\mathcal{P}(U)$ is the set of all primes.

Nothing so precise may be said about the companion Lucas sequence $V = (V_n)_{n \geq 1}$. From $U_{2n} = U_n V_n$ ($n \geq 1$) it follows that $\mathcal{P}(V)$

is a subset of $\mathcal{P}(U)$. From (IV.18), $2 \in \mathcal{P}(V)$ if and only if Q is odd. Also, from (IV.24) and (IV.6), if $p \neq 2$ and if $p \mid DQ$, then $p \notin \mathcal{P}(V)$, while if $p \nmid 2DQ$ and $(Q \mid p) = -1$, then $p \in \mathcal{P}(V)$ [see (IV.23)]; on the other hand, if $p \nmid 2DQ$, $(Q \mid p) = 1$, and $(D \mid p) = -(-1 \mid p)$, then $p \notin \mathcal{P}(V)$. This does not determine, without a further analysis, whether a prime p, such that $p \nmid 2DQ$, $(Q \mid p) = 1$, and $(D \mid p) = (-1 \mid p)$ belongs, or does not belong, to $\mathcal{P}(V)$. At any rate, it shows that $\mathcal{P}(V)$ is also an infinite set.

For the sequence of Lucas numbers, with $P = 1$, $Q = -1$, $D = 5$, the preceding facts may be explicitly stated as follows:

   if $p \equiv 3, 7, 11, 19 \pmod{20}$, then $p \in \mathcal{P}(V)$;

   if $p \equiv 13, 17 \pmod{20}$, then $p \notin \mathcal{P}(V)$.

For $p \equiv 1, 9 \pmod{20}$, no decision may be obtained without a careful study, as, for example, that done by Ward in 1961. Already in 1958 Jarden had shown that there exist infinitely many primes p, $p \equiv 1 \pmod{20}$, such that $p \notin \mathcal{P}(V)$, and, on the other hand, there exist also infinitely many primes p, $p \equiv 1 \pmod{40}$, such that $p \in \mathcal{P}(V)$.

Later, in Chapter 5, Section VIII, I shall return to the study of the sets $\mathcal{P}(U)$, $\mathcal{P}(V)$, asking for their density in the set of all primes.

In analogy with the theorem of Bang and Zsigmondy, Carmichael also considered the primitive prime factors of the Lucas sequences, with parameters (P, Q): p is a primitive prime factor of U_k (resp. V_k) if $p \mid U_k$ (resp. $p \mid V_k$), but p does not divide any preceding number in the sequence in question.

The proof of Zsigmondy's theorem is not too simple; here it is somewhat more delicate.

Carmichael showed that if the discriminant D is positive, then for every $n \neq 1, 2, 6$, U_n has a primitive prime factor, except if $n = 12$ and $P = \pm 1$, $Q = -1$.

Moreover, if D is a square, then it is better: for every n, U_n has a primitive prime factor, except if $n = 6$, $P = \pm 3$, $Q = 2$.

Do you recognize that this second statement includes Zsigmondy's theorem? Also, if $P = 1$, $Q = -1$ the exception is the Fibonacci number $U_{12} = 144$.

For the companion sequence, if $D > 0$, then for every $n \neq 1, 3$, V_n has a primitive prime factor, except if $n = 6$, $P = \pm 1$, $Q = -1$ (the Lucas number $V_6 = 18$). Moreover, if D is a square, then the only exception is $n = 3$, $P = \pm 3$, $Q = 2$, also contained in Zsigmondy's theorem.

If, however, $D < 0$, the result indicated is no longer true. Thus, as Carmichael already noted, if $P = 1$, $Q = 2$, then for $n = 1, 2, 3,$ 5, 8, 12, 13, 18, U_n has no primitive prime factors.

Schinzel showed the following in 1962:

Let $(U_n)_{n \geq 0}$ be the Lucas sequence with relatively prime parameters (P, Q) and assume that the discriminant is $D < 0$. Assume that α/β is not a root of unity. Then there exists n_0 (depending on P, Q), effectively computable, such that if $n > n_0$, then U_n has a primitive prime factor.

Later, in 1974, Schinzel proved the same result with an absolute constant n_0—independent of the Lucas sequence. This was a remarkable result.

Making use of the methods of Baker, Stewart determined in 1977 that if $n > e^{452} 2^{67}$, then U_n has a primitive prime factor. Moreover, Stewart also showed that if n is given ($n \neq 6$, $n > 4$), there are only finitely many Lucas sequences, which may be determined explicitly (so says Stewart, without doing it), for which U_n has no primitive prime factor.

It is interesting to consider the primitive part U_n^* of U_n:

$$U_n = U_n^* U_n' \quad \text{with} \quad \gcd(U_n^*, U_n') = 1$$

and p divides U_n^* if and only if p is a primitive prime factor of U_n.

In 1963, Schinzel indicated conditions for the existence of two (or even $e > 2$) distinct primitive prime factors. It follows that if $D > 0$ or $D < 0$ and α/β is not a root of unity, there exist infinitely many n such that the primitive part U_n^* is composite.

Can one say anything about U_n^* being square-free? This is a very deep question. Just think of the special case when $P = 3$, $Q = 2$, which gives the sequence $2^n - 1$ (see my comments in Section II).

Table 2.   Fibonacci and Lucas numbers

$$P = 1, \ Q = -1$$

Fibonacci numbers	Lucas numbers
$U(0) = 0 \quad U(1) = 1$	$V(0) = 2 \quad V(1) = 1$
$U(2) = 1$	$V(2) = 3$
$U(3) = 2$	$V(3) = 4$
$U(4) = 3$	$V(4) = 7$
$U(5) = 5$	$V(5) = 11$
$U(6) = 8$	$V(6) = 18$
$U(7) = 13$	$V(7) = 29$
$U(8) = 21$	$V(8) = 47$
$U(9) = 34$	$V(9) = 76$
$U(10) = 55$	$V(10) = 123$
$U(11) = 89$	$V(11) = 199$
$U(12) = 144$	$V(12) = 322$
$U(13) = 233$	$V(13) = 521$
$U(14) = 377$	$V(14) = 843$
$U(15) = 610$	$V(15) = 1364$
$U(16) = 987$	$V(16) = 2207$
$U(17) = 1597$	$V(17) = 3571$
$U(18) = 2584$	$V(18) = 5778$
$U(19) = 4181$	$V(19) = 9349$
$U(20) = 6765$	$V(20) = 15127$
$U(21) = 10946$	$V(21) = 24476$
$U(22) = 17711$	$V(22) = 39603$
$U(23) = 28657$	$V(23) = 64079$
$U(24) = 46368$	$V(24) = 103682$
$U(25) = 75025$	$V(25) = 167761$
$U(26) = 121393$	$V(26) = 271443$
$U(27) = 196418$	$V(27) = 439204$
$U(28) = 317811$	$V(28) = 710647$
$U(29) = 514229$	$V(29) = 1149851$
$U(30) = 832040$	$V(30) = 1860498$
$U(31) = 1346269$	$V(31) = 3010349$
$U(32) = 2178309$	$V(32) = 4870847$
$U(33) = 3524578$	$V(33) = 7881196$
$U(34) = 5702887$	$V(34) = 12752043$
$U(35) = 9227465$	$V(35) = 20633239$
$U(36) = 14930352$	$V(36) = 33385282$
$U(37) = 24157817$	$V(37) = 54018521$
$U(38) = 39088169$	$V(38) = 87403803$
$U(39) = 63245986$	$V(39) = 141422324$
$U(40) = 102334155$	$V(40) = 228826127$

Table 3.   Numbers $2^n - 1$ and $2^n + 1$

$$P = 3, \ Q = 2$$

Numbers $2^n - 1$	Numbers $2^n + 1$
$U(0) = 0 \quad U(1) = 1$	$V(0) = 2 \quad V(1) = 3$
$U(2) = 3$	$V(2) = 5$
$U(3) = 7$	$V(3) = 9$
$U(4) = 15$	$V(4) = 17$
$U(5) = 31$	$V(5) = 33$
$U(6) = 63$	$V(6) = 65$
$U(7) = 127$	$V(7) = 129$
$U(8) = 255$	$V(8) = 257$
$U(9) = 511$	$V(9) = 513$
$U(10) = 1023$	$V(10) = 1025$
$U(11) = 2047$	$V(11) = 2049$
$U(12) = 4095$	$V(12) = 4097$
$U(13) = 8191$	$V(13) = 8193$
$U(14) = 16383$	$V(14) = 16385$
$U(15) = 32767$	$V(15) = 32769$
$U(16) = 65535$	$V(16) = 65537$
$U(17) = 131071$	$V(17) = 131073$
$U(18) = 262143$	$V(18) = 262145$
$U(19) = 524287$	$V(19) = 524289$
$U(20) = 1048575$	$V(20) = 1048577$
$U(21) = 2097151$	$V(21) = 2097153$
$U(22) = 4194303$	$V(22) = 4194305$
$U(23) = 8388607$	$V(23) = 8388609$
$U(24) = 16777215$	$V(24) = 16777217$
$U(25) = 33554431$	$V(25) = 33554433$
$U(26) = 67108863$	$V(26) = 67108865$
$U(27) = 134217727$	$V(27) = 134217729$
$U(28) = 268435455$	$V(28) = 268435457$
$U(29) = 536870911$	$V(29) = 536870913$
$U(30) = 1073741823$	$V(30) = 1073741825$
$U(31) = 2147483647$	$V(31) = 2147483649$
$U(32) = 4294967295$	$V(32) = 4294967297$
$U(33) = 8589934591$	$V(33) = 8589934593$
$U(34) = 17179869183$	$V(34) = 17179869185$
$U(35) = 34359738367$	$V(35) = 34359738369$
$U(36) = 68719476735$	$V(36) = 68719476737$
$U(37) = 137438953471$	$V(37) = 137438953473$
$U(38) = 274877906943$	$V(38) = 274877906945$
$U(39) = 549755813887$	$V(39) = 549755813889$
$U(40) = 1099511627775$	$V(40) = 1099511627777$

Table 4.   Pell numbers

$$P = 2, \ Q = -1$$

Pell numbers	Companion Pell numbers
$U(0) = 0 \quad U(1) = 1$	$V(0) = 2 \quad V(1) = 2$
$U(2) = 2$	$V(2) = 6$
$U(3) = 5$	$V(3) = 14$
$U(4) = 12$	$V(4) = 34$
$U(5) = 29$	$V(5) = 82$
$U(6) = 70$	$V(6) = 198$
$U(7) = 169$	$V(7) = 478$
$U(8) = 408$	$V(8) = 1154$
$U(9) = 985$	$V(9) = 2786$
$U(10) = 2378$	$V(10) = 6726$
$U(11) = 5741$	$V(11) = 16238$
$U(12) = 13860$	$V(12) = 39202$
$U(13) = 33461$	$V(13) = 94642$
$U(14) = 80782$	$V(14) = 228486$
$U(15) = 195025$	$V(15) = 551614$
$U(16) = 470832$	$V(16) = 1331714$
$U(17) = 1136689$	$V(17) = 3215042$
$U(18) = 2744210$	$V(18) = 7761798$
$U(19) = 6625109$	$V(19) = 18738638$
$U(20) = 15994428$	$V(20) = 45239074$
$U(21) = 38613965$	$V(21) = 109216786$
$U(22) = 93222358$	$V(22) = 263672646$
$U(23) = 225058681$	$V(23) = 636562078$
$U(24) = 543339720$	$V(24) = 1536796802$
$U(25) = 1311738121$	$V(25) = 3710155682$
$U(26) = 3166815962$	$V(26) = 8957108166$
$U(27) = 7645370045$	$V(27) = 21624372014$
$U(28) = 1845756052$	$V(28) = 52205852194$
$U(29) = 44560482149$	$V(29) = 126036076402$
$U(30) = 107578520350$	$V(30) = 304278004998$
$U(31) = 259717522849$	$V(31) = 734592086398$
$U(32) = 627013566048$	$V(32) = 1773462177794$
$U(33) = 1513744654945$	$V(33) = 4281516441986$
$U(34) = 3654502875938$	$V(34) = 10336495061766$
$U(35) = 8822750406821$	$V(35) = 24954506565518$
$U(36) = 21300003689580$	$V(36) = 60245508192802$
$U(37) = 51422757785981$	$V(37) = 145445522951122$
$U(38) = 124145519261542$	$V(38) = 351136554095046$
$U(39) = 299713796309065$	$V(39) = 847718631141214$
$U(40) = 723573111879672$	$V(40) = 2046573816377474$

Table 5.   Numbers $U(4,3)$ and $V(4,3)$

$$P = 4, \; Q = 3$$

Numbers	Companion numbers
$U(0) = 0 \quad U(1) = 1$	$V(0) = 2 \quad V(1) = 4$
$U(2) = 4$	$V(2) = 10$
$U(3) = 13$	$V(3) = 28$
$U(4) = 40$	$V(4) = 82$
$U(5) = 121$	$V(5) = 244$
$U(6) = 364$	$V(6) = 730$
$U(7) = 1093$	$V(7) = 2188$
$U(8) = 3280$	$V(8) = 6562$
$U(9) = 9841$	$V(9) = 19684$
$U(10) = 29524$	$V(10) = 59050$
$U(11) = 88573$	$V(11) = 177148$
$U(12) = 265720$	$V(12) = 531442$
$U(13) = 797161$	$V(13) = 1594324$
$U(14) = 2391484$	$V(14) = 4782970$
$U(15) = 7174453$	$V(15) = 14348908$
$U(16) = 21523360$	$V(16) = 43046722$
$U(17) = 64570081$	$V(17) = 129140164$
$U(18) = 193710244$	$V(18) = 387420490$
$U(19) = 581130733$	$V(19) = 1162261468$
$U(20) = 1743392200$	$V(20) = 3486784402$
$U(21) = 5230176601$	$V(21) = 10460353204$
$U(22) = 15690529804$	$V(22) = 31381059610$
$U(23) = 47071589413$	$V(23) = 94143178828$
$U(24) = 141214768240$	$V(24) = 282429536482$
$U(25) = 423644304721$	$V(25) = 847288609444$
$U(26) = 1270932914164$	$V(26) = 2541865828330$
$U(27) = 3812798742493$	$V(27) = 7625597484988$
$U(28) = 11438396227480$	$V(28) = 22876792454962$
$U(29) = 34315188682441$	$V(29) = 68630377364884$
$U(30) = 102945566047324$	$V(30) = 205891132094650$
$U(31) = 308836698141973$	$V(31) = 617673396283948$
$U(32) = 926510094425920$	$V(32) = 1853020188851842$
$U(33) = 2779530283277761$	$V(33) = 5559060566555524$
$U(34) = 8338590849833284$	$V(34) = 16677181699666570$
$U(35) = 25015772549499853$	$V(35) = 50031545098999708$
$U(36) = 75047317648499560$	$V(36) = 150094635296999122$
$U(37) = 225141952945498681$	$V(37) = 450283905890997364$
$U(38) = 675425858836496044$	$V(38) = 1350851717672992090$
$U(39) = 2026277576509488133$	$V(39) = 4052555153018976268$
$U(40) = 6078832729528464400$	$V(40) = 12157665459056928802$

V   Primality Tests Based on Lucas Sequences

Lucas began, Lehmer continued, others refined. The primality tests of N, to be presented now, require the knowledge of prime factors of $N+1$, and they complement the tests indicated in Section III, which needed the prime factors of $N - 1$. Now, the tool will be the Lucas sequences. By (IV.18), if N is an odd prime, if $U = (U_n)_{n \geq 0}$ is a Lucas sequence with discriminant D and $N \nmid DPQ$, then N divides $U_{N-(D|N)}$. So, if the Jacobi symbol $(D \mid N) = -1$, then N divides U_{N+1}.

However, I note right away (as I did in Section III) that a crude converse does not hold, because there exist composite integers N, and Lucas sequences $(U_n)_{n \geq 0}$ with discriminant D, such that N divides $U_{N-(D|N)}$. Such numbers will be studied in Section X.

It will be convenient to introduce for every integer $D > 1$ the function ψ_D, defined as follows:

If $N = \prod_{i=1}^{s} p_i^{e_i}$, let

$$\psi_D(N) = \frac{1}{2^{s-1}} \prod_{i=1}^{s} p_i^{e_i-1} \left(p_i - \left(\frac{D}{p_i} \right) \right).$$

Note that if $(U_n)_{n \geq 0}$ is a Lucas sequence with discriminant D, if α, β are the roots of the associated polynomial, then the function $\psi_{\alpha,\beta}$ considered in Section IV is related to ψ_D as follows:

$$\psi_{\alpha,\beta}(N) = 2^{s-1}\psi_D(N).$$

As it will be necessary to consider simultaneously several Lucas sequences with the same discriminant D, it is preferable to work with ψ_D, and not with the functions $\psi_{\alpha,\beta}$ corresponding to the various sequences.

Note, for example, that if $U(P,Q)$ has discriminant D, if $P' = P + 2$, $Q' = P + Q + 1$, then also $U(P',Q')$ has discriminant D.

It is good to start with some preparatory and easy results.

(V.1)   If N is odd, $\gcd(N, D) = 1$, then $\psi_D(N) = N - (D \mid N)$ if and only if N is a prime.

Proof. If N is a prime, by definition $\psi_D(N) = N - (D \mid N)$. If $N = p^e$ with p prime, $e \geq 2$, then $\psi_D(N)$ is a multiple of p, while $N = (D \mid N)$ is not.

If $N = \prod_{i=1}^{s} p_i^{e_i}$, with $s \geq 2$, then

$$\psi_D(N) \leq \frac{1}{2^{s-1}} \prod_{i=1}^{s} p_i^{e_i - 1}(p_i + 1) = 2N \prod_{i=1}^{s} \frac{1}{2}\left(1 + \frac{1}{p_i}\right)$$

$$\leq 2N \times \frac{2}{3} \times \frac{3}{5} \times \cdots \leq \frac{4N}{5} < N - 1,$$

since $N > 5$. $\square$

(V.2)   If N is odd, $\gcd(N, D) = 1$, and $N - (D \mid N)$ divides $\psi_D(N)$, then N is a prime.

Proof. Assume that N is composite. First, let $N = p^e$, with p prime, $e \geq 2$; then $\psi_D(N) = p^e - p^{e-1}(D \mid p)$. Hence,

$$p^e - p^{e-1} < p^e - 1 \leq N - (D \mid N) \leq \psi_D(N) = p^e - p^{e-1}(D \mid p),$$

so $(D \mid p) = -1$ and $N - (D \mid N) = p^e \pm 1$ divides $\psi_D(N) = p^e + p^{e-1} = p^e \pm 1 + (p^{e-1} \mp 1)$, which is impossible.

If N has at least two distinct prime factors, it was seen in (V.1) that $\psi_D(N) < N - 1 \leq N - (D \mid N)$, which is contrary to the hypothesis. So N must be a prime. $\square$

(V.3)   If N is odd, $U = U(P, Q)$ is a Lucas sequence with discriminant D, and $\gcd(N, QD) = 1$, then $N \mid U_{\psi_D(N)}$.

Proof. Since $\gcd(N, Q) = 1$, then by (IV.12), N divides $\lambda_{\alpha,\beta}(N)$, where α, β are the roots of $X^2 - PX + Q$. If $N = \prod_{i=1}^{s} p_i^{e_i}$, then

$$\lambda_{\alpha,\beta}(N) = \gcd\left\{p_i^{e_i - 1}\left(p_i - \left(\frac{D}{p_i}\right)\right)\right\}$$

$$= 2\gcd\left\{\frac{1}{2} p_i^{e_i - 1}\left(p_i - \left(\frac{D}{p_i}\right)\right)\right\}$$

and $\lambda_{\alpha,\beta}(N)$ divides

$$2\prod_{i=1}^{s} \frac{1}{2} p_i^{e_i - 1}\left(p_i - \left(\frac{D}{p_i}\right)\right) = \psi_D(N).$$

By (IV.15), N divides $U_{\psi_D(N)}$. $\square$

(V.4)   If N is odd, $U = U(P, Q)$ is a Lucas sequence with discriminant D such that $(D \mid N) = -1$, and N divides U_{N+1}, then $\gcd(N, QD) = 1$.

Proof. Since $(D \mid N) \neq 0$, then $\gcd(N, D) = 1$. If there exists a prime p such that $p \mid N$ and $p \mid Q$, since $p \nmid D = P^2 - 4Q$, then $p \nmid P$. By (IV.18) $p \nmid U_n$ for every $n \geq 1$, which is contrary to the hypothesis. So $\gcd(N, Q) = 1$.   □

One more result which will be needed is the following:

(V.5)   Let N be odd and q be any prime factor of $N + 1$. Assume that $U = U(P, Q)$ and $V = V(P, Q)$ are the Lucas sequences associated with the integers P, Q, having discriminant $D \neq 0$. Assume $\gcd(P, Q) = 1$ or $\gcd(N, Q) = 1$. If N divides $U_{(N+1)/q}$ and $V_{(N+1)/2}$, then N divides $V_{(N+1)/2q}$.

Proof.

$$\frac{N+1}{2} = \frac{N+1}{2q} + \frac{N+1}{q}u \quad \text{with} \quad u = \frac{q-1}{2}.$$

By (IV.4):

$$2V_{(N+1)/2} = V_{(N+1)/2q}V_{[(N+1)/q]u} + DU_{(N+1)/2q}U_{[(N+1)/q]u}.$$

By (IV.15), N divides $U_{[(N+1)/q]u}$ so N divides $V_{(N+1)/2q}V_{[(N+1)/q]u}$.
  If $\gcd(P, Q) = 1$, by (IV.21) $\gcd(U_{[(N+1)/q]u}, V_{[(N+1)/q]u}) = 1$ or 2, hence $\gcd(N, V_{[(N+1)/q]u}) = 1$, so N divides $V_{(N+1)/2q}$.
  If $\gcd(N, Q) = 1$ and if there exists a prime p dividing N and $V_{[(N+1)/q]u}$, then by (IV.6) p also divides $4Q$; since p is odd, then $p \mid Q$, which is a contradiction.   □

Before indicating primality tests, it is easy to give sufficient conditions for a number to be composite:
  Let $N > 1$ be an odd integer. Assume that there exists a Lucas sequence $(U_n)_{n \geq 0}$ with parameters (P, Q), discriminant D, such that $\gcd(N, QD) = 1$, $(Q \mid N) = 1$, and $N \nmid U_{\frac{1}{2}[N-(D/N)]}$. Then N is composite.
  Similarly, assume that there exists a companion Lucas sequence $(V_n)_{n \geq 0}$, with parameters (P, Q), discriminant D, such that $N \nmid QD$, $(Q \mid N) = -1$ and $N \nmid V_{\frac{1}{2}[N-(D/N)]}$. Then N is composite.

Proof. Indeed, if $N = p$ is an odd prime not dividing QD, and if $(Q \mid p) = 1$, then $p \mid U_{\psi(p)/2}$, and similarly, if $(Q \mid p) = -1$, then $p \mid V_{\psi(p)/2}$, as stated in (IV.23). In both cases there is a contradiction. $\square$

Now I am ready to present several tests; each one better than the preceding one.

Test 1. Let $N > 1$ be an odd integer and $N + 1 = \prod_{i=1}^{s} q_i^{f_i}$. Assume that there exists an integer D such that $(D \mid N) = -1$, and for every prime factor q_i of $N+1$, there exists a Lucas sequence $(U_n^{(i)})_{n \geq 0}$ with discriminant $D = P_i^2 - 4Q_i$, where $\gcd(P_i, Q_i) = 1$, or $\gcd(N, Q_i) = 1$ and such that $N \mid U_{N+1}^{(i)}$ and $N \nmid U_{(N+1)/q_i}^{(i)}$. Then N is a prime.

Defect of this test: it requires the knowledge of all the prime factors of $N + 1$ and the calculation of $U_n^{(i)}$ for $n = 1, 2, \ldots, N + 1$.

Proof. By (V.3), (V.4), $N \mid U_{\psi_D(N)}^{(i)}$ for every $i = 1, \ldots, s$. Let $\rho^{(i)}(N)$ be the smallest integer r such that $N \mid U_r^{(i)}$. By (IV.29) or (IV.22) and the hypothesis, $\rho^{(i)}(N) \mid (N + 1)$, $\rho^{(i)}(N) \nmid (N + 1)/q_i$, and also $\rho^{(i)}(N) \mid \psi_D(N)$. Hence $q_i^{f_i} \mid \rho^{(i)}(N)$ for every $i = 1, \ldots, s$. Therefore, $(N + 1) \mid \psi_D(N)$ and by (V.2), N is a prime. $\square$

The following test needs only half of the computations:

Test 2. Let $N > 1$ be an odd integer and $N + 1 = \prod_{i=1}^{s} q_i^{f_i}$. Assume that there exists an integer D such that $(D \mid N) = -1$, and for every prime factor q_i of $N+1$, there exists a Lucas sequence $(V_n^{(i)})_{n \geq 0}$ with discriminant $D = P_i^2 - 4Q_i$, where $\gcd(P_i, Q_i) = 1$ or $\gcd(N, Q_i) = 1$, and such that $N \mid V_{(N+1)/2}^{(i)}$ and $N \nmid V_{(N+1)/2q_i}^{(i)}$. Then N is a prime.

Proof. By (IV.2), $N \mid U_{N+1}^{(i)}$. By (V.5), $N \nmid U_{(N+1)/q_i}^{(i)}$. By the test 1, N is a prime. $\square$

The following tests will require only a partial factorization of $N+1$.

Test 3. Let $N > 1$ be an odd integer, let q be a prime factor of $N+1$ such that $2q > \sqrt{N} + 1$. Assume that there exists a Lucas sequence $(V_n)_{n \geq 0}$, with discriminant $D = P^2 - 4Q$, where $\gcd(P, Q) = 1$ or

$\gcd(N, Q) = 1$, and such that $(D \mid N) = -1$, and $N \mid V_{(N+1)/2}$, $N \nmid V_{(N+1)/2q}$. Then N is a prime.

Defect of this test: it needs the knowledge of a fairly large prime factor of $N + 1$.

Proof. Let $N = \prod_{i=1}^{s} p_i^{e_i}$. By (IV.2), $N \mid U_{N+1}$, so by (IV.29) or (IV.22), $\rho(N) \mid (N + 1)$. By (V.5), $N \nmid U_{(N+1)/q}$; hence, $\rho(N) \nmid (N + 1)/q$, therefore $q \mid \rho(N)$. By (V.4) and (V.3), $N \mid U_{\psi_D(N)}$, so $\rho(N)$ divides $\psi_D(N)$, which in turn divides $N \prod_{i=1}^{s} (p_i - (D \mid p_i))$.

Since $q \nmid N$, then there exists p_i such that q divides $p_i - (D \mid p_i)$, thus $p_i \equiv (D \mid p_i) \pmod{2q}$. In conclusion, $p_i \geq 2q - 1 > \sqrt{N}$ and $1 \leq N/p_i < \sqrt{N} < 2q - 1$, and this implies that $N/p_i = 1$, that is, N is a prime. $\qquad\square$

The next test, which was proposed by Morrison in 1975, may be viewed as the analogue of Pocklington's test indicated in Section III:

Test 4. Let $N > 1$ be an odd integer and $N + 1 = FR$, where $\gcd(F, R) = 1$ and the factorization of F is known. Assume that there exists D such that $(D \mid N) = -1$ and, for every prime q_i dividing F, there exists a Lucas sequence $(U_n^{(i)})_{n \geq 0}$ with discriminant $D = P_i^2 - 4Q_i$, where $\gcd(P_i, Q_i) = 1$ or $\gcd(N, Q_i) = 1$ and such that $N \mid U_{N+1}^{(i)}$ and $\gcd(U_{(N+1)/q_i}^{(i)}, N) = 1$. Then each prime factor p of N satisfies $p \equiv (D \mid p) \pmod{F}$. If, moreover, $F > \sqrt{N} + 1$, then N is a prime.

Proof. From the hypothesis, $\rho^{(i)}(N) \mid (N + 1)$; a fortiori, $\rho^{(i)}(p) \mid (N + 1)$. But $p \nmid U_{(N+1)/q}^{(i)}$, so $\rho^{(i)}(p) \mid (N + 1)/q_i$, by (IV.29) or (IV.22). If $q_i^{f_i}$ is the exact power of q_i dividing F, then $q_i^{f_i} \mid \rho^{(i)}(p)$, so by (IV.18), $q_i^{f_i}$ divides $p - (D \mid p)$, and this implies that F divides $p - (D \mid p)$.

Finally, if $F > \sqrt{N} + 1$, then $p + 1 \geq p - (D \mid p) \geq F > \sqrt{N} + 1$; hence, $p > \sqrt{N}$. This implies that N itself is a prime. $\qquad\square$

The next result tells more about the possible prime factors of N.

(V.6)   Let N be an odd integer, $N + 1 = FR$, where $\gcd(F, R) = 1$ and the factorization of F is known. Assume that there exists a

Lucas sequence $(U_n)_{n\geq 0}$ with discriminant $D = P^2 - 4Q$, where $\gcd(P, Q) = 1$ or $\gcd(N, Q) = 1$ and such that $(D \mid N) = -1$, $N \mid U_{N+1}$, and $\gcd(U_F, N) = 1$. If p is a prime factor of N, then there exists a prime factor q of R such that $p \equiv (D \mid p) \pmod{q}$.

Proof. $\rho(p) \mid (p - (D \mid p))$ by (IV.18) and $\rho(p) \mid (N + 1)$. But $p \nmid U_F$, so $\rho(p) \nmid F$. Hence, $\gcd(\rho(p), R) \neq 1$ and there exists a prime q such that $q \mid R$ and $q \mid \rho(p)$; in particular, $p \equiv (D \mid p) \pmod{q}$.   □

This result is used in the following test:

Test 5. Let $N > 1$ be an odd integer and $N + 1 = FR$, where $\gcd(F, R) = 1$, the factorization of F is known, R has no prime factor less than B, where $BF > \sqrt{N} + 1$. Assume that there exists D such that $(D \mid N) = -1$ and the following conditions are satisfied:

(i) For every prime q_i dividing F, there exists a Lucas sequence $(U_n^{(i)})_{n\geq 0}$, with discriminant $D = P_i^2 - 4Q_i$, where $\gcd(P_i, Q_i) = 1$ or $\gcd(N, Q_i) = 1$ and such that $N \mid U_{N+1}^{(i)}$ and $\gcd(U_{(N+1)/q_i}^{(i)}, N) = 1$.

(ii) There exists a Lucas sequence $(U_n')_{n\geq 0}$, with discriminant $D = P'^2 - 4Q'$, where $\gcd(P', Q') = 1$ or $\gcd(N, Q') = 1$ and such that $N \mid U_{N+1}'$ and $\gcd(U_F', N) = 1$.

Then N is a prime.

Proof. Let p be a prime factor of N. By Test 4, $p \equiv (D \mid p) \pmod{F}$ and by the preceding result, there exists a prime factor q of R such that $p \equiv (D \mid p) \pmod{q}$. Hence, $p \equiv (D \mid p) \pmod{qF}$ and so,

$$p + 1 \geq p - (D \mid p) \geq qF \geq BF > \sqrt{N} + 1.$$

Therefore, $p > \sqrt{N}$ and N is a prime number.   □

The preceding test is more flexible than the others, since it requires only a partial factorization of $N + 1$ up to a point where it may be assured that the nonfactored part of $N + 1$ has no factors less than B.

Now I want to indicate, in a very succinct way, how to quickly calculate the terms of Lucas sequences with large indices. One of the methods is similar to that used in the calculations of high powers, which was indicated in Section III.

Write $n = n_0 2^k + n_1 2^{k-1} + \cdots + n_k$, with $n_i = 0$ or 1 and $n_0 = 1$; so $k = [(\log n)/(\log 2)]$. To calculate U_n (or V_n) it is necessary to perform the simultaneous calculation of U_m, V_m for various values of m. The following formulas are needed:

$$\begin{cases} U_{2j} = U_j V_j, \\ V_{2j} = V_j^2 - 2Q^j, \end{cases} \qquad \text{[see formulas (IV.2)]}$$

$$\begin{cases} 2U_{2j+1} = V_{2j} + PU_{2j}, \\ 2V_{2j+1} = PV_{2j} + DU_{2j}. \end{cases} \qquad \text{[see formulas (IV.5)]}$$

Put $s_0 = n_0 = 1$, and $s_{j+1} = 2s_j + n_{j+1}$. Then $s_k = n$. So, it suffices to calculate U_{s_j}, V_{s_j} for $j \le k$; note that

$$U_{s_{j+1}} = U_{2s_j + n_{j+1}} = \begin{cases} U_{2s_j} & \text{or} \\ U_{2s_j+1}, & \end{cases}$$

$$V_{s_{j+1}} = V_{2s_j + n_{j+1}} = \begin{cases} V_{2s_j} & \text{or} \\ V_{2s_j+1}. & \end{cases}$$

Thus, it is sufficient to compute $2k$ numbers U_i and $2k$ numbers V_i, that is, only $4k$ numbers.

If it is needed to know U_n modulo N, then in all steps the numbers may be replaced by their least positive residues modulo N.

The second method is also very quick. For $j \ge 1$,

$$\begin{pmatrix} U_{j+1} & V_{j+1} \\ U_j & V_j \end{pmatrix} = \begin{pmatrix} P & -Q \\ 1 & 0 \end{pmatrix} \begin{pmatrix} U_j & V_j \\ U_{j-1} & V_{j-1} \end{pmatrix}.$$

If

$$M = \begin{pmatrix} P & -Q \\ 1 & 0 \end{pmatrix},$$

then

$$\begin{pmatrix} U_n & V_n \\ U_{n-1} & V_{n-1} \end{pmatrix} = M^{n-1} \begin{pmatrix} U_1 & V_1 \\ 0 & 2 \end{pmatrix}.$$

To find the powers of M, say M^m, write m in binary form and proceed in the manner followed to calculate a power of a number.

If U_n modulo N is to be determined, all the numbers appearing in the above calculation should be replaced by their least positive residues modulo N.

To conclude this section, I would like to stress that there are many other primality tests of the same family, which are appropriate for numbers of certain forms, and use either Lucas sequences or other similar sequences.

Sometimes it is practical to combine tests involving Lucas sequences with the tests discussed in Section III; see the paper of Brillhart, Lehmer & Selfridge (1975). As a comment, I add (half-jokingly) the following rule of thumb: the longer the statement of the testing procedure, the quicker it leads to a decision about the primality.

The tests indicated so far are applicable to numbers of the form $2^n - 1$ (see Section VII on Mersenne numbers, where the test will be given explicitly), but also to numbers of the form $k \times 2^n - 1$ (see, for example, Inkeri's paper of 1960 or Riesel's book, 1985).

In 1998, H.C. Williams published a book dedicated to a historical and mathematical study of the work of Lucas. His authoritative and thorough treatment is recommended to anyone who wants to learn more than I could include in my succinct presentation.

VI   Fermat Numbers

For numbers having a special form, there are more suitable methods to test whether they are prime or composite.

The numbers of the form $2^m + 1$ were considered long ago.

If $2^m + 1$ is a prime, then m must be of the form $m = 2^n$, so it is a Fermat number, $F_n = 2^{2^n} + 1$.

The Fermat numbers $F_0 = 3$, $F_1 = 5$, $F_2 = 17$, $F_3 = 257$, $F_4 = 65537$ are primes. Fermat believed, and tried to prove, that all Fermat numbers are primes. Since F_5 has 10 digits, in order to test its primality, it would be necessary to have a table of primes up to $100\,000$ (which was unavailable to him) or to derive and use some criterion for a number to be a factor of a Fermat number. This, Fermat failed to do.

Euler showed that every factor of F_n (with $n \geq 2$) must be of the form $k \times 2^{n+2} + 1$ and thus he discovered that 641 divides F_5:

$$F_5 = 641 \times 6700417.$$

Proof. It suffices to show that every prime factor p of F_n is of the form indicated. Since $2^{2^n} \equiv -1 \pmod{p}$, then $2^{2^{n+1}} \equiv 1 \pmod{p}$, so 2^{n+1} is the order of 2 modulo p. By Fermat's little theorem 2^{n+1} divides $p - 1$; in particular, 8 divides $p - 1$. Therefore the Legendre symbol is $2^{(p-1)/2} \equiv (2 \mid p) \equiv 1 \pmod{p}$, and so 2^{n+1} divides $(p - 1)/2$; this shows that $p = k \times 2^{n+2} + 1$. $\qquad\square$

Since the numbers F_n increase very rapidly with n, it becomes laborious to check their primality.

Using the converse of Fermat's little theorem, as given by Lucas, Pepin obtained in 1877 a test for the primality of Fermat numbers. Namely:

Pepin's Test. Let $F_n = 2^{2^n} + 1$ (with $n \geq 2$) and $k \geq 2$. Then, the following conditions are equivalent:

(i) F_n is prime and $(k \mid F_n) = -1$.

(ii) $k^{(F_n-1)/2} \equiv -1 \pmod{F_n}$.

Proof. If (i) is assumed, then by Euler's criterion for the Legendre symbol

$$k^{(F_n-1)/2} \equiv \left(\frac{k}{F_n}\right) \equiv -1 \pmod{F_n}.$$

If, conversely, (ii) is supposed true, let a, $1 \leq a < F_n$, be such that $a \equiv k \pmod{F_n}$. Since $a^{(F_n-1)/2} \equiv -1 \pmod{F_n}$, then $a^{F_n-1} \equiv 1 \pmod{F_n}$. By Test 3 in Section III, F_n is prime. Hence

$$\left(\frac{k}{F_n}\right) \equiv k^{(F_n-1)/2} \equiv -1 \pmod{F_n}. \qquad\square$$

Possible choices of k are $k = 3, 5, 10$, because $F_n \equiv 2 \pmod{3}$, $F_n \equiv 2 \pmod{5}$, $F_n \equiv 1 \pmod 8$; hence, by Jacobi's reciprocity law

$$\left(\frac{3}{F_n}\right) = \left(\frac{F_n}{3}\right) = \left(\frac{2}{3}\right) = -1,$$

$$\left(\frac{5}{F_n}\right) = \left(\frac{F_n}{5}\right) = \left(\frac{2}{5}\right) = -1,$$

$$\left(\frac{10}{F_n}\right) = \left(\frac{2}{F_n}\right)\left(\frac{5}{F_n}\right) = -1.$$

This test is very practical in application. However, if F_n is composite, the test does not indicate any factor of F_n.

Lucas used it to show that F_6 is composite, and in 1880, at the age of 82, Landry showed that

$$F_6 = 274177 \times 67280421310721.$$

Landry never described how he factored F_6. In a historical reconstitution, Williams (1993) gives indications, obtained from clues in Landry's letters and work, of the method used by Landry.

But the best of the story is a recent "coup de théâtre". In a biography of Clausen by Biermann (1964), it is stated that in a letter to Gauss of January 1, 1855, Clausen (who was known as an able calculator and an important astronomer) already gave the complete factorization of F_6. In this letter, which remains in the library of the University of Göttingen, Clausen also expressed his belief that the larger of the two factors was the largest prime number known at that time. Curiously, the corresponding remark in Biermann's biography remained widely unnoticed for many years.

Generally, the factorization of Fermat numbers known to be composite has been the object of intensive research. In the following table we give the current state of this matter. The notation Pn indicates a prime number of n digits, while Cn denotes a composite number having n digits.

Table 6.   Completely factored Fermat numbers

$F_5 = 641 \times 6700417$

$F_6 = 274177 \times 67280421310721$

$F_7 = 59649589127497217 \times 5704689200685129054721$

$F_8 = 1238926361552897 \times P62$

$F_9 = 2424833 \times$
          $7455602825647884208337395736200454918783366342657 \times P99$

$F_{10} = 45592577 \times 6487031809 \times$
          $4659775785220018543264560743076778192897 \times P252$

$F_{11} = 319489 \times 974849 \times 167988556341760475137 \times$
                    $3560841906445833920513 \times P564$

Notes.

F_5 : Euler (1732)

F_6 : factor 1 Clausen (unpublished, 1855), Landry and Le Lasseur (1880)

F_7 : Morrison and Brillhart (1970)

F_8 : factor 1 Brent and Pollard (1980)

F_9 : factor 1 Western (1903),
　　other factors A.K. Lenstra and Manasse (1990)

F_{10} : factor 1 Selfridge (1953), factor 2 Brillhart (1962),
　　other factors Brent (1995)

F_{11} : factors 1 and 2 Cunningham (1899), other factors Brent (1988),
　　primality of factor 5 Morain (1988)

It is quite difficult to keep track of all the new results that accumulate rapidly, but also to remain acquainted with the most recent methods developed for the factorization of such numbers. In this regard, the articles of Brent (1999), and of Brent, Crandall, Dilcher & van Halewyn (2000) are very informative. I thank W. Keller for keeping me up-to-date on developments concerning the Fermat numbers.

Table 7.   Incomplete factorizations of Fermat numbers

$$F_{12} = 114689 \times 26017793 \times 63766529 \times 190274191361 \times$$
$$1256132134125569 \times C1187$$
$$F_{13} = 2710954639361 \times 2663848877152141313 \times$$
$$3603109844542291969 \times 319546020820551643220672513 \times C2391$$
$$F_{15} = 1214251009 \times 2327042503868417 \times$$
$$168768817029516972383024127016961 \times C9808$$
$$F_{16} = 825753601 \times 188981757975021318420037633 \times C19694$$
$$F_{17} = 31065037602817 \times C39444$$
$$F_{18} = 13631489 \times 81274690703860512587777 \times C78884$$
$$F_{19} = 70525124609 \times 646730219521 \times C157804$$
$$F_{21} = 4485296422913 \times C631294$$
$$F_{23} = 167772161 \times C2525215$$

Table 8.   Composite Fermat numbers without known factor

F_{14} :	Selfridge and Hurwitz (1963)
F_{20} :	Buell and Young (1987)
F_{22} :	Crandall, Doenias, Norrie and Young (1993), independently by Carvalho and Trevisan (1993)
F_{24} :	Mayer, Papadopoulos and Crandall (1999)

The smallest Fermat numbers of unknown character are: F_{33}, F_{34}, F_{35}, F_{40}, F_{41}, F_{44}, ...

RECORDS

A. The largest known Fermat prime is $F_4 = 65537$.

B. The largest known composite Fermat number is $F_{2145351}$, which has the factor $3 \times 2^{2145353} + 1$. This 645817-digit factor was discovered by J.B. Cosgrave and his Proth-Gallot Group at St. Patrick's College (Dublin, Ireland) on February 21, 2003. Programs of P. Jobling, G. Woltman and Y. Gallot were essential for the discovery.

C. As of the end of May 2003, there was a total of 214 Fermat numbers known to be composite.

Here are some open problems:

(1) Are there infinitely many prime Fermat numbers?

This question became significant with the famous result of Gauss (see *Disquisitiones Arithmeticae*, articles 365, 366—the last ones in the book—as a crowning result for much of the theory previously developed). He showed that if $n \geq 3$ is an integer, and if the regular polygon with n sides may be constructed by ruler and compass, then $n = 2^k p_1 p_2 \cdots p_h$, where $k \geq 0$, $h \geq 0$ and $p_1, \ldots, p_h$ are distinct odd primes, each being a Fermat number.

In 1844, Eisenstein proposed, as a problem, to prove that there are indeed infinitely many prime Fermat numbers. I should add, that already in 1828, an anonymous writer stated that

$$2 + 1, \ 2^2 + 1, \ 2^{2^2} + 1, \ 2^{2^{2^2}} + 1, \ 2^{2^{2^{2^2}}} + 1, \ \ldots$$

are all primes, and added that they are the only prime Fermat numbers (apart from $2^{2^3} + 1$). However, Selfridge discovered in 1953 a factor of F_{16}, which therefore is not a prime, and this fact disproved that conjecture.

(2) Are there infinitely many composite Fermat numbers?

Questions (1) and (2) seem beyond the reach of present-day methods and, side by side, they show how little is known on this matter.

(3) Is every Fermat number square-free (i.e., without square factors)?

It has been conjectured, for example by Lehmer and by Schinzel, that there exist infinitely many square-free Fermat numbers.

It is not difficult to show that if p is a prime number and p^2 divides some Fermat number, then $2^{p-1} \equiv 1 \pmod{p^2}$—this will be proved in detail in Chapter 5, Section III. Since Fermat numbers are pairwise relatively prime, if there exist infinitely many Fermat numbers with a square factor, then there exist infinitely many primes p satisfying the above congruence.

I shall discuss this congruence in Chapter 5. Let it be said here that it is very rarely satisfied. In particular, it is not known whether it holds infinitely often.

Sierpiński considered in 1958 the numbers of the form $S_n = n^n + 1$, with $n \geq 2$. He proved that if S_n is a prime, then there exists $m \geq 0$ such that $n = 2^{2^m}$, so S_n is a Fermat number:

$$S_n = F_{m+2^m}.$$

It follows that the only numbers S_n which are primes and have less than 3×10^{20} digits, are 5 and 257. Indeed, if $m = 0, 1$ one has $F_1 = 5$, $F_3 = 257$; if $m = 2, 3, 4$ or 5, we have F_6, F_{11}, F_{20} and F_{37}, which are composite numbers. For $m = 5$, one obtains F_{70}, which is not known to be prime or composite. Since $2^{10} > 10^3$, then

$$F_{70} > 2^{2^{70}} > 2^{10^{21}} = (2^{10})^{10^{20}} > 10^{3 \times 10^{20}}.$$

The primes of the form $n^n + 1$ are very rare. Are there only finitely many such primes? If so, there are infinitely many composite Fermat numbers. But all this is pure speculation, with no basis for any reasonable conjecture.

The recent book by 3 authors (Křížek, Luca & Somer), entitled *17 Lectures on Fermat's Last* (oops) *Numbers*, contains 257 pages of very interesting facts around the Fermat numbers. With the rapid progress in the study of these numbers, I ask to my readers: How many pages will have the next book on Fermat numbers?

VII   Mersenne Numbers

If a number of the form $2^m - 1$ is a prime, then $m = q$ is a prime. Even more, it is not a difficult exercise to show that if $2^m - 1$ is a prime power, it must be a prime, and so m is a prime. [If you cannot do it alone, look at the paper of Ligh & Neal (1974).]

The numbers $M_q = 2^q - 1$ (with q prime) are called Mersenne numbers, and their consideration was motivated by the study of perfect numbers (see the addendum to this section).

Already at Mersenne's time, it was known that some Mersenne numbers were prime, others composite. For example, $M_2 = 3$, $M_3 = 7$, $M_5 = 31$, $M_7 = 127$ are primes, while $M_{11} = 23 \times 89$. In 1640, Mersenne stated that M_q is also a prime for $q = 13$, 17, 19, 31, 67, 127, 257; he was wrong about 67 and 257, and he did not include 61, 89, 107 (among those less than 257), which also produce Mersenne primes. Yet, his statement was quite astonishing, in view of the size of the numbers involved.

The obvious problem is to recognize if a Mersenne number is a prime, and if not, to determine its factors.

A classical result about factors was stated by Euler in 1750 and proved by Lagrange (1775) and again by Lucas (1878):

If q is a prime $q \equiv 3 \pmod 4$, then $2q + 1$ divides M_q if and only if $2q + 1$ is a prime; in this case, if $q > 3$, then M_q is composite.

Proof. Let $n = 2q + 1$ be a factor of M_q. Since $2^2 \not\equiv 1 \pmod n$, $2^q \not\equiv 1 \pmod n$, $2^{2q} - 1 = (2^q + 1)M_q \equiv 0 \pmod n$, then by Lucas test 3 (see Section III), n is a prime.

Conversely, let $p = 2q + 1$ be a prime. Since $p \equiv 7 \pmod 8$, then $(2 \mid p) = 1$, so there exists m such that $2 \equiv m^2 \pmod p$. It follows that $2^q \equiv 2^{(p-1)/2} \equiv m^{p-1} \equiv 1 \pmod p$, so p divides M_q.

If, moreover, $q > 3$, then $M_q = 2^q - 1 > 2q + 1 = p$, so M_q is composite. $\square$

Thus if $q = 11$, 23, 83, 131, 179, 191, 239, 251, then M_q has the factor 23, 47, 167, 263, 359, 383, 479, 503, respectively.

Around 1825, Sophie Germain considered, in connection with Fermat's last theorem, the primes q such that $2q + 1$ is also a prime. These primes are now called *Sophie Germain primes*, and I shall return to them in Chapter 5.

It is also very easy to determine the form of the factors of Mersenne numbers:

If n divides M_q ($q > 2$), then $n \equiv \pm 1 \pmod 8$ and $n \equiv 1 \pmod q$.

Proof. It suffices to show that each prime factor p of M_q is of the form indicated.

If p divides $M_q = 2^q - 1$, then $2^q \equiv 1 \pmod{q}$; so by Fermat's little theorem, q divides $p - 1$, that is, $p - 1 = 2kq$ (since $p \neq 2$). So

$$\left(\frac{2}{p}\right) \equiv 2^{(p-1)/2} \equiv 2^{qk} \equiv 1 \pmod{p},$$

therefore $p \equiv \pm 1 \pmod{8}$, by the property of the Legendre symbol already indicated in Section II. $\qquad\square$

The primality of M_{13} and M_{17} was determined by Cataldi using trial division. Euler also used trial division to show that M_{31} is a prime, but he could spare many calculations, in view of the above mentioned form of factors of Mersenne numbers. In this respect, see Williams & Shallit (1994).

The best method presently known to find out whether M_q is a prime or a composite number is based on the computation of a recurring sequence, indicated by Lucas (1878), and Lehmer (1930, 1935); see also Western (1932), Hardy & Wright (1938, p. 223), and Kaplansky (1945). However, explicit factors cannot be found in this manner.

If n is odd, $n \geq 3$, then $M_n = 2^n - 1 \equiv 7 \pmod{12}$. Also, if $N \equiv 7 \pmod{12}$, then the Jacobi symbol

$$\left(\frac{3}{N}\right) = \left(\frac{N}{3}\right)(-1)^{(N-1)/2} = -1.$$

Primality test for Mersenne numbers. Let $P = 2$, $Q = -2$, and consider the associated Lucas sequences $(U_m)_{m \geq 0}$, $(V_m)_{m \geq 0}$, which have discriminant $D = 12$. Then $N = M_n$ is a prime if and only if N divides $V_{(N+1)/2}$.

Proof. Let N be a prime. By (IV.2)

$$V_{(N+1)/2}^2 = V_{N+1} + 2Q^{(N+1)/2} = V_{N+1} - 4(-2)^{(N-1)/2}$$

$$\equiv V_{N+1} - 4\left(\frac{-2}{N}\right) \equiv V_{N+1} + 4 \pmod{N},$$

because

$$\left(\frac{-2}{N}\right) = \left(\frac{-1}{N}\right)\left(\frac{2}{N}\right) = -1,$$

since $N \equiv 3 \pmod 4$ and $N \equiv 7 \pmod 8$. Thus it suffices to show that $V_{N+1} \equiv -4 \pmod N$.

By (IV.4), $2V_{N+1} = V_N V_1 + DU_N U_1 = 2V_N + 12U_N$; hence, by (IV.14) and (IV.13):

$$V_{N+1} = V_N + 6U_N \equiv 2 + 6(12 \mid N) \equiv 2 - 6 \equiv -4 \pmod N.$$

Conversely, assume that N divides $V_{(N+1)/2}$. Then N divides U_{N+1} [by (IV.2)]. Also, by (IV.6) $V_{(N+1)/2}^2 - 12U_{(N+1)/2}^2 = 4(-1)^{(N+1)/2}$; hence $\gcd(N, U_{(N+1)/2}) = 1$. Since $\gcd(N, 2) = 1$, then by the Test 1 (Section V), N is a prime.   □

For the purpose of calculation, it is convenient to replace the Lucas sequence $(V_m)_{m \geq 0}$ by the following sequence $(S_k)_{k \geq 0}$, defined recursively as follows:

$$S_0 = 4, \qquad S_{k+1} = S_k^2 - 2;$$

thus, the sequence begins with 4, 14, 194, ... Then the test is phrased as follows:

$M_n = 2^n - 1$ *is prime if and only if* M_n *divides* S_{n-2}.

Proof. $S_0 = 4 = V_2/2$. Assume that $S_{k-1} = V_{2^k}/2^{2^{k-1}}$; then

$$S_k = S_{k-1}^2 - 2 = \frac{V_{2^k}^2}{2^{2^k}} - 2 = \frac{V_{2^{k+1}} + 2^{2^{k+1}}}{2^{2^k}} - 2 = \frac{V_{2^{k+1}}}{2^{2^k}}.$$

By the test, M_n is prime if and only if M_n divides

$$V_{(M_n+1)/2} = V_{2^{n-1}} = 2^{2^{n-2}} S_{n-2},$$

or equivalently, M_n divides S_{n-2}.   □

The repetitive nature of the computations makes this test quite suitable. In this way, all examples of large Mersenne primes have been discovered. Lucas himself showed, in 1876, that M_{127} is a prime, while M_{67} is composite. Not much later, Pervushin showed that M_{61} is also a prime. Finally, in 1927 (published in 1932) Lehmer showed that M_{257} is also composite, settling one way or another, what Mersenne had asserted. Note that M_{127} has 39 digits and was the largest prime known before the age of computers. In this competition this was the longest lasting record!

The Mersenne primes with $q \leq 127$ were discovered before the computer age. A. Turing made, in 1951, the first attempt to find Mersenne primes using an electronic computer; however, he was unsuccessful. In 1952, Robinson carried out Lucas' test using a computer SWAC (from the National Bureau of Standards in Los Angeles), with the assistance of D.H. and E. Lehmer. He discovered the Mersenne primes M_{521}, M_{607} on January 30, 1952—the first such discoveries with a computer. The primes M_{1279}, M_{2203}, M_{2281} were found later in the same year.

The Lucas-Lehmer primality test for Mersenne numbers M_q, when q is large, requires much calculation. To face this situation, the work has to be done by teams, using very powerful computers. Moreover, one uses programs especially created for the purpose. A great role is played by multiplication done via the fast Fourier transform, invented by Schönhage & Strassen in 1971. The programs of Crandall and Woltman have been determinant in the discovery of large primes.

The GIMPS ("Great Internet Mersenne Prime Search"), organized by Woltman, has as its aim to discover large Mersenne primes. Anyone, so willing, may participate with his personal computer. He will receive the software and an interval of prime exponents as his territory for search. Presently the project has recruited several thousands participants.

In a not so distant past the gold and diamond prospectors sacrificed family and friends going to inhospitable places, jungles with snakes, disease infested marshes, or high mountains with cliffs and snow, all this in search of the precious discovery which would make them rich. The modern searcher of Mersenne primes lives a transposed but similar adventure. The location of his findings cannot be anticipated; lucky the one who first finds IT. No riches, but fame. My metaphor is not so different from reality. I suggest you learn the ways to the 38th Mersenne prime in Woltman's own description (1999)—the captain explorer tells ...

RECORD

The first 38 Mersenne primes are shown in Table 9. The largest known Mersenne prime, with $q = 13466917$, has 4053946 digits. Its discovery, which occurred on November 14, 2001, is credited to

M. Cameron, G.F. Woltman, S. Kurowski, and to GIMPS. The fact is that Cameron found that prime working on a segment assigned to him by GIMPS.

Note that this Mersenne prime is currently the largest known prime, and only the second *megaprime* known, i.e., a prime with one million digits at least.

It should be remarked that the prime M_{110503} was found only after M_{132049} and M_{216091} were known. So it may happen that the next Mersenne prime to be found has $q < 13466917$, since not all of the primes q below this limit have been tested to see if M_q is a prime.

On the other hand, the search for Sophie Germain primes q of the form $q = k \times 2^N - 1$ (so, $2q + 1$ is also a prime) yields, as already indicated, composite Mersenne numbers M_q.

RECORD

The largest Mersenne number M_q known to be composite has $q = 2540041185 \times 2^{114729} - 1$ and was found by D. Underbakke, G.F. Woltman and Y. Gallot in January 2003. The prime q is the largest known Sophie Germain prime (see Chapter 5, Section II).

Riesel's book (1985) has a table of complete factorization of all numbers $M_n = 2^n - 1$, with n odd, $n \leq 257$. A more extensive table is in the book of Brillhart et al. (1983, 1988; see also the third edition, 2002).

Just as for Fermat numbers, there are many open problems about Mersenne numbers:

(1) Are there infinitely many Mersenne primes?

(2) Are there infinitely many composite Mersenne numbers?

The answer to both questions ought to be "yes", as I will try to justify. For example, I will indicate in Chapter 6, Section A, after (D5), that some sequences, similar to the sequence of Mersenne numbers, contain infinitely many composite numbers.

(3) Is every Mersenne number square-free?

Table 9.   Mersenne primes M_q with $q < 7000000$

q	Year	Discoverer
2	–	–
3	–	–
5	–	–
7	–	–
13	1461	Anonymous*
17	1588	P.A. Cataldi
19	1588	P.A. Cataldi
31	1750	L. Euler
61	1883	I.M. Pervushin
89	1911	R.E. Powers
107	1913	E. Fauquembergue
127	1876	E. Lucas
521	1952	R.M. Robinson
607	1952	R.M. Robinson
1279	1952	R.M. Robinson
2203	1952	R.M. Robinson
2281	1952	R.M. Robinson
3217	1957	H. Riesel
4253	1961	A. Hurwitz
4423	1961	A. Hurwitz
9689	1963	D.B. Gillies
9941	1963	D.B. Gillies
11213	1963	D.B. Gillies
19937	1971	B. Tuckerman
21701	1978	L.C. Noll and L. Nickel
23209	1979	L.C. Noll
44497	1979	H. Nelson and D. Slowinski
86243	1982	D. Slowinski
110503	1988	W.N. Colquitt and L. Welsh, Jr.
132049	1983	D. Slowinski
216091	1985	D. Slowinski
756839	1992	D. Slowinski and P. Gage
859433	1993	D. Slowinski and P. Gage
1257787	1996	D. Slowinski and P. Gage
1398269	1996	J. Armengaud, G.F. Woltman and GIMPS
2976221	1997	G. Spence, G.F. Woltman and GIMPS
3021377	1998	R. Clarkson, G.F. Woltman, S. Kurowski and GIMPS
6972593	1999	N. Hajratwala, G.F. Woltman, S. Kurowski and GIMPS

*See Dickson's *History of the Theory of Numbers*, Vol. I, p. 6.

Rotkiewicz showed in 1965 that if p is a prime and p^2 divides some Mersenne number, then $2^{p-1} \equiv 1 \pmod{p^2}$, the same congruence which already appeared in connection with Fermat numbers having a square factor.

I wish to mention two other problems involving Mersenne numbers, one of which has been solved, while the other one is still open.

Is it true that if M_q is a Mersenne prime, then M_{M_q} is also a prime number?

The answer is negative, since despite M_{13} being prime, $M_{M_{13}} = 2^{8191} - 1$ is composite; this was shown by Wheeler, see Robinson (1954). Note that $M_{M_{13}}$ has more than 2400 digits. In 1976, Keller discovered the prime factor

$$p = 2 \times 20644229 \times M_{13} + 1 = 338193759479$$

of the Mersenne number $M_{M_{13}}$, thus providing an easier proof that it is composite; only 13 squarings modulo p are needed to verify that $2^{2^{13}} \equiv 2 \pmod{p}$. This has been communicated to me by Keller in a letter.

The second problem, proposed by Catalan in 1876 and reported in Dickson's *History of the Theory Numbers*, Vol. I, p. 22, is the following. Consider the sequence of numbers

$$C_1 = 2^2 - 1 = 3 = M_2\,,$$
$$C_2 = 2^{C_1} - 1 = 7 = M_3\,,$$
$$C_3 = 2^{C_2} - 1 = 127 = M_7\,,$$
$$C_4 = 2^{C_3} - 1 = 2^{127} - 1 = M_{127}\,,$$

. . . .

$$C_{n+1} = 2^{C_n} - 1$$

. . . .

Are all numbers C_n primes? Are there infinitely many which are prime? At present, it is impossible to test C_5, which has more than 10^{37} digits!

I conclude with the interesting conjecture of Bateman, Selfridge & Wagstaff (1989), concerning the Mersenne primes.

Conjecture. Let p be an odd natural number (not necessarily a prime). If two of the following conditions are satisfied, so is the third one:

(a) p is equal to $2^k \pm 1$ or to $4^k \pm 3$ (for some $k \geq 1$).

(b) M_p is a prime.

(c) $(2^p + 1)/3$ is a prime.

In a private communication, H. and R. Lifchitz informed that the conjecture holds for all $p < 720000$. In this range, the only primes satisfying the three conditions are $p = 3, 5, 7, 13, 17, 19, 31, 61, 127$. It is conceivable that these are the only primes for which the above three conditions hold.

ADDENDUM ON PERFECT NUMBERS

I shall now consider perfect numbers and tell how they are related to Mersenne numbers.

A natural number $n > 1$ is said to be *perfect* if it is equal to the sum of all its aliquot parts, that is, its divisors d, with $d < n$. For example, $n = 6, 28, 496, 8128$ are the perfect numbers smaller than 10000.

Perfect numbers were already known in ancient times. The first perfect number 6 was connected, by mystic and religious writers, to perfection, thus explaining that the Creation required 6 days, so PERFECT is the world.

Euclid showed, in his *Elements*, Book IX, Proposition 36, that if q is a prime and $M_q = 2^q - 1$ is a prime, then $N = 2^{q-1}(2^q - 1)$ is a perfect number.

In a posthumous paper, Euler proved the converse: any even perfect number is of the form indicated by Euclid. Thus, the knowledge of even perfect numbers is equivalent to the knowledge of Mersenne primes.

And what about odd perfect numbers? Do they exist? Not even one has ever been found! This is a question which has been extensively searched, but its answer is still unknown.

Quick information on the progress made toward the solution of the problem may be found in Guy's book (new edition 1994), quoted in General References. More recent facts are also mentioned below.

The methods to tackle the problem have been legion. I believe it is useful to describe them so the reader will get a feeling of what to do when nothing seems reasonable. The idea is to assume that there exists an odd perfect number N and to derive various consequences, concerning the number $\omega(N)$ of its distinct prime factors, the size of N, the multiplicative form, and the additive form of N, etc. I shall review what has been proved in each count.

(a) Number of distinct prime factors $\omega(N)$

Hagis (1980, announced in 1975) proved that $\omega(N) \geq 8$. The same result was also obtained by Chein (1979) in his thesis.

In 1983, Hagis and, independently, Kishore proved that if $3 \nmid N$, then $\omega(N) \geq 11$.

Another result in this line was given by Dickson in 1913: for every $k \geq 1$ there are at most finitely many odd perfect numbers N, such that $\omega(N) = k$. In 1949, Shapiro gave a simpler proof.

Dickson's theorem was generalized in 1956 by Kanold, for numbers N satisfying the condition $\sigma(N)/N = \alpha$ (α is a given rational number and $\sigma(N)$ denotes the sum of all divisors of N). The proof involved the fact that the equation $aX^3 - bY^3 = c$ has at most finitely many solutions in integers x, y. Since an effective estimate for the number of solutions was given by Baker, with his celebrated method of linear forms in logarithms, it became possible for Pomerance to show in 1977 (taking $\alpha = 2$), for every $k \geq 1$: If the odd perfect number N has k distinct prime factors, then

$$N < (4k)^{(4k)^{2^{k^2}}}.$$

In 1994, Heath-Brown sharpened substantially the result of Pomerance: If an odd perfect number N has k distinct prime factors, then

$$N < 4^{4^k}.$$

Improving further, Cook (1999) showed that the base 4 may be replaced by $195^{1/7} = 2.123\ldots$.

(b) Lower bound for N

Brent, Cohen & te Riele (1991) have established that if N is an odd perfect number, then $N > 10^{300}$. Previously, in 1989, Brent & Cohen showed that $N > 10^{160}$, and in 1973 Hagis proved that $N > 10^{50}$.

In 1976, Buxton & Elmore claimed that $N > 10^{200}$, but this statement has not been substantiated in detail, so it should not be accepted. In 1999, Grytczuk & Wojtowicz published a far larger lower bound for N, but F. Saidak found a flaw in the proof, and this was acknowledged by the authors in 2000.

(c) Multiplicative structure of N

The first result is by Euler: $N = p^e k^2$, where p is a prime not dividing k, and $p \equiv e \equiv 1 \pmod 4$.

There have been numerous results on the kind of number k. For example, in 1972 Hagis & McDaniel showed that k is not a cube.

(d) Largest prime factor of N

In 1998, Hagis & Cohen showed that N must have a prime factor greater than 10^6. Earlier, in 1975, Hagis & McDaniel had proved that the largest prime factor of N should be greater than 100110.

For prime-power factors, Muskat showed in 1966 that N must have one which is greater than 10^{12}.

(e) Other prime factors of N

In 1975, Pomerance showed that the second largest prime factor of N should be at least 139. That limit was raised to 10^3 by Hagis (1981) and to 10^4 by Iannucci (1999). In 2000, Iannucci also showed that the third largest prime factor of N exceeds 100.

In 1952, Grün showed that the smallest prime factor p_1 of N should satisfy the relation $p_1 < \frac{2}{3}\omega(N) + 2$.

In his thesis, Kishore (1977) showed that if $i = 2, 3, 4, 5, 6$, the ith smallest prime factor of N is less than $2^{2^{i-1}}(\omega(N) - i + 1)$.

In 1958, Perisastri proved that

$$\frac{1}{2} < \sum_{p \mid N} \frac{1}{p} < 2 \log \frac{\pi}{2}.$$

This has been sharpened by Suryanarayana (1963), Suryanarayana & Hagis (1970), and Cohen (1978).

(f) Additive structure of N

In 1953, Touchard proved that $N \equiv 1 \pmod{12}$ or $N \equiv 9 \pmod{36}$. An easier proof was later given by Satyanarayana (1959).

(g) Ore's conjecture

In 1948, Ore considered the harmonic mean of the divisors of N, namely,

$$H(N) = \frac{\tau(N)}{\sum\limits_{d|N}(1/d)},$$

where $\tau(N)$ denotes the number of divisors of N.

If N is a perfect number, then $H(N)$ is an integer; indeed, whether N is even or odd, this follows from Euler's results.

Actually, Laborde noted in 1955, that N is an even perfect number if and only if

$$N = 2^{H(N)-1}\left(2^{H(N)} - 1\right),$$

hence $H(N)$ is an integer, and in fact a prime.

Ore conjectured that if N is odd, then $H(N)$ is not an integer. The truth of this conjecture would imply, therefore, that there do not exist odd perfect numbers.

Ore verified that the conjecture is true if N is a prime-power or if $N < 10^4$. Since 1954 (published only in 1972), Mills checked its truth for $N < 10^7$, as well as for numbers of special form, in particular, if all prime-power factors of N are smaller than 65551^2.

Pomerance (unpublished) verified Ore's conjecture when $\omega(N) \leq 2$, by showing that if $\omega(N) \leq 2$ and $H(N)$ is an integer, then N is an even perfect number (kindly communicated to me by letter).

The next results do not distinguish between even or odd perfect numbers. They concern the distribution of perfect numbers. The idea is to define, for every $x \geq 1$, the function $V(x)$, which counts the perfect numbers less or equal to x:

$$V(x) = \#\{N \text{ perfect} \mid N \leq x\}.$$

The limit $\lim_{x\to\infty} V(x)/x$ represents a natural density for the set of perfect numbers. In 1954, Kanold showed the $\lim_{x\to\infty} V(x)/x = 0$. Thus, $V(x)$ grows to infinity slower than x does.

The following more precise result of Wirsing (1959) tells how slowly $V(x)$ grows: there exist x_0 and $C > 0$ such that if $x \geq x_0$ then

$$V(x) \leq e^{(C\log x)/(\log\log x)}.$$

Earlier work was done by Hornfeck (1955, 1956), Kanold (1957), and Hornfeck & Wirsing (1957), who had established that for every $\varepsilon > 0$ there exists a positive constant C such that $V(x) < Cx^\varepsilon$.

All the results that I have indicated about the problem of the existence of odd perfect numbers represent a considerable amount of work, sometimes difficult and delicate. Yet I believe the problem stands like an unconquerable fortress. For all that is known, it would be almost by luck that an odd perfect number would be found. On the other hand, nothing that has been proved is promising to show that odd perfect numbers do not exist. New ideas are required.

I wish to conclude this overview of perfect numbers with the following results of Sinha (1974)—the proof is elementary and should be an amusing exercise (just get your pencil ready!): 28 is the only even perfect number that is of the form $a^n + b^n$ with $n \geq 2$, and $\gcd(a, b) = 1$. It is also the only even perfect number of the form $a^n + 1$, with $n \geq 2$. And finally, there is no even perfect number of the form

$$a^{n^{n^{\cdot^{\cdot^{\cdot^{n}}}}}} + 1$$

with $n \geq 2$ and at least two exponents n.

Looking back, perfect numbers are defined by comparing N with $\sigma(N)$, the sum of its divisors. Demanding just that N divides $\sigma(N)$ leads to the *multiply perfect numbers*. Numbers N with $2N < \sigma(N)$ are called *abundant*, while those with $2N \geq \sigma(N)$ are called *deficient*.

Let $s(N) = \sigma(N) - N$, the sum of aliquot parts of N, that is, the sum of proper divisors of N. Since some numbers are abundant and others are deficient, it is natural to iterate the process of getting $s(N)$, namely, to build the sequence $s(N)$, $s^2(N)$, $s^3(N), \dots$, where $s^k(N) = s\big(s^{k-1}(N)\big)$. This leads to many fascinating questions, as they are described in Guy's book. Because of space limitations, I am forced to abstain from discussing these matters.

VIII   Pseudoprimes

In this section I shall consider composite numbers having a property which one would think that only prime numbers possess.

A   PSEUDOPRIMES IN BASE 2 (psp)

A problem, commonly attributed to the ancient Chinese, was to ascertain whether a natural number n must be a prime if it satisfies the congruence

$$2^n \equiv 2 \pmod{n}.$$

On this subject, there are legends and speculations. One should be prudent before making preemptory statements. In view of what one believes to be the knowledge about numbers in ancient China, it seems difficult to conceive that such a question could even be formulated. Siu Man-Keung, a mathematician from Hong Kong interested in the history of mathematics, wrote to me:

> This myth originated in a paper by J.H. Jeans, in the *Messenger of Mathematics*, 27, 1897/8, who wrote that "a paper found among those of the late Sir Thomas Wade and dating from the time of Confucius" contained the theorem that $2^n \equiv 2 \pmod{n}$ holds if and only if n is a prime number. However, in a footnote to his monumental work *Science and Civilisation in China*, Vol. 3, Chap. 19 (Mathematics), J. Needham dispels Jeans' assertion, which is due to an erroneous translation of a passage of the famous book *The Nine Chapters of Mathematical Art*.

This mistake has been perpetuated by several Western scholars. In Dickson's *History of the Theory of Numbers*, Vol. I, p. 91, it is quoted that Leibniz believed to have proved that the so-called Chinese congruence indicated above implies that n is prime. The story is also repeated, for example, in Honsberger's very nicely written chapter "An Old Chinese Theorem and Pierre de Fermat" in his book *Mathematical Gems*, Vol. I, (1973).

There is now a better founded version of the events. In a more recent letter (February 1992), Siu wrote:

I have just seen the doctoral thesis, written in Chinese, of Han Qi, on the mathematics in the Qing period, entitled *Transmission of Western Mathematics during the Kangxi Kingdom and its Influence Over Chinese Mathematics*, Beijing, 1991. The author points out new evidence concerning "the old Chinese theorem". According to Han, this "theorem" is due to Li Shan-Lan (1811–1882), a well-known mathematician of the Qing period (thus the statement is not so old). Li mentioned his criterion to Alexander Wylie, who was his collaborator in the translation of Western texts. Wylie, who probably did not understand mathematics, presented Li's criterion in a note "A Chinese theorem" to the journal *Notes and Queries on China*, Hong Kong, 1869 (1873).

In the succeeding months, at least four readers have written comments on the work of Li; one of the readers pointed out that Li's statement was wrong. Among the readers there was a certain J. von Gumpach, a German who later became a colleague of Li in Beijing. Apparently, Gumpach told Li of his mistake. As a result, in a later publication on number theory (1872), Li Shan-Lan deleted any reference to his criterion. However, in 1882, Hua Heng-Fang, another well-known mathematician of the Qing period, published a treatise on numbers in which he included Li's criterion as if it were correct. This might help to explain why the Western historians of Chinese mathematics were led to think that the criterion might be an old Chinese theorem. Han Qi has announced that he will publish an article on this question, with more details.

I take this opportunity to thank Siu Man-Keung for this well-founded and interesting information.

Concerning the works of Li Shan-Lan you may wish to consult the book of Li Yan and Du Shiran, in an English translation of 1987.

After these comments of historical character, I return to the problem concerning the congruence $2^n \equiv 2 \pmod{n}$, which might be appropriately called, if not as a joke, the "pseudo-Chinese congruence on pseudoprimes".

The first counterexample to the conjecture was obtained in 1819, so much earlier than the events in China. Sarrus showed that $2^{341} \equiv 2$ (mod 341), yet $341 = 11 \times 31$ is a composite number. In particular, a crude converse of Fermat's little theorem is false.

Other composite numbers with this property are, for example: 561, 645, 1105, 1387, 1729, 1905.

A composite number n satisfying the congruence $2^{n-1} \equiv 1$ (mod n) is called a *pseudoprime*, or also a *Poulet number* since that was the focus of his attention. In particular, Poulet computed, as early as 1926, a table of pseudoprimes up to 5×10^7, and in 1938 up to 10^8; see references in Chapter 4.

Every pseudoprime n is odd and also satisfies the congruence $2^n \equiv 2$ (mod n); conversely, every odd composite number satisfying this congruence is a pseudoprime.

Clearly, every odd prime number satisfies the above congruence, so if $2^{n-1} \not\equiv 1$ (mod n), then n must be composite. This is useful as a first step in testing primality.

In order to know more about primes, it is natural to study the integers for which $2^{n-1} \equiv 1$ (mod n).

Suppose I would like to write a chapter about pseudoprimes for the *Guinness Book of Records*. How would I organize it?

The natural questions should be basically the same as those for prime numbers. For example: How many pseudoprimes are there? Can one tell whether a number is a pseudoprime? Are there ways of generating pseudoprimes? How are the pseudoprimes distributed?

As it turns out, not surprisingly, there are infinitely many pseudoprimes, and there are many ways to generate infinite sequences of pseudoprimes.

The simplest proof was given in 1903 by Malo, who showed that if n is a pseudoprime, and if $n' = 2^n - 1$, then n' is also a pseudoprime. Indeed, n' is obviously composite, because if $n = ab$ with $1 < a$, $b < n$, then

$$2^n - 1 = (2^a - 1)(2^{a(b-1)} + 2^{a(b-2)} + \cdots + 2^a + 1).$$

Also n divides $2^{n-1} - 1$, hence n divides $2^n - 2 = n' - 1$; so $n' = 2^n - 1$ divides $2^{n'-1} - 1$.

In 1904, Cipolla gave another proof, using the Fermat numbers:

If $m > n > \cdots > s > 1$ are integers and N is the product of the Fermat numbers $N = F_m F_n \cdots F_s$, then N is a pseudoprime if and only if $2^s > m$. Indeed, the order of 2 modulo N is 2^{m+1}, which is equal to the least common multiple of the orders 2^{m+1}, $2^{n+1}, \ldots, 2^{s+1}$ of 2 modulo each factor $F_m, F_n, \ldots, F_s$ of N. Thus $2^{N-1} \equiv 1 \pmod{N}$ if and only if $N - 1$ is divisible by 2^{m+1}. But $N - 1 = F_m F_n \cdots F_s - 1 = 2^{2^s} Q$, where Q is an odd integer. Thus, the required condition is $2^s > m$.   □

As it was indicated in Chapter 1, the Fermat numbers are pairwise relatively prime, so the above method leads to pairwise relatively prime pseudoprimes. One can also obtain pseudoprimes having an arbitrarily large number of prime factors.

Cipolla presented another method that will be described below.

In 1936, Lehmer found a very simple method to generate infinitely many pseudoprimes, each one being the product of two distinct primes p, q. Namely, let $k \geq 5$ be an arbitrary odd integer, let p be a primitive prime factor of $2^k - 1$, and let q be a primitive prime factor of $2^k + 1$. Then pq is a pseudoprime. Thus, for every $m \geq 1$ there exist at least m pseudoprimes $n = pq$ such that

$$n \leq \left(2^{2m+3} - 1\right)\left(\frac{2^{2m+3} + 1}{3}\right) = \frac{4^{2m+3} - 1}{3}.$$

There also exist even composite integers satisfying the congruence $2^n \equiv 2 \pmod{n}$—they may be called *even pseudoprimes*. The smallest one is $m = 2 \times 73 \times 1103 = 161038$, discovered by Lehmer in 1950. In 1951, Beeger showed the existence of infinitely many even pseudoprimes; each one must have at least two odd prime factors.

How "far" are pseudoprimes from being primes? From Cipolla's result, there are pseudoprimes with arbitrarily many prime factors. This is not an accident. In fact, in 1949 Erdös proved that for every $k \geq 2$ there exist infinitely many pseudoprimes, which are the product of exactly k distinct primes.

In 1936, Lehmer gave criteria for the product of two or three distinct odd primes to be a pseudoprime: $p_1 p_2$ is a pseudoprime if and only if the order of 2 modulo p_2 divides $p_1 - 1$ and the order of 2 modulo p_1 divides $p_2 - 1$. If $p_1 p_2 p_3$ is a pseudoprime, then the least common multiple of $\mathrm{ord}(2 \bmod p_1)$ and $\mathrm{ord}(2 \bmod p_2)$ divides $p_3(p_1 + p_2 - 1) - 1$.

Here is an open question: Are there infinitely many integers $n > 1$ such that $2^{n-1} \equiv 1 \pmod{n^2}$? This is equivalent to each of the following problems (see Rotkiewicz, 1965):

Are there infinitely many pseudoprimes that are squares?

Are there infinitely many primes p such that $2^{p-1} \equiv 1 \pmod{p^2}$?

This congruence was already encountered in the question of square factors of Fermat numbers and Mersenne numbers. I shall return to primes of this kind in Chapter 5, Section III.

On the other hand, a pseudoprime need not be square-free. The smallest such examples are $1\,194\,649 = 1093^2$, $12\,327\,121 = 3511^2$, $3\,914\,864\,773 = 29 \times 113 \times 1093^2$.

B   Pseudoprimes in Base a (psp(a))

It is also useful to consider the congruence $a^{n-1} \equiv 1 \pmod{n}$, for $a > 2$. If n is a prime and $1 < a < n$, then the above congruence holds necessarily. So, if, for example, $2^{n-1} \equiv 1 \pmod{n}$, but, say, $3^{n-1} \not\equiv 1 \pmod{n}$, then n is not a prime.

This leads to the more general study of the *pseudoprimes in base a* (or *a-pseudoprimes*) which are the composite integers $n > a$ such that $a^{n-1} \equiv 1 \pmod{n}$.

In 1904, Cipolla also indicated how to obtain a-pseudoprimes. Let $a \geq 2$, let p be any odd prime such that p does not divide $a(a^2 - 1)$. Let

$$n_1 = \frac{a^p - 1}{a - 1}, \quad n_2 = \frac{a^p + 1}{a + 1}, \quad n = n_1 n_2;$$

then n_1 and n_2 are odd and n is composite. Since $n_1 \equiv 1 \pmod{2p}$ and $n_2 \equiv 1 \pmod{2p}$, then $n \equiv 1 \pmod{2p}$. From $a^{2p} \equiv 1 \pmod{n}$ it follows that $a^{n-1} \equiv 1 \pmod{n}$, so n is an a-pseudoprime.

Since there exist infinitely many primes, then there also exist infinitely many a-pseudoprimes (also when $a > 2$).

There are other methods in the literature to produce very quickly increasing sequences of a-pseudoprimes.

For example, Crocker proceeded as follows in 1962. Let a be even, but not of the form 2^{2^r}, with $r \geq 0$. Then, for every $n \geq 1$, the number $a^{a^n} + 1$ is an a-pseudoprime.

In 1948, Steuerwald established the following infinite sequence of a-pseudoprimes. Let n be an a-pseudoprime, which is prime to $a - 1$. For example, for a prime q, put $a = q + 1$ and let p be a prime such that $p > a^2 - 1$; as in the Cipolla construction, let

$$n_1 = \frac{a^p - 1}{a - 1} \equiv a^{p-1} + a^{p-2} + \cdots + a + 1 \equiv p \pmod{q},$$

$$n_2 = \frac{a^p + 1}{a + 1} \equiv a^{p-1} - a^{p-2} + \cdots + a^2 - a + 1 \equiv 1 \pmod{q},$$

so $n = n_1 n_2 \equiv p \pmod{q}$. Let now $f(n) = (a^n - 1)/(a - 1) > n$. Then $f(n)$ is also an a-pseudoprime. Indeed,

$$f(n) = \frac{a^{n_1 n_2} - 1}{a^{n_2} - 1} \times \frac{a^{n_2} - 1}{a - 1}$$

is composite. Since n is prime to $a - 1$ and $a^{n-1} \equiv 1 \pmod{n}$, then n divides $(a^n - a)/(a - 1) = f(n) - 1$. Thus $f(n)$ divides $a^n - 1$, which divides $a^{f(n)-1} - 1$, hence $f(n)$ is an a-pseudoprime. The process may be iterated, noting that $f(n)$ is prime to $a - 1$:

$$f(n) = \frac{\left[(a - 1) + 1\right]^n - 1}{a - 1} = (a - 1)^{n-1} + \binom{n}{1}(a - 1)^{n-2}$$

$$+ \cdots + \binom{n}{n - 2}(a - 1) + n \equiv n \pmod{a - 1},$$

so $f(n)$ is an a-pseudoprime that is prime to $a - 1$. This process leads to an infinite increasing sequence of a-pseudoprimes $n < f(n) < f(f(n)) < f(f(f(n))) < \cdots$, which grows as n, a^n, a^{a^n} $a^{a^{a^n}}$,.... . The method of Lehmer indicated above, applied to binomials $a^k - 1$ and $a^k + 1$, produces a-pseudoprimes which are the product of two distinct prime factors.

From these considerations it follows that it is futile to wish to discover the largest a-pseudoprime.

In 1958, Schinzel showed that for every $a \geq 2$, there exist infinitely many pseudoprimes in base a that are products of two distinct primes.

In 1971, in his thesis, Lieuwens extended simultaneously this result of Schinzel and Erdös' result about pseudoprimes in base 2: for every $k \geq 2$ and $a > 1$, there exist infinitely many pseudoprimes in base a, which are products of exactly k distinct primes.

In 1972, Rotkiewicz showed that if $p \geq 2$ is a prime not dividing $a \geq 2$, then there exist infinitely many pseudoprimes in base a that are multiples of p; the special case when $p = 2$ dates back to 1959, also by Rotkiewicz.

It may occur that a number is a pseudoprime for different bases, like 561 for the bases 2, 5, 7. Indeed, Baillie & Wagstaff and Monier showed independently, in 1980, the following result: Let n be a composite number, and let $B_{\mathrm{psp}}(n)$ be the number of bases a, $1 < a < n$, with $\gcd(a, n) = 1$, for which n is an a-pseudoprime. Then

$$B_{\mathrm{psp}}(n) = \left\{ \prod_{p \mid n} \gcd(n-1, p-1) \right\} - 1.$$

It follows that if n is an odd composite number, which is not a power of 3, then n is a pseudoprime for at least two bases a, $1 < a \leq n - 1$.

It will be seen in Section IX that there exist composite numbers n, which are pseudoprimes for all bases a, $1 < a < n$, with $\gcd(a, n) = 1$.

Here is a table, from the paper by Pomerance, Selfridge & Wagstaff (1980), which gives the smallest pseudoprimes for various bases, or simultaneous bases.

Table 10.   Smallest pseudoprimes for several bases

Bases	Smallest psp
2	$341 = 11 \times 31$
3	$91 = 7 \times 13$
5	$217 = 7 \times 31$
7	$25 = 5 \times 5$
2, 3	$1105 = 5 \times 13 \times 17$
2, 5	$561 = 3 \times 11 \times 17$
2, 7	$561 = 3 \times 11 \times 17$
3, 5	$1541 = 23 \times 67$
3, 7	$703 = 19 \times 37$
5, 7	$561 = 3 \times 11 \times 17$
2, 3, 5	$1729 = 7 \times 13 \times 19$
2, 3, 7	$1105 = 5 \times 13 \times 17$
2, 5, 7	$561 = 3 \times 11 \times 17$
3, 5, 7	$29341 = 13 \times 37 \times 61$
2, 3, 5, 7	$29341 = 13 \times 37 \times 61$

As I have said, if there exists a such that $1 < a < n$ and $a^{n-1} \not\equiv 1$ (mod n), then n is composite, but not conversely. This gives therefore a very practical way to ascertain that many numbers are composite. There are other congruence properties, similar to the above, which give also easy methods to discover that certain numbers are composite.

I shall describe several of these properties; their study has been justified by the problem of primality testing. As a matter of fact, without saying it explicitly, I have already considered these properties in Sections III and V. First, there are properties about the congruence $a^m \equiv 1 \pmod{n}$, which lead to the Euler a-pseudoprimes and strong a-pseudoprimes. In another section, I will examine the Lucas pseudoprimes, which concern congruence properties satisfied by terms of Lucas sequences.

C   EULER PSEUDOPRIMES IN BASE a (epsp(a))

According to Euler's congruence for the Legendre symbol, if $a \geq 2$, p is a prime and p does not divide a, then

$$\left(\frac{a}{p}\right) \equiv a^{(p-1)/2} \pmod{p}.$$

This leads to the notion of an *Euler pseudoprime in base a* (epsp(a)), proposed by Shanks in 1962. These are odd composite numbers n, such that $\gcd(a, n) = 1$ and the Jacobi symbol satisfies the congruence

$$\left(\frac{a}{n}\right) \equiv a^{(n-1)/2} \pmod{n}.$$

Clearly, every epsp(a) is an a-pseudoprime.

There are many natural questions about epsp(a) which I enumerate now:

(e1) Are there infinitely many epsp(a), for each a?

(e2) Are there epsp(a) with arbitrary large number of distinct prime factors, for each a?

(e3) For every $k \geq 2$ and base a, are there infinitely many epsp(a), which are equal to the product of exactly k distinct prime factors?

(e4) Can an odd composite number be an epsp(a) for every possible a, $1 < a < n$, $\gcd(a, n) = 1$?

(e5) For how many bases a, $1 < a < n$, $\gcd(a, n) = 1$, can the number n be an $\mathrm{epsp}(a)$?

In 1986, Kiss, Phong & Lieuwens showed that given $a \geq 2$, $k \geq 2$, and $d \geq 2$, there exist infinitely many $\mathrm{epsp}(a)$, which are the product of k distinct primes and are congruent to 1 modulo d.

This gives a strong affirmative answer to (e3), and therefore also to (e2) and (e1).

In 1976, Lehmer showed that if n is odd composite, then it cannot be an $\mathrm{epsp}(a)$, for every a, $1 < a < n$, $\gcd(a, n) = 1$. So the answer to (e4) is negative.

In fact, more is true, as shown by Solovay & Strassen in 1977: a composite integer n can be an Euler pseudoprime for at most $\frac{1}{2}\varphi(n)$ bases a, $1 < a < n$, $\gcd(a, n) = 1$. This gives an answer to question (e5). The proof is immediate, noting that the residue classes $a \bmod n$, for which $(a \mid n) \equiv a^{(n-1)/2} \pmod{n}$ form a subgroup of $(\mathbb{Z}/n)^{\times}$ (group of invertible residue classes modulo n), which is a proper subgroup (by Lehmer's result); hence it has at most $\frac{1}{2}\varphi(n)$ elements—by dear old Lagrange's theorem.

Let n be an odd composite integer. Denote by $B_{\mathrm{epsp}}(n)$ the number of bases a, $1 < a < n$, $\gcd(a, n) = 1$, such that n is an $\mathrm{epsp}(a)$. Monier showed in 1980 that

$$B_{\mathrm{epsp}}(n) = \delta(n) \prod_{p \mid n} \gcd\left(\frac{n-1}{2}, p-1\right) - 1.$$

Here

$$\delta(n) = \begin{cases} 2 & \text{if } v_2(n) - 1 = \min_{p \mid n}\{v_2(p-1)\}, \\ \frac{1}{2} & \text{if there exists a prime } p \text{ dividing } n \text{ such that} \\ & v_p(n) \text{ is odd and } v_2(p-1) < v_2(n-1), \\ 1 & \text{otherwise,} \end{cases}$$

and for any integer m and prime p, $v_p(m)$ denotes the exponent of p in the factorization of m, that is, the p-adic value of m.

D   STRONG PSEUDOPRIMES IN BASE a (spsp(a))

A related property is the following: Let n be an odd composite integer, let $n - 1 = 2^s d$, with d odd and $s \geq 1$; let a be such that $1 < a < n$, $\gcd(a, n) = 1$.

Then n is called a *strong pseudoprime in base a* (spsp(a)) if $a^d \equiv 1$ (mod n) or $a^{2^r d} \equiv -1$ (mod n) for some r, $0 \le r < s$.

Note that if n is a prime, then it satisfies the above condition for every a, $1 < a < n$, $\gcd(a, n) = 1$.

Selfridge showed (see the proof in Williams' paper, 1978) that every spsp(a) is an epsp(a). There are partial converses.

By Malm (1977): if $n \equiv 3$ (mod 4) and n is an epsp(a), then n is a spsp(a).

By Pomerance, Selfridge & Wagstaff (1980): if n is odd, $(a \mid n) = -1$ and n is an epsp(a), then n is also a spsp(a). In particular, if $n \equiv 5$ (mod 8) and n is an epsp(2), then it is a spsp(2).

Concerning the strong pseudoprimes, I may ask questions (s1)–(s5), analogous to the questions about Euler pseudoprimes posed in Section VIII, C.

In 1980, Pomerance, Selfridge & Wagstaff proved that for every base $a > 1$, there exist infinitely many spsp(a), and this answers in the affirmative question (s1), as well as (e1). I shall say more about this in the study of the distribution of pseudoprimes (Chapter 4, Section VI).

For base 2, it is possible to give infinitely many spsp(2) explicitly, as I indicate now.

If n is a psp(2), then $2^n - 1$ is a spsp(2). Since there are infinitely many psp(2), this gives explicitly infinitely many spsp(2); among these are all composite Mersenne numbers. It is also easy to see that if a Fermat number is composite, then it is a spsp(2).

Similarly, since there exist pseudoprimes with arbitrarily large numbers of distinct prime factors, then (s2), as well as (e2), have a positive answer; just note that if $p_1, p_2, \ldots, p_k$ divide the pseudoprime n, then $2^{p_i} - 1$ ($i = 1, \ldots, k$) divides the spsp(2) $2^n - 1$.

In virtue of Lehmer's negative answer to (e4) and Selfridge's result, then clearly (s4) has also a negative answer. Very important—as I shall indicate later, in connection with the Monte Carlo primality testing methods—is the next theorem by Rabin, corresponding to Solovay & Strassen's result for Euler pseudoprimes. And it is tricky to prove:

If $n > 4$ is composite, there are at least $3(n - 1)/4$ integers a, $1 < a < n$, for which n is not a spsp(a). So, the number of bases a, $1 < a < n$, $\gcd(a, n) = 1$, for which an odd composite integer is spsp(a), is at most $(n - 1)/4$. This answers question (s5).

Monier (1980) has also determined a formula for the number $B_{\text{spsp}}(n)$, of bases a, $1 < a < n$, $\gcd(a, n) = 1$, for which the odd composite integer n is spsp(a). Namely:

$$B_{\text{spsp}}(n) = \left(1 + \frac{2^{\omega(n)\nu(n)} - 1}{2^{\omega(n)} - 1}\right) \left(\prod_{p|n} \gcd(n^*, p^*)\right) - 1,$$

where

$$\omega(n) = \text{ number of distinct prime factors of } n,$$
$$\nu(n) = \min_{p|n} \left\{v_2(p - 1)\right\},$$
$$v_p(m) = \text{ exponent of } p \text{ in the factorization of } m$$
$$\text{(any natural number)},$$
$$m^* = \text{ largest odd divisor of } m - 1.$$

Just for the record, the smallest spsp(2) is $2047 = 23 \times 89$. It is interesting and also useful to know the smallest strong pseudoprimes to several bases simultaneously. Their knowledge is used in strong primality testing.

Given $k \geq 1$, denote by t_k the smallest integer which is a strong pseudoprime for the bases $p_1 = 2$, $p_2 = 3$, ... , p_k, simultaneously. Then the calculations of Pomerance, Selfridge & Wagstaff (1980), extended by Jaeschke (1993), provide the following values:

$$t_2 = 1\,373\,653 = 829 \times 1657,$$
$$t_3 = 25\,326\,001 = 2251 \times 11251,$$
$$t_4 = 3\,215\,031\,751 = 151 \times 751 \times 28351,$$
$$t_5 = 2\,152\,302\,898\,747 = 6763 \times 10627 \times 29947,$$
$$t_6 = 3\,474\,749\,660\,383 = 1303 \times 16927 \times 157543,$$
$$t_7 = t_8 = 341\,550\,071\,728\,321 = 10670053 \times 32010157.$$

Jaeschke's work also showed that there are only 101 numbers below 10^{12} which are strong pseudoprimes for the bases 2, 3, and 5, simultaneously. Since their complete list is fairly large, I reproduce only the one published by the three Knights of Numerology, which is restricted to numbers less than 25×10^9.

Table 11.

Numbers less than 25×10^9, which are spsp in bases 2, 3, 5

Number	psp to base			Factorization
	7	11	13	
25 326 001	no	no	no	2251×11251
161 304 001	no	spsp	no	7333×21997
960 946 321	no	no	no	11717×82013
1 157 839 381	no	no	no	24061×48121
3 215 031 751	spsp	psp	psp	$151 \times 751 \times 28351$
3 697 278 427	no	no	no	30403×121609
5 764 643 587	no	no	spsp	37963×151849
6 770 862 367	no	no	no	41143×164569
14 386 156 093	psp	psp	psp	$397 \times 4357 \times 8317$
15 579 919 981	psp	spsp	no	88261×176521
18 459 366 157	no	no	no	67933×271729
19 887 974 881	psp	no	no	81421×244261
21 276 028 621	no	psp	psp	103141×206281

To this table, I add the list of pseudoprimes up to 25×10^9 which are not square-free and their factorizations:

$$1\,194\,649 = 1093^2,$$
$$12\,327\,121 = 3511^2,$$
$$3\,914\,864\,773 = 29 \times 113 \times 1093^2,$$
$$5\,654\,273\,717 = 1093^2 \times 4733,$$
$$6\,523\,978\,189 = 43 \times 127 \times 1093^2,$$
$$22\,178\,658\,685 = 5 \times 47 \times 79 \times 1093^2.$$

With the exception of the last two, the numbers in the above list are strong pseudoprimes.

Note that the only prime factors to the square are 1093 and 3511. The occurrence of these numbers will be explained in Chapter 5, Section III.

IX   Carmichael Numbers

In a short article which remained unnoticed, Korselt considered in 1899 a more rare kind of numbers; they were also introduced independently by Carmichael in 1912, who first studied their properties. Since his article was noted, such numbers came to be called *Carmichael numbers*. By definition, they are the composite numbers n such that $a^{n-1} \equiv 1 \pmod{n}$ for every integer a, $1 < a < n$, such that a is relatively prime to n. The smallest Carmichael number is $561 = 3 \times 11 \times 17$.

I shall now indicate a characterization of Carmichael numbers. Recall that I have introduced, in Section II, Carmichael's function $\lambda(n)$, which is the maximum of the orders of a mod n, for every a, $1 \le a < n$, $\gcd(a, n) = 1$; in particular, $\lambda(n)$ divides $\varphi(n)$.

Carmichael showed that n is a Carmichael number if and only if n is composite and $\lambda(n)$ divides $n - 1$. (It is the same as saying that if p is any prime dividing n, then $p - 1$ divides $n - 1$.)

It follows that every Carmichael number is odd and is the product of three or more distinct prime numbers.

Explicitly, if $n = p_1 p_2 \cdots p_r$ (product of distinct primes), then n is a Carmichael number if and only if $p_i - 1$ divides $(n/p_i) - 1$ (for $i = 1, 2, \ldots, r$). Therefore, if n is a Carmichael number, then also $a^n \equiv a \pmod{n}$, for every integer $a \ge 1$.

Schinzel noted in 1959 that for every $a \ge 2$ the smallest pseudoprime m_a in base a satisfies necessarily $m_a \le 561$. Moreover, there exists a such that $m_a = 561$. Explicitly, let p_i ($i = 1, \ldots, s$) be the primes such that $2 < p_i < 561$; for each p_i let e_i be such that $p_i^{e_i} < 561 < p_i^{e_i+1}$; let g_i be a primitive root modulo $p_i^{e_i}$, and by the Chinese remainder theorem, let a be such that $a \equiv 3 \pmod{4}$ and $a \equiv g_i \pmod{p_i^{e_i}}$ for $i = 1, \ldots, s$. Then $m_a = 561$.

Carmichael and Lehmer determined the smallest Carmichael numbers:

$561 = 3 \times 11 \times 17$	$15841 = 7 \times 31 \times 73$	$101101 = 7 \times 11 \times 13 \times 101$
$1105 = 5 \times 13 \times 17$	$29341 = 13 \times 37 \times 61$	$115921 = 13 \times 37 \times 241$
$1729 = 7 \times 13 \times 19$	$41041 = 7 \times 11 \times 13 \times 41$	$126217 = 7 \times 13 \times 19 \times 73$
$2465 = 5 \times 17 \times 29$	$46657 = 13 \times 37 \times 97$	$162401 = 17 \times 41 \times 233$
$2821 = 7 \times 13 \times 31$	$52633 = 7 \times 73 \times 103$	$172081 = 7 \times 13 \times 31 \times 61$
$6601 = 7 \times 23 \times 41$	$62745 = 3 \times 5 \times 47 \times 89$	$188461 = 7 \times 13 \times 19 \times 109$
$8911 = 7 \times 19 \times 67$	$63973 = 7 \times 13 \times 19 \times 37$	$252601 = 41 \times 61 \times 101$
$10585 = 5 \times 29 \times 73$	$75361 = 11 \times 13 \times 17 \times 31$	

I consider now the following questions, which are of course closely related:

(1) Are there infinitely many Carmichael numbers?

(2) Given $k \geq 3$, are there infinitely many Carmichael numbers having exactly k prime factors?

The first problem was solved in 1992, in the affirmative, in a brilliant paper by Alford, Granville & Pomerance that appeared in 1994; see also the expository paper by Pomerance (1993).

It is believed that the answer to the second question is also affirmative, but this has yet to be established. For example, it is not even known if there exist infinitely many Carmichael numbers, which are products of exactly three primes. In this respect, there is a result of Duparc (1952) (see also Beeger, 1950):

For every $r \geq 3$, there exist only finitely many Carmichael numbers with r prime factors, of which the smallest $r - 2$ factors are given in advance. I shall return to these questions in Chapter 4.

In 1939, Chernick gave the following method to obtain Carmichael numbers. Let $m \geq 1$ and

$$M_3(m) = (6m + 1)(12m + 1)(18m + 1).$$

If m is such that all three factors above are prime, then $M_3(m)$ is a Carmichael number. This yields Carmichael numbers with three prime factors. But obviously we do not know if there exist infinitely many integers m having that property.

Similarly, if $k \geq 4$, $m \geq 1$, let

$$M_k(m) = (6m + 1)(12m + 1) \prod_{i=1}^{k-2} (9 \times 2^i m + 1).$$

If m is such that all k factors are prime numbers and, moreover, 2^{k-4} divides m, then $M_k(m)$ is a Carmichael number with k prime factors.

This method, or variants of it, have been used to produce Carmichael numbers which are large or have many prime factors.

I note: Wagstaff in 1980 (321 digits), Atkin in 1980 (370 digits), Woods & Huenemann in 1982 (432 digits), Dubner in 1985 (1057 digits), Dubner in 1989 (3710 digits).

While these examples have only a few prime factors, Yorinaga (1978) determined Carmichael numbers with up to 15 prime factors.

The search for large Carmichael numbers with many prime factors continued. In 1994 (published in 1996), Löh & Niebuhr constructed a Carmichael number with 16142049 digits and 1101518 prime factors.

RECORD

The largest known Carmichael number was determined by W.R. Alford and J. Grantham in 1998; it has 20163700 digits and 1371497 prime factors. Also, this number has the following additional property: for every k with $62 \leq k \leq 1371435$ it is divisible by a Carmichael number having exactly k prime factors.

This unpublished record was kindly communicated to me by the authors.

Stimulated by a deeper understanding of this kind of computations, Alford, Granville & Pomerance (1994) established the thruth of this old conjecture: There exist infinitely many Carmichael numbers.

Concerning the calculation of Carmichael numbers, Pinch has produced, in 1998, the complete list of these numbers up to 10^{16}. I shall discuss his findings in Chapter 4, Section VI, B.

The distribution of Carmichael numbers will be studied in Chapter 4, Section VIII.

ADDENDUM ON KNÖDEL NUMBERS

For every $k \geq 1$, let C_k be the set of all composite integers $n > k$ such that if $1 < a < n$ and $\gcd(a, n) = 1$, then $a^{n-k} \equiv 1 \pmod{n}$.

Thus, C_1 is the set of Carmichael numbers. For $k \geq 2$, the numbers C_k were considered by Knödel in 1953. Even before it was proved that there exist infinitely many Carmichael numbers, Mąkowski proved in 1962:

For each $k \geq 2$, the set C_k is infinite.

Proof. For every a, $1 < a < k$, $\gcd(a, k) = 1$, let r_a be the order of a modulo k. Let $r = \prod r_a$ (product for all a as above). So $a^r \equiv 1 \pmod{k}$.

There exist infinitely many primes p such that $p \equiv 1 \pmod{r}$; see Chapter 4, Section IV, for a proof of this very useful theorem. For each such $p > k$, write $p - 1 = hr$, and let $n = kp$. Then $n \in C_k$.

Indeed, let $1 \le a < n$, $\gcd(a, n) = 1$, so $\gcd(a, k) = 1$; hence

$$a^{n-k} = a^{k(p-1)} = a^{khr} \equiv 1 \pmod{k},$$
$$a^{n-k} = a^{k(p-1)} \equiv 1 \pmod{p}.$$

Since $p \nmid k$, then $a^{n-k} \equiv 1 \pmod{n}$, showing that $n = kp$ is in C_k. □

It follows from the above proof that if $k = 2$, then $2p \in C_2$ for every prime $p > 2$. If $k = 3$, then $3p \in C_3$ for every prime $p > 3$; this last fact was proved by Morrow in 1951.

X   Lucas Pseudoprimes

In view of the analogy between sequences of binomials $a^n - 1$ $(n \ge 1)$ and Lucas sequences, it is no surprise that pseudoprimes should have a counterpart involving Lucas sequences. For each parameter $a \ge 2$, there were the a-pseudoprimes and their cohort of Euler pseudoprimes and strong pseudoprimes in base a. In this section, to all pairs (P, Q) of nonzero integers will be associated the corresponding Lucas pseudoprimes, the Euler-Lucas pseudoprimes, and the strong Lucas pseudoprimes. Their use will parallel that of pseudoprimes.

Let P, Q be nonzero integers, $D = P^2 - 4Q$ and consider the associated Lucas sequences $(U_n)_{n \ge 0}$, $(V_n)_{n \ge 0}$.

Recall (from Section IV) that if n is an odd prime, then:

(X.1)   If $\gcd(n, D) = 1$, then $U_{n-(D|n)} \equiv 0 \pmod{n}$.

(X.2) $U_n \equiv (D \mid n) \pmod{n}$.

(X.3) $V_n \equiv P \pmod{n}$.

(X.4)   If $\gcd(n, D) = 1$, then $V_{n-(D|n)} \equiv 2Q^{(1-(D/n))/2} \pmod{n}$.

If n is an odd composite number and the congruence (X.1) holds, then n is called a *Lucas pseudoprime* (with the parameters (P, Q)), abbreviated lpsp(P, Q).

It is alright to make such a definition, but do these numbers exist? If so, are they worthwhile to study?

A   FIBONACCI PSEUDOPRIMES

To begin, it is interesting to look at the special case of Fibonacci numbers, where $P = 1$, $Q = -1$, $D = 5$. In this situation, it is more appropriate to call *Fibonacci pseudoprimes* the lpsp$(1, -1)$.

The smallest Fibonacci pseudoprimes are $323 = 17 \times 19$ and $377 = 13 \times 29$; indeed, $(5 \mid 323) = (5 \mid 377) = -1$ and it may be calculated that $U_{324} \equiv 0 \pmod{323}$, $U_{378} \equiv 0 \pmod{377}$.

E. Lehmer showed in 1964 that there exist infinitely many Fibonacci pseudoprimes; more precisely, if p is any prime greater than 5, then U_{2p} is a Fibonacci pseudoprime.

Property (X.2) was investigated by Parberry (in 1970) and later by Yorinaga (1976).

Among his several results, Parberry showed that if $\gcd(h, 30) = 1$ and condition (X.2) is satisfied by h, then it is also satisfied by $k = U_h$; moreover, $\gcd(k, 30) = 1$ and, if h is composite, clearly U_h is also composite. This shows that if there exists one composite Fibonacci number U_n such that $U_n \equiv (5 \mid n) \pmod{n}$, then there exist infinitely many such numbers. As I shall say (in a short while) there do exist such Fibonacci numbers.

Actually, this also follows from another result of Parberry: If p is prime and $p \equiv 1$ or $4 \pmod{15}$, then $n = U_{2p}$ is odd composite and it satisfies both properties (X.1) and (X.2). In particular, there are infinitely many Fibonacci pseudoprimes which, moreover, satisfy (X.2). (Here I use the fact—to be indicated later in Chapter 4, Section IV—that there exist infinitely many primes p such that $p \equiv 1 \pmod{15}$, resp. $p \equiv 4 \pmod{15}$.)

If $p \not\equiv 1$ or $4 \pmod{15}$, then (X.2) is not satisfied, as follows from various divisibility properties and congruences indicated in Section IV.

Yorinaga considered the primitive part of the Fibonacci number U_n. If you remember, I have indicated in Section IV that every Fibonacci number U_n (with $n \neq 1, 2, 6, 12$) admits a primitive prime factor p—these are the primes that divide U_n, but do not divide U_d, for every d, $1 < d < n$, d dividing n. Thus $U_n = U_n^* \times U_n'$, where $\gcd(U_n^*, U_n') = 1$ and p divides U_n^* if and only if p is a primitive prime factor of U_n.

Yorinaga showed that if m divides U_n^* (with $n > 5$) then $U_m \equiv (5 \mid m) \pmod{m}$.

According to Schinzel's result (1963), discussed in Section IV, there exist infinitely many integers n such that U_n^* is not a prime. So, Yorinaga's result implies that there exist infinitely many odd composite n such that the congruence (X.2) is satisfied.

Yorinaga published a table of all 109 composite numbers n up to 707000, such that $U_n \equiv (5 \mid n) \pmod{n}$. Some of these numbers also give Fibonacci pseudoprimes, like $n = 4181 = 37 \times 113$, $n = 5777 = 53 \times 109$, and many more. Four of the numbers in the table give pseudoprimes in base 2:

$$219781 = 271 \times 811,$$
$$252601 = 41 \times 61 \times 101,$$
$$399001 = 31 \times 61 \times 211,$$
$$512461 = 31 \times 61 \times 271.$$

Another result of Parberry, later generalized by Baillie & Wagstaff, is the following:

If n is an odd composite number, not a multiple of 5, if congruences (X.1) and (X.2) are satisfied, then

$$\begin{cases} U_{(n-(5|n))/2} \equiv 0 \pmod{n} & \text{if} \quad n \equiv 1 \pmod{4}, \\ V_{(n-(5|n))/2} \equiv 0 \pmod{n} & \text{if} \quad n \equiv 3 \pmod{4}. \end{cases}$$

In particular, since there are infinitely many composite integers n such that $n \equiv 1 \pmod{4}$, then there are infinitely many odd composite integers n satisfying the congruence $U_{(n-(5|n))/2} \equiv 0 \pmod{n}$.

The composite integers n such that $V_n \equiv 1 \pmod{n}$ (where $(V_k)_{k \geq 0}$ is the sequence of Lucas numbers) have also been studied. They have been called *Lucas pseudoprimes*, but this name is used here with a different meaning.

In 1983, Singmaster found the following 25 composite numbers $n < 10^5$ with the above property:

705, 2465, 2737, 3745, 4181, 5777, 6721,
10877, 13201, 15251, 24465, 29281, 34561,
35785, 51841, 54705, 64079, 64681, 67861,
68251, 75077, 80189, 90061, 96049, 97921.

B   LUCAS PSEUDOPRIMES (lpsp(P,Q))

I shall now consider lpsp(P,Q) associated to arbitrary pairs of parameters (P,Q). To stress the analogy with the pseudoprimes in base a, the discussion should follow the same lines, but it will be clear that much less is known about these numbers. For example, there is no explicit mention of any algorithm to generate infinitely many lpsp(P,Q), when P, Q are given—except the results mentioned for Fibonacci pseudoprimes.

However, in his thesis in 1971, Lieuwens stated that for every $k \geq 2$, there exist infinitely many Lucas pseudoprimes with given parameters (P,Q), which are the product of exactly k distinct primes.

It is quite normal for an odd integer n to be a Lucas pseudoprime with respect to many different sets of parameters. Let $D \equiv 0$ or 1 (mod 4), let $B_{\mathrm{lpsp}}(n,D)$ denote the number of integers P, $1 \leq P \leq n$, such that there exists Q, with $P^2 - 4Q \equiv D$ (mod n) and n is a lpsp(P,Q). Baillie & Wagstaff showed in 1980 that

$$B_{\mathrm{lpsp}}(n,D) = \prod_{p|n} \left\{ \gcd\left(n - \left(\frac{D}{n} \right), p - \left(\frac{D}{p} \right) \right) - 1 \right\}.$$

In particular, if n is odd and composite, there exists D and, correspondingly, at least three pairs (P,Q), with $P^2 - 4Q = D$ and distinct values of P modulo n, such that n is a lpsp(P,Q).

Another question is the following: If n is odd, for how many distinct D modulo n, do there exist (P,Q) with $P^2 - 4Q \equiv D$ (mod n), $P \not\equiv 0$ (mod n), and n is a lpsp(P,Q)? Baillie & Wagstaff also discussed this matter when $n = p_1 p_2$, where p_1, p_2 are distinct primes.

C   EULER-LUCAS PSEUDOPRIMES (elpsp(P,Q)) AND
STRONG LUCAS PSEUDOPRIMES (slpsp(P,Q))

Let P, Q be given, $D = P^2 - 4Q$, as before. Let n be an odd prime number. If $\gcd(n, QD) = 1$, it was seen in Section V that

$$\text{(el)} \quad \begin{cases} U_{(n-(D|n))/2} \equiv 0 \pmod{n} & \text{when} \quad (Q \mid n) = 1, \\ V_{(n-(D|n))/2} \equiv D \pmod{n} & \text{when} \quad (Q \mid n) = -1. \end{cases}$$

This leads to the following definition. An odd composite integer n, such that $\gcd(n, QD) = 1$, satisfying the above condition is called a

Euler–Lucas pseudoprime with parameters (P, Q), abbreviated elpsp(P, Q).

Let n be an odd composite integer, with $\gcd(n, D) = 1$, let $n - (D \mid n) = 2^s d$, with d odd, $s \geq 0$. If

$$
\text{(sl)} \quad \begin{cases} U_d \equiv 0 \pmod{n}, & \text{or} \\ V_{2^r d} \equiv 0 \pmod{n} & \text{for some } r, \ 0 \leq r < s, \end{cases}
$$

then n is called a *strong Lucas pseudoprime* with parameters (P, Q), abbreviated slpsp(P, Q). In this case, necessarily, $\gcd(n, Q) = 1$.

If n is an odd prime, and $\gcd(n, QD) = 1$, then n satisfies the congruences (el) and (sl) above. It is also clear that if n is an elpsp(P, Q) and $\gcd(n, Q) = 1$, then n is a lpsp(P, Q).

What are the relations between elpsp(P, Q) and slpsp(P, Q)? Just as in the case of Euler and strong pseudoprimes in base a, Baillie & Wagstaff showed that if n is a slpsp(P, Q), then n is an elpsp(P, Q)—this is the analogue of Selfridge's result.

Conversely, if n is an elpsp(P, Q) and either $(Q \mid n) = -1$ or $n - (D \mid n) \equiv 2 \pmod 4$, then n is a slpsp(P, Q)—this is the analogue of Malm's result.

If $\gcd(n, Q) = 1$, n is a lpsp(P, Q), $U_n \equiv (D \mid n) \pmod{n}$ and if, moreover, n is an elpsp(P, Q), then n is also a slpsp(P, Q). The special case for Fibonacci numbers was proved by Parberry, as already indicated.

Previously, I mentioned the result of Lehmer, saying that no odd composite number can be an epsp(a), for all possible bases. Here is the analogous result of Williams (1977): Given $D \equiv 0$ or $1 \pmod 4$, if n is an odd composite integer, and $\gcd(n, D) = 1$, there exist P, Q, nonzero integers, with $P^2 - 4Q = D$, $\gcd(P, Q) = 1$, $\gcd(n, Q) = 1$, and such that n is not an elpsp(P, Q).

With the present terminology, I have mentioned already that Parberry had shown, for the Fibonacci sequence, that there exist infinitely many elpsp$(1, -1)$.

This has been improved by Kiss, Phong & Lieuwens (1986): Given (P, Q) such that the sequence $(U_n)_{n \geq 0}$ is nondegenerate (that is, $U_n \neq 0$ for every $n \geq 0$), given $k \geq 2$, there exist infinitely many elpsp(P, Q), each being the product of k distinct primes. Moreover, given also $d \geq 2$, if $D = P^2 - 4Q > 0$, then the prime factors may all be chosen to be of the form $dm + 1$ $(m \geq 1)$.

As for Fibonacci numbers, I now consider the congruences (X.2) and also (X.3), (X.4). It may be shown that if $\gcd(n, 2PQD) = 1$ and if n satisfies any two of the congruences (X.1) to (X.4), then it satisfies the other two.

In 1986, Kiss, Phong & Lieuwens extended a result of Rotkiewicz (1973) and proved: Given $P, Q = \pm 1$ (but $(P, Q) \neq (1, 1)$), given $k \geq 2$, $d \geq 2$, there exist infinitely many integers n, which are Euler pseudoprimes in base 2, and which satisfy the congruences (X.1) to (X.4); moreover, each such number n is the product of exactly k distinct primes, all of the form $dm + 1$ (with $m \geq 1$).

D   CARMICHAEL–LUCAS NUMBERS

Following the same line of thought that led from pseudoprimes to Carmichael numbers, it is natural to consider the following numbers.

Given $D \equiv 0$ or $1 \pmod 4$, the integer n is called a *Carmichael–Lucas number* (associated to D), if $\gcd(n, D) = 1$ and for all nonzero relatively prime integers P, Q with $P^2 - 4Q = D$ and $\gcd(n, Q) = 1$, the number is an $\mathrm{lpsp}(P, Q)$.

Do such numbers exist? A priori, this is not clear. Of course, if n is a Carmichael–Lucas number associated to $D = 1$, then n is a Carmichael number.

Williams, who began the consideration of Carmichael–Lucas numbers, showed in 1977:

If n is a Carmichael–Lucas number associated to D, then n is the product of $k \geq 2$ distinct primes p_i such that $p_i - (D \mid p_i)$ divides $n - (D \mid n)$.

Note that $323 = 17 \times 19$ is a Carmichael–Lucas number (with $D = 5$); but it cannot be a Carmichael number, because it is the product of only two distinct primes.

Adapting the method of Chernick, it is possible to generate many Carmichael–Lucas numbers. Thus, for example, $1649339 = 67 \times 103 \times 239$ is such a number (with $D = 8$).

XI   Primality Testing and Factorization

I reserve the last section to treat a burning topic, full of tantalizing ideas and the object of intense research, in view of immediate direct applications.

Immediate direct applications of number theory! Who would dream of it, even some 40 years ago? Von Neumann yes, not me, not many people. Poor number theory, the Queen relegated (or raised?) to be the object of a courtship inspired by necessity not by awe.

In recent years, progress on the problems of primality testing and factorization have been swift. More and more deep results of number theory have been invoked. Brilliant brains devised clever procedures, not less brilliant technicians invented tricks, shortcuts to implement the methods in a reasonable time—and thus, a whole new branch of number theory is evolving.

In previous sections of this chapter, I have attempted to develop the foundations needed to present in a lucid way the main procedures for primality testing. But this was doomed to failure. Indeed, with the latest developments I would need, for example, to use facts about the theory of Jacobi sums, algebraic number theory, elliptic curves, abelian varieties, etc. This is far beyond what I intend to discuss. It is more reasonable to assign supplementary reading for those who are avidly interested in the problem. Happily enough, there are now many excellent expository articles and books, which I will recommend at the right moment.

Despite the shortcomings just mentioned, I feel that presenting an overview of the question, even one with gaps, will still be useful. Having apologized, I may now proceed with my incomplete treatment.

First, money: how much it costs to see the magic. Then, I shall discuss more amply primality tests, indicate some noteworthy recent factorizations, to conclude with a quick description of applications to public key cryptography. I will be happy if the presentation which follows will make my reader thirsty. Thirsty to know more about what he has read here, and for this purpose, I recommend the books of Williams (1998) and of Crandall & Pomerance (2001).

A   THE COST OF TESTING

The cost of applying an algorithm to a number N is proportional to the time required and, in turn, it depends on the machine, the program, and the size of the number.

The operations should be counted in an appropriate way, since it is clear that addition or multiplication of very large numbers is more time consuming than if the numbers were small. So, in the last analysis, the cost is proportional to the number of operations with digits—such indivisible operations are called bit operations. Thus, for the calculation, the input is not the integer N, but the number of its digits in some base system, which is then proportional to $\log N$.

The algorithm runs in *polynomial time* if there exists a polynomial $f(X)$ such that, for every N, the time required to perform the algorithm on the number N is bounded by $f(\log N)$. An algorithm, not of polynomial time, whose running time is bounded by $f(N)$ (for every N) where $f(X)$ is a polynomial, is said to have an *exponential* running time, since $N = e^{\log N}$. An algorithm can only be economically justified if it runs in polynomial time.

The theory of complexity of algorithms deals specifically with the determination of bounds for the running time. It is a very elaborate sort of bookkeeping, which requires a careful analysis of the methods involved. Through the discovery of clever tricks, algorithms may sometimes be simplified into others requiring only a polynomial running time.

It may be said that the main problem faced in respect to primality testing (and many other problems) is the following:

Does there exist an algorithm to perform the test, which runs in polynomial time?

This problem has just been solved in the affirmative, as I shall discuss soon at the appropriate place. But first, I will consider other tests for primality, which do not run in polynomial time, and yet are very practical for actual testing.

All this should not be confused with the following.

If a number N is known to be composite, this fact may be proved with only one operation. Indeed, it is enough to produce two numbers a, b, such that $N = ab$, so the number of bit operations required is at most $(\log N)^2$. Paraphrasing Lenstra, it is irrelevant whether a, b were found after consulting a clairvoyant, or after three years of

Sundays, like Cole's factorization of the Mersenne number M_{67}:

$$2^{67} - 1 = 193707721 \times 761838257287.$$

If p is known to be a prime, what is the number of bit operations required to prove it? This is not so easy to answer. In 1975, Pratt showed that it suffices a $C(\log p)^4$ bit operations (where C is a positive constant).

In 1987, Pomerance applied the Hasse-Weil theorem on the number of points on elliptic curves defined modulo some integer n. He was able to show that if p is known to be a prime, then a proof of this fact may be done involving at most $C \log p$ multiplications modulo p. This was better than all the other earlier certification proofs.

B   MORE PRIMALITY TESTS

I return once more to primality testing. There are many kinds of tests, and according to the point of view, they may be classified as follows:

$$\begin{cases} \text{Tests for numbers of special forms} \\ \text{Tests for generic numbers} \end{cases}$$

or

$$\begin{cases} \text{Tests with full justification} \\ \text{Tests with justification based on conjectures} \end{cases}$$

or

$$\begin{cases} \text{Deterministic tests} \\ \text{Probabilistic or Monte Carlo tests.} \end{cases}$$

In the sequel, I shall encounter tests of each of the above kinds.

If sufficiently many prime factors of $N-1$ or $N+1$ are known, the tests indicated in Sections III and V run in polynomial time on the number of digits of the input. These are *special purpose* primality tests, each one being very effective for numbers of appropriate form.

In contrast, a *general purpose* primality test is applicable to any number and is not specifically designed to handle more effectively any one kind of number.

The justification of a primality test ought to be based on theorems of number theory. But there are cases where no justification is known without appealing to unproved conjectures, like some form of Riemann's hypothesis.

Many of the tests are deterministic and the steps are all prescribed in advance. In other tests, there are random choices made in some steps during the testing.

When a number N is submitted to a primality test, the desired output is one of the following two answers: "N is a prime," or "N is composite." However, there are tests leading to the following outputs: "N is composite," or "N satisfies a property shared by prime numbers." Since there are measures of probability attached to the test, these are called probabilistic or Monte Carlo tests.

If it has been ascertained that a number N has a high probability of being a prime, it is customary to call such a number a *probable prime*. Of course, it should be borne in mind that a number $N > 1$ is either prime or composite. The designation of "probable prime" reflects the lack of knowledge, at a given moment, of the exact kind of number, prime, or composite.

Once a test is performed and the number is designated to be a prime, often after extensive calculations, usually subjected to the hazards of human or machine errors, it is of the utmost importance to ratify the result obtained. A second or third repetition of the test, preferably performed with different programs and on different machines, giving the same output is reassuring enough—but not a proof that the output is correctly given.

In this respect, the most desirable feature is a certificate of primality, when the number is declared a prime; this certificate would be a proof of primality for the number. — Now I wish to discuss a few—very few—of the methods to test primality.

Trial division

For numbers that are not of a special form, the very naive primality test is by trial division of N by all primes $p < \sqrt{N}$. It will be seen in Chapter 4 that, for any large integer N, the number of primes less than $\sqrt{N}$ is about $2\sqrt{N}/\log N$ (this statement will be made much more precise later on); thus there will be at most $C\sqrt{N}/\log N$ operations (where $C > 0$ is a constant), which tells that the running time could be $C\sqrt{N}/\log N$. So this procedure does not run in polynomial time on the input.

Miller's test

In 1976, Miller proposed a primality test, which was justified using a generalized form of Riemann's hypothesis. I will not explain the exact meaning of this hypothesis or conjecture, but in Chapter 4, I shall discuss the classical Riemann's hypothesis.

To formulate Miller's test, which involves the congruences used in the definition of strong pseudoprimes, it is convenient to use the terminology introduced by Rabin.

Let N be an integer, $N-1 = 2^s d$, with $s \geq 0$, d odd. Let $1 < a < N$ with $\gcd(a, N) = 1$. Then a is said to be a *witness* for N when $a^d \not\equiv 1$ (mod N) and $a^{2^r d} \not\equiv -1$ (mod N) for every r, $0 \leq r < s$.

If N has a witness, it is composite. If N is composite, if $1 < a < N$, $\gcd(a, N) = 1$, and a is not a witness, then N is a spsp(a). Conversely, if N is odd and N is a spsp(a) then a is not a witness for N.

In this terminology, it suffices to show that no integer a, $1 < a < N$, $\gcd(a, N) = 1$, is a witness, in order to deduce that N is prime. Since N is assumed to be very large, this task is overwhelming! It would be wonderful just to settle the matter by considering small integers a, and checking whether any one is a witness for N. Here is where the generalized Riemann's hypothesis is needed. It was used to show:

Miller's test. *Let N be an odd integer. If there exists a, such that $\gcd(a, N) = 1$, $1 < a < 2(\log N)^2$, which is a witness for N, then N is composite. Otherwise, N is a prime.*

I should add here that for numbers up to 25×10^9, because of the calculations reported in Section VIII, the only composite integer N that is a strong pseudoprime simultaneously to the bases 2, 3, 5, 7, is the number $3\,215\,031\,751$. So if $N < 25 \times 10^9$ is not this number, and 2, 3, 5, 7 are not witnesses, then N is a prime. As shown by Jaeschke (1993), this is also true up to $N < 118\,670\,087\,467$.

This test may be easily implemented on a pocket calculator.

The number of bit operations for testing whether a number is a witness for N is at most $C(\log N)^5$, where C is a positive constant. So, this test runs in polynomial time on the input, provided the generalized Riemann's hypothesis is assumed true.

In 1979, Lenstra published a streamlined version of Miller's test, which he discussed again in his paper of 1982. See also the nice expository paper by Wagon (1986).

The APR test

The primality test devised by Adleman, Pomerance & Rumely (1983), usually called the APR test, represents a breakthrough. To wit:

(i) It is a deterministic general purpose primality test; thus, it is applicable to arbitrary natural numbers N, without requiring the knowledge of factors of $N - 1$ or $N + 1$.

(ii) The running time $t(N)$ is almost polynomial; more precisely, there exist effectively computable constants $0 < C' < C$, such that

$$(\log N)^{C' \log \log \log N} \leq t(N) \leq (\log N)^{C \log \log \log N}.$$

(iii) The test is justified rigorously, and for the first time ever in this domain, it was necessary to appeal to deep results in the theory of algebraic numbers. The test involves calculations with roots of unity and the general reciprocity law for the power residue symbol. (Did you notice that I have not explained these concepts? It is far beyond what I plan to treat.)

Up to 2002, the APR test had the best running time among all deterministic general purpose primality tests.

Soon after its publication, Cohen & Lenstra (1984) modified the APR test, making it more flexible, using Jacobi sums in the proof (instead of the reciprocity law), and having the new test programmed for practical applications. It was the first primality test in existence that could routinely handle numbers of up to 200 decimal digits, the test being executed in about ten minutes, while numbers of up to 100 digits were treated in about 45 seconds.

In 1987, Cohen & Lenstra, Br. (Brother, not Junior), tested a number of 247 digits (a prime factor of $2^{892} + 1$), in about 15 minutes.

A presentation of the APR test was made by Lenstra in the Séminaire Bourbaki, Exposé 576 (1981). It was also discussed in papers of Lenstra (1982) and Nicolas (1984), as well as in the important book by Cohen (1993).

Tests with elliptic curves

In 1986, Atkin presented his own new primality test which used elliptic curves over finite fields, the first test of this kind. It runs in random polynomial time, it is fully justified, and if the output is "prime", it comes assorted with a list of numbers from which it is easily verified, without performing all the calculations again, that the number is indeed a prime. Such a list of intermediate results is called just a *certificate* for the prime number.

Atkin & Morain (1993) published a long paper devoted to their method, called ECPP ("elliptic curve primality proving"), which is described in its various aspects. The algorithm has been refined by Morain, who succeeded to prove, and to certify, the primality of various interesting numbers having more than 1000 digits. Other, most effective implementations of the test are currently being used.

Due to its complexity, I shall not even try to indicate the basic steps of the ECPP algorithm.

RECORD

The largest number proved prime by using a general purpose primality test (rigorously justified and applicable to an arbitrary number), is a 5878 digit number $16282536\ldots36478311$, which has the special property that it is preceded by a row of 233821 composite numbers.

The certification of this prime, completed in February 2003, was accomplished by J.L. Gómez Pardo, using the ECPP implementation of M. Martin. The computations required 3581 hours (about 21 weeks) on one of the fastest available PCs. The produced certificate is a text file containing nearly 5 800 000 characters (please count how many books, more boring than this one, would be needed to contain them). Using the existing certificate, primality of the number can be verified within less than two days.

To illustrate the extraordinary progress that has been achieved in the performance of the ECPP method during the past years, here are the previous records:

Prime number	Digits	Date
$10^{5019} + (3^2 \times 7^5 \times 11^{11})$	5020	September 2001
$10^{3999} + 4771$	4000	May 2001
$(348^{1223} - 1)/347$	3106	January 2001
$(30^{1789} - 1)/29$	2642	October 2000
$(2^{7331} - 1)/458072843161$	2196	October 1997

Except for the last one, these records were due to the brothers G. and M. La Barbera and to Martin. The last prime, which is the second and largest factor of the Mersenne number

$$M_{7331} = 458072843161 \times P2196,$$

was verified by E. Mayer and F. Morain using Morain's ECPP program.

To feel how well a general purpose primality test performs, it is a good idea to apply the test to random numbers, namely, numbers whose digits were obtained by repeatedly spinning a wheel with ten possible positions. Some numbers which appear in nature, like the ubiquitous constant π, seem to have randomly distributed digits in their decimal part.

Indeed, in September 1999 more than 206 billion decimal digits of π were calculated by Y. Kanada and his coworkers. A statistical analysis confirms that any given succession of digits appears as often as it should be expected from randomness. In particular, Caldwell & Dubner (2000) analysed the occurrence of primes made out of a sequence of successive digits of π, obtaining a remarkable agreement.

More recently, in December 2002, Kanada announced that he had calculated 1.2411 trillion digits of π; for details, see Bailey (2003). This brings to a true story, not to be forgotten. Ludolph van Ceulen became famous for having calculated 35 correct digits of π (published posthumously in 1615). These digits were inscribed in his epitaph. I wish long life to Kanada—his epitaph will create problems.

Monte Carlo methods

Early in this century, the casino in Monte Carlo attracted the aristocracy and adventurers, who were addicted to gambling. Tragedy and fortune were determined by the spinning wheel.

I read with particular pleasure the novel by Luigi Pirandello, telling how the life of Mattia Pascal was changed when luck favored him, both at Monte Carlo and in his own Sicilian village. But Monte Carlo is not always so good. More often, total ruin, followed by suicide, is the price paid!

As you enter into the Monte Carlo primality game, and if your Monte Carlo testing will be unsuccessful, I sincerely hope that you will not be driven to suicide.

I wish to mention three Monte Carlo tests, due to Baillie & Wagstaff (1980), Solovay & Strassen (1977) and Rabin (1976, 1980). In each of these tests a number of witnesses a are used, in connection with congruences like those satisfied by $\text{psp}(a)$, $\text{epsp}(a)$, $\text{spsp}(a)$ numbers.

I describe briefly Rabin's test, which is very similar to Miller's. Based on the same idea of Solovay & Strassen, Rabin proposed the following test:

Step 1. Choose, at random, $k > 1$ small numbers a, such that $1 < a < N$ and $\gcd(a, N) = 1$.

Step 2. Test, in succession, for each chosen basis a, whether N satisfies the condition in the definition of a strong pseudoprime in base a; writing $N - 1 = 2^s d$, with d odd, $s \geq 0$, either $a^d \equiv 1 \pmod{N}$ or $a^{2^r d} \equiv -1 \pmod{N}$ for some r, $0 \leq r < s$.

If an a is found for which the above condition does not hold, then declare N to be composite. In the other case, the probability that N is a prime, when certified prime, is at least $1 - 1/4^k$. So, for $k = 30$, the likely error is at most one in 10^{18} tests.

You may wish to sell prime numbers—yes, I say sell—to be used in public key cryptography (be patient, I will soon come to this application of primality and factorization). And you wish to be sure, or sure with only a negligible margin of error, that you are really selling a prime number, so that you may advertise: "Satisfaction guaranteed or money back."

On the basis of Rabin's test, you can safely develop a business and honestly back the product sold.

The recent AKS test

In August 2002, Agrawal, Kayal & Saxena posted in their website a paper containing an algorithm for primality testing which is for general purpose, deterministic, fully justified *and* runs in polynomial time. This solved the long-standing problem mentioned earlier in this subsection.

The theoretical basis of the test is a proposition which, except at one step, involves only arguments dealing with simple polynomials with coefficients in integers modulo N, and a binomial. The crucial step, presently required, is a deep theorem of Fouvry pertaining to sieve theory. I like to state this theorem (not in the stronger original form):

Let $\theta = 0.6687\ldots > 2/3$. For every $x > 2$ there exists a prime p such that $x^\theta < p < x$, and there exists k, not a multiple of 3, such that $2kp + 1 \leq x$ and $2kp + 1$ is a prime.

It is reasonable to hope that the test will be suitably modified and perhaps become dependent on a less profound theorem than Fouvry's.

As for the running time (with fast multiplication), it was originally evaluated as essentially $(\log N)^{12}$, and lately lowered to $(\log N)^{7.5}$. An analysis of the running time may also be found in Morain's preprint (2002).

I have asked Agrawal to prepare a short presentation of the AKS algorithm, which I reproduce here. I am thankful for his collaboration.

The central idea in the new primality testing algorithm is the following identity characterizing primes:

N is prime if and only if $(1 - X)^N \equiv 1 - X^N \pmod{N}$.

The simplest way of verifying this identity efficiently is to choose a random small degree polynomial $Q(X)$ and check the identity modulo $Q(X)$. With high probability the result will be correct. This gives a very simple randomized polynomial time algorithm.

To get a deterministic algorithm, one way is to show that if the identity is false, then modulo only a "few" small degree polynomials $Q(X)$ the check will fail. And one of the simplest sets of such polynomials is $Q(X) = X^r - 1$ for small degrees r.

In what follows, let $P_1(X) \equiv P_2(X) \pmod{X^r - 1, n}$ denote the identity of the remainders of $P_1(X)$ and $P_2(X)$ after division by $X^r - 1$ and after dividing the coefficients by n. Then the following weaker version of the above statement is proved:

$N = p^k$ (where p is a prime) if and only if $(a - X)^N \equiv a - X^N$ $\pmod{X^r - 1, p}$ for a "few" values of a and r.

In fact, r can be fixed to be a specific value. The characterization immediately gives a deterministic and efficient primality test as the identity can be verified modulo N (but not modulo p, of course), and the standard method can be used to handle the case when N is a non-trivial power of p.

One direction of the equivalence is trivial to show. To prove the other direction use is made of the following facts:

(i) If $(a - X)^N \equiv a - X^N \pmod{X^r - 1, p}$ for several values of a, then for any polynomial $g(X)$ in the multiplicative group generated by the corresponding linear polynomials $(a - X)$, the following property holds:

$$g(X)^N \equiv g(X^N) \pmod{X^r - 1, p}.$$

This gives exponentially many polynomials $g(X)$ satisfying the identity, provided the order of p modulo r is large, and this can be ensured using existing results in sieve theory.

(ii) If $g(X)^N \equiv g(X^N) \pmod{X^r - 1, p}$, as above, and $g(X)^p \equiv g(X^p) \pmod{X^r - 1, p}$ (trivially), then for any $s = n^i p^j$,

$$g(X)^s \equiv g(X^s) \pmod{X^r - 1, p}.$$

(iii) Since powers of X are reduced modulo $X^r - 1$, there exist s and t, $s \neq t$, such that

$$g(X)^s \equiv g(X^t) \pmod{X^r - 1, p}.$$

This is not possible when both s and t are smaller than the size of the group in (i), but this is ensured, as noted above, by known results in sieve theory.

C TITANIC AND CURIOUS PRIMES

In an article of 1983/84, Yates coined the expression "titanic prime" to name any prime with at least 1000 digits. In the paper with the suggestive title *Sinkers of the Titanics* (1984/85), Yates compiled a list of the largest known titanic primes. By January 1, 1985, he knew 581 titanic primes, of which 170 had more than 2000 digits. These were listed in the paper.

In September 1988, Yates' list comprised already 876 titanic primes. The *Six of Amdahl* (J. Brown, L.C. Noll, B. Parady, G. Smith, J. Smith & S. Zarantonello) announced at the beginning of 1990 the discovery of 550 new titanic primes.

It is not surprising that these primes have special forms, a few being Mersenne primes, others being of the form $k \times 2^n \pm 1$, or $k \times$

$b^n + 1$ ($b > 2$). The reason is simply that there are more efficient primality testing algorithms for numbers of these forms.

In 1992, Yates called *gigantic* all primes with at least 10000 digits. For primes with 1 000 000 or more digits, we use the expression *megaprimes*; as it was mentioned, the largest Mersenne primes are megaprimes. After Yates' death, C. Caldwell became the keeper of the titanic primes, gigantic primes, and other jewels. But he is also the author and manager of a very informative and up-to-date Internet site about "matters primes". I benefited from visiting this site—it is not less interesting than the San Diego Zoo.

The rapid progress of primality testing increased these lists, almost every day. At the end of 2002, the 5000 largest known primes (the only ones displayed in Caldwell's list) had more than 30000 digits. It would be futile to try to report these numbers. Since there are already more known titanic, gigantic and megaprimes than the total number of lines of this book, I do not have bad conscience with this omission. However, it would be unforgivable to hide the following curiosities from you.

A *palindromic* number (in base 10) is an integer $N = a_1 a_2 \ldots a_{n-1} a_n$ with decimal digits a_i ($0 \leq a_i \leq 9$) such that $a_1 = a_n$, $a_2 = a_{n-1}$, ... Due to the survival of the old mysticism attached so often to numbers (perfect numbers, amicable numbers, abundant numbers, etc.), the palindromic numbers still command the attention of numerologists.

For many years, Dubner has been finding larger and larger palindromic prime numbers, keeping safe his title of record man until 2001, when he found the prime $10^{39026} + 4538354 \times 10^{19510} + 1$, with 39027 digits.

RECORD

The largest known palindromic prime is $10^{104281} - 10^{52140} - 1$, a number with 104281 digits. It was found in January 2003 by D. Heuer using a program called PrimeForm, whose developers include C. Nash, Y. Gallot and G. Woltman.

An earlier record by Dubner, was a number that might be called a triply palindromic prime: $10^{35352} + 2049402 \times 10^{17673} + 1$; it has 35353 digits—a number which is again a palindromic prime, with 5 digits, and where 5 is again a palindomic prime!

We may consider the following, apparently silly problem: Given $k \geq 4$, to determine a sequence $N_1, N_2, \ldots, N_k$, where each N_i is a palindromic prime and N_{i+1} is the number of digits of N_i (for $i = 1, \ldots, k - 1$).

For the description of the subsequent pearls, the following notation is useful: $(23)_4$, for example, means 23232323, and $(1)_{15}$ means that the digit 1 is repeated 15 times; and so on.

RECORDS

A. The largest known prime, all of whose digits are prime numbers (2, 3, 5, 7), is

$$72323252323272325252 \times \frac{10^{3120} - 1}{10^{20} - 1} + 1$$

$$= (72323252323272325252)_{156} + 1.$$

It has 3120 digits and was discovered by Dubner in 1992.

B. The largest known prime with all digits equal to 0 or 1 is $1(0)_{15397}1110111(0)_{15397}1$, with 30803 digits. It is also a palindrome and was discovered by Dubner in 1999.

C. The largest known primes with initial digit d (of course, not divisible by 3), followed by n digits equal to 9, are:

d	n	Year
1	55347	2002
2	49314	2002
4	21456	2001
5	34936	2001
7	49808	2002
8	48051	2000

Most of these primes were discovered by E.J. Sorensen. Only the last one was found by Dubner. In each case Gallot's program was used.

D. The largest known prime with all digits odd is the number $1(9)_{55347}$ listed in the previous topic.

E. The largest known prime number with the largest number of digits equal to 0 is $105994 \times 10^{105994} + 1$ and was discovered by G. Löh and Y. Gallot in 2000.

F.   The most exotic curious prime is

$(1)_{1000}(2)_{1000}(3)_{1000}(4)_{1000}(5)_{1000}(6)_{1000}(7)_{1000}(8)_{1000}(9)_{1000}(0)_{6645}1.$

This prime has 15646 digits and was discovered, of course, by Dubner (in 2000).

G.   And last (but surely least): The smallest prime with 1000 digits is $10^{999} + 7$. Its primality was verified by P. Mihăilescu in 1998.

D   FACTORIZATION

The factorization of large integers is a hard problem: there is no known algorithm that runs in polynomial time. It is also an important problem, because it has found a notorious application to public key cryptography.

Nevertheless, I shall not discuss here the methods of factorization—this would once again lead me too far from the subject of records on prime numbers. The best I can do is to quote some books and research papers, which may serve as an Ariadne thread in the labyrinth. Recommended books are, in chronologocal order, the following.

The volume by Brillhart, Lehmer, Selfridge, Tuckerman & Wagstaff (1983) contains tables of known factors of $b^n \pm 1$ ($b = 2$, 3, 5, 6, 7, 10, 11, 12) for various ranges of n. For example, the table of factors of $2^n - 1$ extends for $n < 1200$; for larger bases b, the range is smaller. The second edition of the book, which appeared in 1988, contains 2045 new factorizations, reflecting the important progress accomplished in those few years, both in the methods and in the technology. The recent third edition includes another 2332 new factorizations.

This collective work, also dubbed "the Cunningham project", was originally undertaken to extend the tables published by Cunningham & Woodall in 1925. It is likely that this activity will go on unabated. Heaven is the limit!

The book of Riesel (1985) discusses factorization (and primality) at length. It also contains tables of factors of Fermat numbers, of Mersenne numbers, of numbers of the forms 2^n+1, 10^n+1, of repunits $(10^n - 1)/9$, and many more. It is a good place to study techniques of factorization, which are exposed in a coherent and unified way. Due to its deserved success, a second updated edition has appeared in 1994, which also contains a description of the elliptic curve factoring method.

In 1989, Bressoud published an undergraduate text on factorization and primality. It contains not only the standard background, but also the quadratic sieve and elliptic curve methods.

Among the expository papers, the following deserve attention: Guy (1975) discusses the methods now considered classical; Williams (1984) covers about the same ground, being naturally more up to date—it is pleasant reading. Dixon (1984) writes about factorization as well as on primality. The lecture notes of a short course by Pomerance (1984) contain an annotated bibliography.

To quote more technical papers, the use of elliptic curves in factoring may be read, first hand, in the paper by Lenstra (1987); the paper of the brothers Lenstra of 1990 is also of fundamental importance. More recently, I indicate a paper on the number field sieve, by the brothers Lenstra, Manasse & Pollard (1993).

Just as an illustration, and for the delight of lovers of large numbers, I will give now explicit factorizations of some Mersenne, Fermat, and other numbers; for the older references, see Dickson's *History of the Theory of Numbers*, Vol. I, pp. 22, 29, 377, and Archibald (1935):

$$M_{59} = 2^{59} - 1 = 179951 \times 3203431780337,$$

by Landry in 1869;

$$M_{67} = 2^{67} - 1 = 193707721 \times 761838257287,$$

by Cole in 1903, already mentioned;

$$M_{73} = 2^{73} - 1 = 439 \times 2298041 \times 9361973132609,$$

the factor 439 by Euler, the other factors by Poulet in 1923;

$$F_6 = 2^{2^6} + 1 = (1071 \times 2^8 + 1) \times (262814145745 \times 2^8 + 1)$$
$$= 274177 \times 67280421310721,$$

by Clausen in 1856.

The above factorizations were obtained before the advent of computers!

More recently, the following factorizations were obtained:

$$M_{113} = 2^{113} - 1 = 3391 \times 23279 \times 65993 \times 1868569$$
$$\times 1066818132868207,$$

the smallest factor by Reuschle in 1856, and the remaining factors by Lehmer in 1947;

$$M_{193} = 2^{193} - 1 = 13821503 \times 61654440233248340616559$$
$$\times 14732265321145317331353282383,$$

by Naur (1983) and, independently, by Pomerance & Wagstaff in 1983. The next factorization has direct historical connection with Mersenne himself (see Section VII):

$$M_{257} = 2^{257} - 1 = 535006138814359$$
$$\times 1155685395246619182673033$$
$$\times 374550598501810936581776630096313181393,$$

by Penk and by Baillie, who found, respectively, the first and the two last factors in 1979, resp. 1980; note that already in 1927, Lehmer had shown that M_{257} is composite, without however finding any factor.

Turning to Fermat numbers, we have:

$$F_7 = 2^{2^7} + 1 = 59649589127497217 \times 5704689200685129054721,$$

by Morrison & Brillhart in 1970 (published in 1971);

$$F_8 = 2^{2^8} + 1 = 1238926361552897$$
$$\times 93461639715357977769163558199606896584051237541638188580280321,$$

by Brent & Pollard in 1980 (published in 1981).

The Fermat number F_{11} has been completely factored in 1988. Two small prime factors were long well-known; two more prime factors were found by Brent (with the elliptic curve method), who indicated that the 564-digit cofactor was probably a prime; this was shown to be the case by F. Morain.

The number F_9 was factored in 1990 by A.K. Lenstra and M.S. Manasse. It could not resist the number field sieve method. The most recently factored Fermat number is F_{10}; the factorization was completed by Brent in 1995.

All this, and much more, was said in the sections dealing with Fermat and Mersenne numbers.

In a paper of 1988, dedicated to Dov Jarden, Brillhart, Montgomery & Silverman gave the known factors of Fibonacci numbers U_n

(for n odd, $n \leq 999$) and of Lucas numbers V_n (for $n \leq 500$). The factorizations were complete to $n \leq 387$ and $n \leq 397$, respectively. In April 2003, Montgomery reported that the factorizations of U_n and V_n had been finished for all $n \leq 1000$. This pushes much further the work which had been done by many other numerologists, among whom Jarden (see the third edition of his book, 1958).

Here are some more noteworthy factorizations, which at their time represented an important step forward:

$$\frac{10^{103} + 1}{11} = 1237 \times 44092859 \times 102860539 \times 984385009$$
$$\times\, 612053256358933 \times 182725114866521155647161$$
$$\times\, 14718654539938553026608876141375211979,$$

factorization completed by Atkin and Rickert in 1984.

A.K. Lenstra and M.S. Manasse were "pleased to announce a first factorization of a 100-digit number by a general purpose factorization algorithm" (October 12, 1988); such an algorithm factors a number N in a deterministic way, based solely on the size of N, and not on any particular property of its factors; in its worst case, the running time for factorization is nearly the same as the average running time.

The happy number was

$$\frac{11^{104} + 1}{11^8 + 1} = 86759222313428390812218077095850708048977$$
$$\times\, 108488104853637470612961399842972948409834611525790577216753.$$

The number field sieve method was used to completely factor the 138-digit number $2^{457} + 1$, which is equal to $3 \times P49 \times P89$, Pn denoting a prime with n digits. This was one of the good successes of the special number field sieve (SNFS) method, achieved by A.K. Lenstra and M.S. Manasse in November 1989; newspapers reported this feat, sometimes at front page!

In 1992, A.K. Lenstra and D. Bernstein factored the 158-digit Mersenne number M_{523} into two prime factors with 69 and 90 digits respectively, using an SNFS implementation on two massively parallel supercomputers.

An extraordinary factorization was announced in April 1999 by a group calling itself *The Cabal*. Using SNFS again, they factored the repunit number $(10^{211} - 1)/9$ into a product $P93 \times P118$, establishing a record for the largest penultimate prime factor ever found. This was the collective effort of S. Cavallar, B. Dodson, A. Lenstra, P. Leyland, W. Lioen, P. Montgomery, H. te Riele and P. Zimmermann.

In the following subsection I shall discuss public key cryptography, where numbers are involved which should be extremely difficult to factorize.

For a deeper understanding of primality and factorization, I warmly recommend the new book by Crandall & Pomerance (2001). It contains the most important methods and proofs and was written by two renowned authorities in the subject.

Anyone interested in primality testing, factorization, or similar calculations with very large numbers needs, of course, access to high-speed sophisticated computers of the latest generation. There is still pioneering work to be done in the development of gadgets adaptable to personal computers. These will allow us to reach substantial results in the comfort of home. If it is snowing outside—as is often the case in Canada—you may test your prime, keeping warm feet.

E   Public Key Cryptography

Owing to the proliferation of means of communication and the need to send messages—like bank transfers, love letters, instructions for buying stocks, secret diplomatic information, as, for example, reports of spying activities—it has become very desirable to develop a safe method of coding messages. In the past, codes have been kept secret, known only to the parties sending and receiving the messages, but it has often been possible to study the intercepted messages and crack the code. In simpler cases, it would be enough to study the frequency of symbols in the message. In war situations, this had disastrous consequences.

Great progress in cryptography came with the advent of public key crypto-systems. The main characteristics of the system are its simplicity, the public key, and the extreme difficulty in cracking it. The idea was proposed in 1976 by Diffie & Hellman, and the effective implementation was proposed in 1978 by Rivest, Shamir, & Adleman. This crypto-system is therefore called the RSA-system. I shall describe it now.

Each letter or sign, including blank space, corresponds to a 3-digit number. In the *American Standard Code for Information Interchange* (ASCII), this correspondence is the following:

—	A	B	C	D	E	F	G	H
032	065	066	067	068	069	070	071	072

I	J	K	L	M	N	O	P	Q
073	074	075	076	077	078	079	080	081

R	S	T	U	V	W	X	Y	Z
082	083	084	085	086	087	088	089	090

Each letter or sign of the message is replaced by its corresponding 3-digit number, giving rise to a number M, which represents the message.

Each user A of the system lists in a public directory his key, which is a pair of positive integers: (n_A, s_A). The first integer n_A is a product of two primes, $n_A = p_A q_A$, which are chosen to be large and are kept secret. Moreover, s_A is chosen to be relatively prime with both $p_A - 1$, $q_A - 1$.

To send a message M to another user B, A encrypts M—the way to encode M depends on who will receive it. Upon receiving the encoded message from A, the user B decodes it using his own secret decoding method.

In detail, the process goes as follows. If the message $M \geq n_B$, it suffices to break M into smaller blocks; so it may be assumed that $M < n_B$. If $\gcd(M, n_B) \neq 1$, a dummy letter is added to the end of M, so that for the new message, $\gcd(M, n_B) = 1$.

A sends to B the encoded message $E_B(M) = M'$, $1 \leq M' < n_B$, where M' is the residue of M^{s_B} modulo n_B: $M' \equiv M^{s_B} \pmod{n_B}$.

In order to decode M', the user B calculates t_B, $1 \leq t_B < (p_B - 1)(q_B - 1) = \varphi(n_B)$, such that $t_B s_B \equiv 1 \pmod{\varphi(n_B)}$; this is done once and for all. Then

$$D_B(M') = M'^{t_B} \equiv M^{s_B t_B} \equiv M \pmod{n_B},$$

so B may read the message M. How simple!

In truth, as it always happens, some technical problems appear. They are discussed in specialized books and numerous articles. Here I adopt a simplistic point of view, illustrated with an example. To make it easier, the message is encoded by groups of two letters— which is not what happens in practice.

Now put your hand in your pocket and pick up your little calculator. Below is an encoded message which a certain person is sending to an individual whose public key is (n, s), where $n = 156287$, $s = 181$:

15147403692507697411796402929902665403692510174310970109517915207006804505517600832900157414996603153311786415459901390703153301398601235306804513375012651013734911786411333812898611786411005204760700157401073800377209664211786407083810914501109811786402860011786405654711786408356704127110914505600611786408356704127110914505600

You don't know the secret prime factors of n. Can you decode the message? The answer is printed somewhere in this book.

I shall now say a little bit on how to crack the crypto-system. It is necessary to discover $\varphi(n_A)$ for each user A. This is equivalent to the factorization of n_A. Indeed, if p_A, q_A are known, then $\varphi(n_A) = (p_A - 1)(q_A - 1)$. Conversely, putting $p = p_A$, $q = q_A$, $n = n_A$, from $\varphi(n) = (p - 1)(q - 1) = n + 1 - (p + q)$, $(p + q)^2 - 4n = (p - q)^2$ (if $p > q$), then

$$p + q = n + 1 - \varphi(n),$$
$$p - q = \sqrt{[n + 1 - \varphi(n)]^2 - 4n},$$

and from this, p, q are expressed in terms of n, $\varphi(n)$.

There is much more to be said about the RSA crypto-system:

(a) how to send "signed" messages, so that the receiver can unmistakably identify the sender;

(b) how to choose well the prime factors of the numbers n_A of the keys, so that the cracking of the system is unfeasible with currently known means.

In relation to (b), it is of foremost importance for the protection of the message that the public key can not be factorized. So, how many digits should the key have in order to make the potential factoring time prohibitive?

To test this point, various keys have been proposed to mathematicians as a factoring challenge. Among them was the following 512-bit number, called RSA-155 to indicate that it has 155 decimal digits:

RSA-155 =
10941738641570527421809707322040357612003732945449205990913842131476349984288934784717997257891267332497625752899781833797076537244027146743531593354333897

This number had carefully been generated as a possible key for the Rivest-Shamir-Adleman method. The challenge to factorize it was broken in August 1999 by a team of scientists from six different countries, led by H. te Riele. They used the general number field sieve to disclose the following two 78-digit prime factors:

> 1026395928297411057720541965739916759007165678080380668033419335217907113077779,
>
> 1066034883801684548209272203600128786792079585759892915222706082371930628808643

This breakthrough showed, much earlier than expected when the practical use of the RSA method was started, that the popular key-size of 512 bits is no longer safe. As a result, 768-bit keys (about 230 digits) are now recommended as the minimum for achieving reliable security. Their two prime factors p, q, chosen at random, should be of equal size.

The current RSA factoring challenge includes, in a notation indicating number of bits, the numbers RSA-576 (174 decimal digits) through RSA-2048 (617 digits). Rewards range from $10,000 to $200,000 (US Dollars).

For all these questions, the reader may consult the original papers of Rivest, Shamir & Adleman (1978), and of Rivest (1978). There are, of course, many expository papers and books on the subject. See the paper by Couvreur & Quisquater (1982) as well as— pardon me the other writers of nice expository papers—the books of Riesel (1985), Koblitz (1987), Bressoud (1989), Coutinho (1999), and Wagstaff (2003). And, for example, the lecture notes of Lemos (1989), which are written in Portuguese—it is like studying cryptography in an encrypted language. Perhaps all this at Copacabana Beach.

3

Are There Functions Defining Prime Numbers?

To determine prime numbers, it is natural to ask for functions $f(n)$ defined for all natural numbers $n \geq 1$, which are computable in practice and produce some or all prime numbers.

For example, one of the following conditions should be satisfied:

(a) $f(n) = p_n$ (the nth prime) for all $n \geq 1$;

(b) $f(n)$ is always a prime number, and if $n \neq m$, then $f(n) \neq f(m)$;

(c) the set of prime numbers is equal to the set of positive values assumed by the function.

Clearly, condition (a) is more demanding than (b) and than (c).

The results reached have been rather disappointing, except for some of theoretical importance, related to condition (c).

I Functions Satisfying Condition (a)

In their famous book, Hardy & Wright asked:

(1) Is there a formula for the nth prime number?

(2) Is there a formula for a prime, in terms of the preceding primes?

What is intended in (1) is to find a closed expression for the nth prime p_n, in terms of n, by means of functions that are computable and, if possible, classical. Intimately related with this problem is to find reasonable expressions for the function counting primes.

For every real number $x > 0$ let $\pi(x)$ denote the number of primes p such that $p \leq x$.

This is a traditional notation for one of the most important functions in the theory of prime numbers. I shall return to it in Chapter 4. Even though the number $\pi = 3.14\ldots$ and the function $\pi(x)$ do occur below in the same formula, this does not lead to any ambiguity.

First I indicate a formula for $\pi(m)$, given by Willans in 1964. It is based on the classical Wilson theorem which I proved in Chapter 2.

For every integer $j \geq 1$ let

$$F(j) = \left[\cos^2 \pi \frac{(j-1)! + 1}{j}\right],$$

where $[x]$ indicates the only integer n such that the real number x verifies $n \leq x < n+1$.

So for any integer $j > 1$, $F(j) = 1$ when j is a prime, while $F(j) = 0$ otherwise. Also $F(1) = 1$.

Thus

$$\pi(m) = -1 + \sum_{j=1}^{m} F(j).$$

Willans also expressed $\pi(m)$ in the form

$$\pi(m) = \sum_{j=2}^{m} H(j) \quad \text{for } m = 2, 3, \ldots,$$

where

$$H(j) = \frac{\sin^2 \pi \dfrac{((j-1)!)^2}{j}}{\sin^2 \dfrac{\pi}{j}}.$$

Mináč gave an alternate (unpublished) expression, which involves neither the cosine nor the sine:

$$\pi(m) = \sum_{j=2}^{m} \left[\frac{(j-1)! + 1}{j} - \left[\frac{(j-1)!}{j}\right]\right].$$

Proof. The proof of Mináč's formula is quite simple, and since it is not published anywhere, I will give it here.

First a remark: if $n \neq 4$ is not a prime, then n divides $(n-1)!$. Indeed, either n is equal to a product $n = ab$, with $2 \leq a, b \leq n-1$, and $a \neq b$, or $n = p^2 \neq 4$. In the first alternative, n divides $(n-1)!$; in the second case, $2 < p \leq n-1 = p^2 - 1$, so $2p \leq p^2 - 1$ and n divides $2p^2 = p \times 2p$, which divides $(n-1)!$.

For any prime j, by Wilson's theorem, $(j-1)! + 1 = kj$ (where k is an integer), so

$$\left[\frac{(j-1)!+1}{j} - \left[\frac{(j-1)!}{j}\right]\right] = \left[k - \left[k - \frac{1}{j}\right]\right] = 1.$$

If j is not a prime and $j \geq 6$, then $(j-1)! = kj$ (where k is an integer), by the above remark. Hence

$$\left[\frac{(j-1)!+1}{j} - \left[\frac{(j-1)!}{j}\right]\right] = \left[k + \frac{1}{j} - k\right] = 0.$$

Finally, if $j = 4$, then

$$\left[\frac{3!+1}{4} - \left[\frac{3!}{4}\right]\right] = 0.$$

This is enough to prove the formula indicated for $\pi(m)$.   □

With the above notations, Willans gave the following formula for the nth prime:

$$p_n = 1 + \sum_{m=1}^{2^n} \left[\left[\frac{n}{\sum_{j=1}^{m} F(j)}\right]^{1/n}\right]$$

or, in terms of the prime counting function,

$$p_n = 1 + \sum_{m=1}^{2^n} \left[\left[\frac{n}{1 + \pi(m)}\right]^{1/n}\right].$$

For the related problem of expressing a prime q in terms of the prime p immediately preceding it, Willans gave the formula:

$$q = 1 + p + F'(p+1) + F'(p+1)F'(p+2) + \cdots + \prod_{j=1}^{p} F'(p+j),$$

where $F'(j) = 1 - F(j)$ and $F(j)$ was defined above.

Another formula for the smallest prime greater than $m \geq 2$ was given by Ernvall, while still a student, and published in 1975: Let

$$d = \gcd((m!)^{m!} - 1, (2m)!),$$

let

$$t = \frac{d^d}{\gcd(d^d, d!)},$$

and let a be the unique integer such that d^a divides t, but d^{a+1} does not divide t. Then the smallest prime larger than m is

$$p = \frac{d}{\gcd(t/d^a, d)}.$$

Taking $m = p_{n-1}$, this yields a formula for p_n.

Despite the nil practical value of these formulas, I tend to believe that they may have some relevance to logicians who wish to understand clearly how various parts of arithmetic may be deduced from different axiomatizations or from fragments of Peano's arithmetic.

In 1971, Gandhi gave a formula for the nth prime p_n. To explain it, I require the *Möbius function*, which is one of the most important arithmetic functions. The Möbius function is defined as follows:

$$\begin{cases} \mu(1) = 1, \\ \text{if } n \text{ is the product of } r \text{ distinct primes, then } \mu(n) = (-1)^r, \\ \text{if the square of a prime divides } n, \text{ then } \mu(n) = 0. \end{cases}$$

Let $P_{n-1} = p_1 p_2 \cdots p_{n-1}$. Then Gandhi showed:

$$p_n = \left[1 - \frac{1}{\log 2} \log \left(-\frac{1}{2} + \sum_{d | P_{n-1}} \frac{\mu(d)}{2^d - 1} \right) \right],$$

or equivalently, p_n is the only integer such that

$$1 < 2^{p_n} \left(-\frac{1}{2} + \sum_{d | P_{n-1}} \frac{\mu(d)}{2^d - 1} \right) < 2.$$

The following simple proof was given by Vanden Eynden in 1972.

Proof. For simplicity of notation, let $Q = P_{n-1}$, $p_n = p$, and

$$S = \sum_{d | Q} \frac{\mu(d)}{2^d - 1}.$$

So

$$(2^Q - 1)S = \sum_{d|Q} \mu(d)\frac{2^Q - 1}{2^d - 1} = \sum_{d|Q} \mu(d)\left(1 + 2^d + 2^{2d} + \cdots + 2^{Q-d}\right).$$

If $0 \le t < Q$, the term $\mu(d)2^t$ occurs exactly when d divides $\gcd(t, Q)$. So the coefficient of 2^t in the last sum is $\sum_{d|\gcd(t,Q)} \mu(d)$; in particular, for $t = 0$ it is equal to $\sum_{d|Q} \mu(d)$.

However, for any integer $m \ge 1$ it is well known, and easy to show, that

$$\sum_{d|m} \mu(d) = \begin{cases} 1 & \text{if } m = 1, \\ 0 & \text{if } m > 1. \end{cases}$$

Writing $\sum'_{0<t<Q}$ for the sum extended over all t, such that $0 < t < Q$ and $\gcd(t, Q) = 1$, then $(2^Q - 1)S = \sum'_{0<t<Q} 2^t$; the largest index t in this summation is $t = Q - 1$. It follows that

$$2(2^Q - 1)\left(-\frac{1}{2} + S\right) = -(2^Q - 1) + \sum_{0<t<Q}{}' 2^{t+1} = 1 + \sum_{0<t<Q-1}{}' 2^{t+1}.$$

If $2 \le j < p_n = p$, there exist some prime q such that $q < p_n = p$ (so $q \mid Q$) and $q \mid Q - j$. Hence, every index t in the above sum satisfies $0 < t \le Q - p$. Thus,

$$\frac{2^{Q-p+1}}{2 \times 2^Q} < -\frac{1}{2} + S = \frac{1 + \displaystyle\sum_{0<t\le Q-p}{}' 2^{t+1}}{2(2^Q - 1)} < \frac{2^{Q-p+2}}{2 \times 2^Q},$$

where the inequalities are easy to establish.

Hence multiplying with 2^p, it follows that

$$1 < 2^p\left(-\frac{1}{2} + S\right) < 2. \qquad \square$$

In 1974, Golomb gave another proof, which I find illuminating. He described it as being the sieve of Eratosthenes (see Chapter 2, Section I) performed on the binary expansion of 1.

To each positive integer n, assign the probability or weight $W(n) = 2^{-n}$. It is clear that $\sum_{n=1}^{\infty} W(n) = 1$.

In this distribution, the probability that a random integer be a multiple of a fixed integer $d \geq 1$, is

$$M(d) = \sum_{n=1}^{\infty} W(nd) = \sum_{n=1}^{\infty} 2^{-nd} = \frac{1}{2^d - 1}.$$

The probability that a random integer be relatively prime to a fixed integer $m \geq 1$ is easily seen to be

$$R(m) = 1 - \sum_{p|m} M(p) + \sum_{pp'|m} M(pp') - \sum_{pp'p''|m} M(pp'p'') + \cdots$$

$$= \sum_{d|m} \mu(d) M(d) = \sum_{d|m} \frac{\mu(d)}{2^d - 1}.$$

As before, let $Q = p_1 p_2 \cdots p_{n-1}$. So

$$R(Q) = \sum_{d|Q} \frac{\mu(d)}{2^d - 1},$$

but, on the other hand, in this distribution, $R(Q)$ is given directly as

$$R(Q) = \sum_{\gcd(m,Q)=1} W(m) = \frac{1}{2} + \frac{1}{2^{p_n}} + \frac{1}{2^{p_{n+1}}} + \alpha,$$

where α is a sum of reciprocals of some higher powers of 2. Thus,

$$R(Q) - \frac{1}{2} = \sum_{d|Q} \frac{\mu(d)}{2^d - 1} - \frac{1}{2} = \frac{1}{2^{p_n}} + \frac{1}{2^{p_{n+1}}} + \alpha.$$

Hence,

$$2^{p_n} \left(\sum_{d|Q} \frac{\mu(d)}{2^d - 1} - \frac{1}{2} \right) = 1 + \theta_n,$$

where $0 < \theta_n < 1$. So p_n is the only integer m such that

$$1 < 2^m \left(\sum_{d|Q} \frac{\mu(d)}{2^d - 1} - \frac{1}{2} \right) < 2,$$

and this is just another way of writing Gandhi's formula. Note that $0 < \theta_n < \frac{1}{2}$, because $p_{n+1} \geq p_n + 2$.

In binary notation, all this becomes even more transparent. Now, $W(n) = 0.000\ldots 1$ (with digit 1 at nth place), so $\sum_{n=1}^{\infty} W(n) = 0.1111\ldots = 1$.

For the even integers,

$$\sum_{n=1}^{\infty} W(2n) = 0.010101\ldots = \frac{1}{2^2 - 1} = \frac{1}{3}.$$

Subtracting, with $P_1 = p_1 = 2$:

$$R(P_1) = \sum_{2|n} W(n) = 0.101010\ldots = 1 - \frac{1}{3}.$$

By subtracting the multiples of 3 and adding back the twice-subtracted multiples of 6, obtain

$$Q(3) = 0.001001001\ldots = \frac{1}{2^3 - 1} = \frac{1}{7},$$

$$Q(6) = 0.000001000001\ldots = \frac{1}{2^6 - 1} = \frac{1}{63},$$

and, with $P_2 = p_1 p_2 = 6$,

$$R(P_2) = R(P_1) - Q(3) + Q(6) = 0.1000101000101000\ldots$$
$$= 1 - \frac{1}{3} - \frac{1}{7} + \frac{1}{63}.$$

Continuing in this way,

$$R(P_{n-1}) = 0.100\ldots 0100\ldots 0100\ldots = \frac{1}{2} + \frac{1}{2^{p_n}} + \frac{1}{2^{p_{n+1}}} + \alpha$$

and $\qquad R(P_{n-1}) - \dfrac{1}{2} = 0.000\ldots 010\ldots,$

where the first digit 1 appears in position p_n.

II Functions Satisfying Condition (b)

The number $f(n) = [\theta^{3^n}]$ is a prime for every $n \geq 1$; here θ is a number which is roughly equal to $1.3064\ldots$ (see Mills, 1947). Similarly,

$$g(n) = \left[2^{2^{2^{\cdot^{\cdot^{\cdot^{2^\omega}}}}}} \right]$$

(a string of n exponents) is a prime for every $n \geq 1$; here ω is a number which is roughly equal to $1.9287800\ldots$ (see Wright, 1951).

The fact that θ and ω are known only approximately and the numbers grow very fast, make these formulas no more than curiosities. For example, $g(1) = 3$, $g(2) = 13$, $g(3) = 16381$, and $g(4)$ has more than 5000 digits. There are many other formulas of a similar kind in the literature, but they are just as useless; see Dudley (1969).

At this point one might wonder: Why not try some polynomial with integral coefficients instead of these weird functions involving exponentials and the largest integer function? I devote the next section to this matter.

III   Prime-Producing Polynomials

Here is the answer to the question at the end of the preceding section:

If $f(X)$ is a non-constant polynomial with integral coefficients, in one indeterminate, then there exist infinitely many integers n such that $|f(n)|$ is not a prime number.

Proof. I may assume that there exists some integer $n_0 \geq 0$ such that $|f(n_0)| = p$ is a prime number. Since the polynomial is not constant, $\lim_{x\to\infty} |f(x)| = \infty$, so there exists $n_1 > n_0$ such that if $n \geq n_1$, then $|f(n)| > p$. For any h such that $n_0 + ph \geq n_1$, $f(n_0 + ph) = f(n_0)$ + (multiple of p) = (multiple of p). Since $|f(n_0 + ph)| > p$, then $|f(n_0 + ph)|$ is a composite integer. $\square$

The above proposition was indicated by Goldbach in a letter to Euler of September 28, 1743.

Now that no polynomial with integral coefficients in one indeterminate is fit for the purpose, could a polynomial with several indeterminates be suitable? Once again, this is excluded by the following strong negative result:

If $f(X_1, X_2, \ldots, X_m)$ is a polynomial with complex coefficients in m indeterminates, such that the values $|f(n_1, n_2, \ldots, n_m)|$ at arbitrary natural numbers $n_1, n_2, \ldots, n_m$ are primes, then f must be a constant.

Even though nonconstant polynomials $f(X)$ with integral coefficients have composite values (in absolute value) at infinitely many natural numbers, Euler discovered in 1772 one such polynomial $f(X)$ with a long string of prime values.

Here is Euler's famous example, communicated in a letter to Bernoulli: $f(X) = X^2 + X + 41$. For $k = 0, 1, 2, 3, \ldots, 39$, all its values are prime numbers, namely 41, 43, 47, 53, 61, 71, 83, 97, 113, 131, 151, 173, 197, 223, 251, 281, 313, 347, 383, 421, 461, 503, 547, 593, 641, 691, 743, 797, 853, 911, 971, 1033, 1097, 1163, 1231, 1301, 1373, 1447, 1523, 1601. For $k = 40$ the value is $1681 = 41^2$.

This example motivated new developments:

(1) To find, in a somewhat systematic way, linear, quadratic, or higher degree polynomials $f(X)$ such that for $k > 0$ as large as possible, $|f(0)|, |f(1)|, \ldots, |f(k)|$ are prime numbers.

(2) To identify polynomials which assume, in absolute value, prime values as often as possible (but not necessarily at consecutive integers).

First, I shall consider linear polynomials. For the case of quadratic polynomials, this study is intimately connected with the arithmetic of quadratic fields. So it is convenient to devote a subsection to this topic. For polynomials of degree higher than 2, very little is known; see Chapter 6, Section II.

A   PRIME VALUES OF LINEAR POLYNOMIALS

Let $f(X) = dX + q$, with $d > 1$, $q > 1$, $\gcd(d, q) = 1$. If $f(0)$ is a prime, then q is a prime. Moreover, $f(q)$ is not a prime. Thus for polynomials of degree one there are at most q sucessive values which are primes. This leads, therefore, to the following open problem: Is it true that for every prime q there exists an integer $d \geq 1$ such that $q, d + q, 2d + q, \ldots, (q - 1)d + q$ are primes? For example,

$q = 3$, $d = 2$ yield the primes 3, 5, 7;

$q = 5$, $d = 6$ yield the primes 5, 11, 17, 23, 29;

$q = 7$, $d = 150$ yield the primes 7, 157, 307, 457, 607, 757, 907.

This question is so difficult that I believe no one will prove it in the near future. Note that Lagrange had shown that if such d exists, then d is a multiple of $\prod_{p<q} p$.

In 1986, G. Löh discovered that for $q = 11$ the smallest value of d is $d = 1536160080$, and for $q = 13$ it is $d = 9918821194590$.

Record

The smallest d for $q = 17$ was found by P. Carmody, who obtained $d = 341976204789992332560$ in November 2001. Another impressive computational feat!

I shall examine this and other questions about primes in arithmetic progressions in Chapter 4, Section V.

B   On Quadratic Fields

In order to understand the problems in the case of quadratic polynomials, it is useful to gather now the facts about quadratic fields which will be required. This material may be found in a variety of books on number theory, so it is not necessary to make here specific recommendations. The classical theory of binary quadratic forms, elaborated by Gauss in his famous masterpiece *Disquisitiones Arithmeticae* (1801) is also presented in many modern books; see, for example, Flath (1989) and a long chapter in my own book *My Numbers, My Friends* (2000). Here I shall only mention what will be required in the sequel.

Let $d \neq 0, 1$ be a square-free integer, which may be positive or negative. The associated (fundamental) *discriminant* is

$$\Delta = \begin{cases} d & \text{if } d \equiv 1 \pmod 4, \\ 4d & \text{if } d \equiv 2 \text{ or } 3 \pmod 4. \end{cases}$$

Then $\mathbb{Q}(\sqrt{d}) = \mathbb{Q}(\sqrt{\Delta})$ is the associated quadratic field. It consists of all numbers $r + s\sqrt{d}$, where $r, s \in \mathbb{Q}$. If $d > 0$, the numbers $r + s\sqrt{d}$ are real and one speaks of a *real quadratic field*; if $d < 0$, then $\mathbb{Q}(\sqrt{d})$ is called an *imaginary quadratic field*.

Each element α of $\mathbb{Q}(\sqrt{d})$ is a root of some polynomial $f(X) \in \mathbb{Q}[X]$ of degree 2 when α is not in $\mathbb{Q}$ and, of course, of degree 1 when α is in $\mathbb{Q}$, having leading coefficient equal to 1. If each coefficient of $f(X)$ is in $\mathbb{Z}$, then α is called an *algebraic integer* of $\mathbb{Q}(\sqrt{d})$. The set A of algebraic integers is a subring of $\mathbb{Q}(\sqrt{d})$. It is easy to characterize the elements of A:

> If $d \equiv 1 \pmod 4$, then the elements of A are those of the form $(m + n\sqrt{d})/2$, where m, n are in $\mathbb{Z}$ and m, n have the same parity.

If $d \equiv 2$ or $3 \pmod 4$, then the elements of A are those of the form $m + n\sqrt{d}$, where m, n are in $\mathbb{Z}$.

For every $n \geq 1$ and $\alpha_1, \alpha_2, \ldots, \alpha_n$ in $\mathbb{Q}(\sqrt{d})$ the set of all elements of the form $\sum_{i=1}^{n} \gamma_i \alpha_i$, where $\gamma_1, \gamma_2, \ldots, \gamma_n$ are in A, is called the *fractional ideal* generated by $\alpha_1, \alpha_2, \ldots, \alpha_n$. If I is a fractional ideal, then $I + I \subseteq I$ and $AI \subseteq I$. Among the fractional ideals there are the *principal fractional ideals* $A\alpha$, for any α in $\mathbb{Q}(\sqrt{d})$. If I, J are fractional ideals, their product IJ is, by definition, the fractional ideal equal to the set of all finite sums $\sum_{i=1}^{n} \alpha_i \beta_i$ with α_i in I, β_i in J for all $i = 1, 2, \ldots, n$, and for all $n \geq 0$. With this multiplication, the non-zero fractional ideals form an abelian group, and the non-zero principal fractional ideals form a subgroup.

Gauss proved (in a different but equivalent language) that the quotient of the group of non-zero fractional ideals by the subgroup of non-zero principal fractional ideals is a finite abelian group. It is denoted by Cl_d (or also by Cl_Δ) and called the *class group* of $\mathbb{Q}(\sqrt{d})$ (or also of Δ). The number of elements of this group is denoted by h_d (or also by h_Δ) and it is called the *class number* of $\mathbb{Q}(\sqrt{d})$ (or also of Δ). To say that the class number is equal to 1 means that every fractional ideal is principal.

If α, β are in $\mathbb{Q}(\sqrt{d})$, by definition α *divides* β if there exists γ in A such that $\alpha\gamma = \beta$. The elements of A which divide 1 are called the *units* of A. A non-zero element $\pi \in A$, not a unit, is said to be a *prime algebraic integer* when the following condition is satisfied: if α, β are in A and $\alpha\beta = \pi$, then α, or β, is a unit.

It turns out that $h_d = 1$ if and only if the fundamental theorem is true in $\mathbb{Q}(\sqrt{d})$: every non-zero algebraic integer is, in a unique way (up to units and order of factors), equal to the product of prime algebraic integers.

The *exponent* e_d of the class group Cl_d is the maximum of the orders of the classes in the group Cl_d. Then the order of every class divides e_d. Of course $e_d = 1$ if and only if $h_d = 1$. Also, e_d is a power of 2 if and only if h_d is a power of 2.

The theory of genera of binary quadratic forms, developed by Gauss, leads to a more precise result. Let

$$h_d^* = \begin{cases} h_d & \text{if} \quad d < 0, \\ 2h_d & \text{if} \quad d > 0. \end{cases}$$

Let $N+1$ be the number of distinct prime factors of Δ. Then Gauss proved that 2^N divides h_d^*. Moreover, if $d < 0$ and $e_d = 1$ or 2, then $h_d = 2^N$.

It follows that if $h_d = 1$ then $N = 0$, so $d = -1, -2$, or $d = -p$, where p is a prime, $p \equiv 3 \pmod 4$. If $h_d = 2$ then $N = 1$, and there are three possible types for d:

(I) $d = -2p$, where p is any odd prime;

(II) $d = -p$, where p is a prime, $p \equiv 1 \pmod 4$;

(III) $d = -pq$, where $p < q$ are primes and $pq \equiv 3 \pmod 4$.

The binary quadratic forms with a given discriminant also form a finite abelian group (under composition of forms); this group is related in a definite way to the group Cl_Δ, which will not be explained here. The forms are organized into genera, each genus containing a certain number of classes. The theory of genera implies that the following two statements are equivalent when $d < 0$:

(1) The class group of binary quadratic forms of discriminant d has exponent 1 or 2.

(2) Each genus of the class group of binary quadratic forms of group discriminant d possesses only one class.

This last property admits an interpretation in terms of Euler's "numeri idonei", which I define now.

Let $n \geq 1$, let $E(n)$ be the set of odd natural numbers q such that there exists at most one pair of integers (x, y) with $x \geq 0$, $y \geq 0$, $q = x^2 + ny^2$, and $\gcd(x, ny) = 1$. The number n is called a *numerus idoneus* (also known as *convenient number*) if $E(n)$ contains no composite integer.

For example, from Fermat's study of integers of the form $x^2 + y^2$ it follows that $n = 1$ is a numerus idoneus. Gauss proved that if $d < 0$ then $e_d = 1$ or 2 if and only if $-d$ is a numerus idoneus. Euler indicated the following 65 numeri idonei:

1, 2, 3, 4, 5, 6, 7, 8, 9, 10, 12, 13, 15, 16, 18, 21, 22, 24,
25, 28, 30, 33, 37, 40, 42, 45, 48, 57, 58, 60, 70, 72, 78, 85,
88, 93, 102, 105, 112, 120, 130, 133, 165, 168, 177, 190,
210, 232, 240, 253, 273, 280, 312, 330, 345, 357, 385, 408,
462, 520, 760, 840, 1320, 1365, 1848.

The complete determination of all numeri idonei is still an open problem. It has been shown that besides the above there may exist at most one more—but, in fact, it is believed that no such number exists and the present list is complete.

Now I discuss the problem of the determination of imaginary quadratic fields $\mathbb{Q}(\sqrt{d})$ for which h_d has a prescribed value.

(1) Gauss showed that $h_d = 1$ when $d = -1, -2, -3, -7, -11, -19, -43, -67, -163$; he also conjectured, in article 303, that these are the only possible integers $d < 0$ for which $\mathbb{Q}(\sqrt{d})$ has $h_d = 1$. (To tell the truth, Gauss did not refer to imaginary quadratic fields, but rather to negative discriminants having only one class of binary quadratic forms.)

In a classical paper, Heilbronn & Linfoot showed, in 1934, that there exists at most one other integer $d < 0$ for which $\mathbb{Q}(\sqrt{d})$ has class number 1, and Lehmer (1933) showed that if such d exists, its absolute value must be quite large: $|d| > 5 \times 10^9$. In 1952, Heegner proved that no other such d could exist, but his proof contained some steps which were unclear, perhaps even incorrect.

Baker reached the same conclusion in 1966, with his method involving effective lower bounds on linear forms of three logarithms; this is also reported in his article of 1971. At about the same time, unaware of Heegner's results but with similar ideas concerning elliptic modular functions, Stark (1967) proved that no further possible integer d exists with the property indicated. So were determined all the imaginary quadratic fields with $h_d = 1$.

It was something of an anticlimax when, in 1968, Deuring was able to straighten out Heegner's proof, and when Stark (1969) also showed that, by 1949, Gelfond & Linnik could have used the theorem on the linear independence of just two logarithms (known at that time) to reach the same conclusion. The technical details involved in these proofs are outside the scope of this book.

(2) With his method of lower bounds of linear forms in logarithms, Baker (1971) proved that all imaginary quadratic fields with class number $h_d = 2$ can be effectively determined, but he gave no explicit bound for the discriminant. In the same volume of the same journal, Stark computed that if $h_d = 2$ then $|d| < 10^{1100}$. Montgomery & Weinberger showed in 1974 that if $10^{12} \leq |d| \leq 10^{1200}$, then h_d is not equal to 2. Earlier, Lehmer had verified that $h_d = 2$ does not occur for $10^6 \leq |d| \leq 10^{12}$.

As a result of these combined efforts, if $d < 0$ and $h_d = 2$, then $d = -5, -6, -10, -13, -15, -22, -35, -37, -51, -58, -91, -115, -123, -187, -235, -267, -403, -427$. Altogether, 18 discriminants.

The determination of all $d < 0$ such that $e_d = 2$ is still incomplete. In 1973, Weinberger determined an effective upper bound for $|d|$ such that $e_d = 2$.

(3) Concerning arbitrary values of h_d, Gauss conjectured that for every $n \geq 1$ there exist at most finitely many fields $\mathbb{Q}(\sqrt{d})$, with $d < 0$, such that $h_d = n$. This conjecture on the class number of imaginary quadratic fields was proved by Gross & Zagier (1983, 1986) and required the previous pathbreaking work of Goldfeld (1977). It follows from the more precise result:

For every $\varepsilon > 0$ there exists $C = C(\varepsilon) > 0$, which is effectively computable, such that for all $d < 0$, $h_d > C(\log |d|)^{1-\varepsilon}$.

The study of the class number of real quadratic fields is rather more difficult. Let it only be said here that Gauss conjectured that there exist infinitely many real quadratic fields with class number $h_d = 1$. It will be a great achievement to prove this conjecture.

C   PRIME-PRODUCING QUADRATIC POLYNOMIALS

It is convenient to say that the quadratic polynomial $f(X) = aX^2 + bX + c$ (with $a \geq 1$, $c \geq 1$) is a *prime-producing polynomial* when there exists $l > 2$ such that $f(0), f(1), \ldots, f(l-1)$ are primes. As it was indicated at the beginning of the section, l is bounded; the maximum value of l is called the *prime-producing length* of $f(X)$.

It will be seen that there is a surprising relation between prime-producing polynomials and class numbers of quadratic fields, which shall be discussed in this subsection. But there is also another relation, of quite a different kind, with the prime k-tuples conjecture. This will be seen in Chapter 4, Section IV.

Consider the polynomials $f_q(X) = X^2 + X + q$, where q is a prime number, and note that $f(q-1) = q^2$. Rabinowitsch proved in 1912 that the following statements are equivalent:

(1) $f_q(X)$ has prime-producing length equal to $q - 1$.

(2) The imaginary quadratic field $\mathbb{Q}(\sqrt{1-4q})$ has class number equal to 1.

In the same year, Frobenius proved that (2) implies (1). I note that Lehmer proved in 1936 once more that (1) implies (2), and more recently, Szekeres (1974), and Ayoub & Chowla (1981) established again the implication (2) $\Rightarrow$ (1). A detailed discussion of this result may be found in Cohn's book (1962) or in my own article (1988), where all the calculations and proofs are given in full detail, almost from first principles.

The complete determination of all imaginary quadratic fields with class number equal to 1 gives the possible values of q. In fact, of the nine values of d listed above, all but -1 and -2 are congruent to 1 modulo 4, and for the remaining ones, with the exclusion of 3, $(1-d)/4$ is a prime q. The primes so obtained are $q = 2, 3, 5, 11, 17, 41$. For these, and only for these values, the polynomial $f_q(X)$ has prime-producing length equal to $q-1$. This gives rise to the following record, which will never be surpassed.

RECORD

For polynomials of the type $X^2 + X + q$, the best possible result is already Euler's: $q = 41$ is the largest prime such that the polynomial assumes prime values for $k = 0, 1, \ldots, q - 2$.

There are many more quadratic polynomials with large prime-producing length. Legendre noted that the polynomials $2X^2 + q$, with $q = 3, 5, 11, 29$ have prime-producing length q, which is the maximal possible. A. Lévy observed in 1914 that $3X^2 + 3X + 23$ has prime-producing length equal to 22, and van der Pol & Speziali mentioned in 1951 that $6X^2 + 6X + 31$ has prime-producing length equal to 29. These examples illustrate the result which I explain now.

According to the three possible types of imaginary quadratic fields with class number 2, let

$f_{\mathrm{I}}(X) = 2X^2 + p$, if p is an odd prime;

$f_{\mathrm{II}}(X) = 2X^2 + 2X + \dfrac{p+1}{2}$, if p is a prime and $p \equiv 1 \pmod 4$;

$f_{\mathrm{III}}(X) = pX^2 + pX + \dfrac{p+q}{4}$, if $p < q$ are primes and $pq \equiv 3 \pmod 4$.

Note that $f_{\mathrm{I}}(p)$, $f_{\mathrm{II}}((p-1)/2)$ and $f_{\mathrm{III}}((p+q)/4-1)$ are composite. The following result is due, in this form, to Louboutin (1991); see also Frobenius (1912) and Hendy (1974).

(I) $h_{-2p} = 2$ if and only if $f_I(X)$ has prime-producing length p;

(II) $h_{-p} = 2$ if and only if $f_{II}(X)$ has prime-producing length $(p-1)/2$;

(III) $h_{-pq} = 2$ if and only if $f_{III}(X)$ has prime-producing length $(p+q)/4 - 1$.

Comparing with the list of $d < 0$ such that $h_d = 2$ this gives in the various cases:

(I) $p = 3, 5, 11, 29$;

(II) $p = 5, 13, 37$;

(III) $(p, q) = (3, 5), (3, 17), (3, 41), (3, 89), (5, 7), (5, 23), (5, 47), (7, 13), (7, 61), (11, 17), (13, 31)$.

In the same paper of 1991, Louboutin obtained the following characterizations of fields with negative discriminant and class number 4. Let $2 < q < p$, where p, q are primes. Then

(1) $h_{-2pq} = 4$ if and only if $2qk^2 + p$ is a prime for all $k = 0, 1, \ldots, p-1$.

(2) Let $pq \equiv 1 \pmod 4$. Then $h_{-pq} = 4$ if and only if $(pq+1)/2$ is a prime and for $k = 0, 1, \ldots, (p+q)/2 - 2$, $k \neq (p-1)/2$, $2qk^2 + 2qk + (p+q)/2$ is a prime.

Mollin developed a more embracing theory relating imaginary quadratic fields of exponent 2 and special prime-producing polynomials. Let $d \neq 0, 1$ be a square-free integer, assumed to be negative. Let Δ be the associated fundamental discriminant. Let $2 \leq q_1 < q_2 < \cdots < q_{N+1} = p$ be the distinct primes dividing Δ, and $q = \prod_{i=1}^N q_i$. For each m such that $1 \leq m$, $m \mid q$, define the following polynomial:

$$f_{\Delta,m}(X) = \begin{cases} mX^2 - \dfrac{\Delta}{4m} & \text{when } 4m \mid \Delta, \\ mX^2 + mX + \dfrac{m^2 - \Delta}{4m} & \text{when } 4m \nmid \Delta \end{cases}$$

(note that in the latter case, $4m$ divides $m^2 - \Delta$). Let $B_{\Delta,m} = [|\Delta|/4m]$. The following notation is used now: if $n = \prod_{i=1}^k p_i^{e_i}$ (where p_i are distinct primes), then $\nu(n) = \sum_{i=1}^k e_i$. Let

$$\Omega(f_{\Delta,m}(X)) = \max \{\nu(f_{\Delta,m}(k)) \mid k = 0, 1, \ldots, B_{\Delta,m} - 1\}.$$

Mollin proved (see his book of 1996 or his expository paper, 1997): With the above notations, and assuming $\Delta < -4$, the following conditions are equivalent:

(1) $e_d \leq 2$.

(2) $h_d = 2^N$ and for every m dividing q, $1 \leq m$,
$\Omega(f_{\Delta,m}(X)) + \nu(m) - 1 = N$.

(3) $h_d = 2^N$ and there exists m dividing q, $1 \leq m$, such that
$\Omega(f_{\Delta,m}(X)) + \nu(m) - 1 = N$.

Taking $m = q$ it follows that $f_{\Delta,q}(k)$ is a prime for $k = 0, 1, \ldots$, $B_{\Delta,q} - 1$. It is easy to verify that the earlier results of Rabinowitsch and Louboutin already mentioned are special cases.

In 1986, Sasaki proved: $h_d = 2$ if and only if $\Omega(f_{\Delta,1}(X)) = 2$. This result may be derived from Mollin's theorem (1996).

Once again, it is easy to verify that the prime-producing behaviour of the polynomials mentioned earlier may be explained in the light of Mollin's theorem: $d = -2 \times 29$, $\Delta = -8 \times 29$, $N = 1$; $-d$ is a numerus idoneus, so $e_d \leq 2$, $m = 2$, $\nu(m) = 1$, $B_{\Delta,m} = 29$, $f_{\Delta,m}(X) = 2X^2 + 29$, so $\Omega(2X^2 + 29) = 1$, thus $f_{\Delta,m}(k)$ is a prime for $k = 0, 1, \ldots, 28$

Hopefully, you may succeed in explaining in the same way the behaviour of the polynomials $3X^2 + 3X + 23$ and $6X^2 + 6X + 31$.

Up to now I have only considered the case of negative discriminant, but there is a similar theory for polynomials with positive discriminant. See, for example, Louboutin (1990), Mollin (1996, 1996a), as well as Sasaki (1986a).

Other prime-producing quadratic polynomials, not justified by any theory, have been found by computer search.

RECORD

The quadratic polynomial $f(X) = 36X^2 - 810X + 2753$, discovered by R. Ruby in 1990, is currently the one yielding the longest string of successive initial values such that $|f(k)|$ is prime, in this case for $k = 0, 1, \ldots, 44$.

The polynomials $103X^2 - 3945X + 34381$ (by R. Ruby) and $47X^2 - 1701X + 10181$ (by G. Fung) give 43 prime absolute values each.

For polynomials of higher degree, Dress & Landreau (2003) found $f(X) = 66x^3 + 83x^2 - 13735x + 30139$, giving 46 successive prime values $|f(k)|$ for $k = -26$ to $k = 19$, and $f(X) = 16x^4 + 28x^3 - 1685x^2 - 23807x + 110647$ with $|f(k)|$ prime for $k = -23$ to $k = 22$.

If the polynomial $f(X)$ is allowed to have rational coefficients such that $f(k)$ is always integer, then the record is

$$f(X) = \frac{1}{4}X^5 + \frac{1}{2}X^4 - \frac{345}{4}X^3 + \frac{879}{2}X^2 + 17500X + 70123.$$

For all integers k from $k = -27$ to $x = 29$, $|f(k)|$ is a prime number. So there are 57 consecutive integers giving 57 prime values. This record, which is also given by Dress & Landreau in their preprint, will not be easy to overcome.

D   THE PRIME VALUES AND PRIME FACTORS RACES

The prime values race

An investigation which has attracted many amateur mathematicians is the following. Let $f(X)$ be a non-constant polynomial with integral coefficients and $N \geq 1$; let

$$\pi^*_{f(X)}(N) = \#\{n \mid 0 \leq n \leq N, \ |f(n)| \text{ is a prime}\}.$$

Note that it is not required that the prime values be distinct.

Given N (usually large) and $d \geq 1$, the problem is to determine a polynomial $f(X)$ of degree d, such that $\pi^*_{f(X)}(N)$ is maximum. If desired, this search may be restricted to monic polynomials, that is, with leading coefficient 1, or even to monic polynomials of special types.

There are many tightly disputed records.

RECORDS

A.   For $N = 1000$, the quadratic polynomial $f(X) = 2X^2 - 1584X + 98621$, indicated by the amateur mathematician S.M. Williams (letter dated October 1993), provides the maximum number of primes, namely, $\pi^*_{f(X)}(1000) = 706$. Previous records, also by Williams, were

$$f_1(X) = 2X^2 - 1904X + 42403,$$
$$f_2(X) = 2X^2 - 1800X - 5749,$$

with $\pi^*_{f_1(X)}(1000) = 693$ and $\pi^*_{f_2(X)}(1000) = 686$, respectively.

B.   For $N = 1000$ and quadratic monic polynomials only, the present champion is

$$g(X) = (X - 499)^2 + (X - 499) + 27941,$$

with $\pi^*_{g(X)}(1000) = 669$, found by N. Boston (private communication).

Note that if $h(X) = X^2 + X + 27941$, then there are 600 distinct primes in the set $\{h(k) \mid 0 \leq k \leq 1000\}$. This may well be a record for the modified race (up to $N = 1000$) of quadratic polynomials with distinct prime values.

In 1973, Karst made the polynomial $f(X) = 2X^2 - 199$, which gives 597 prime values. On the other hand, Euler's polynomial $X^2 + X + 41$ yields 582 prime values.

The race between these two famous polynomials has been extended to much larger bounds N. In a message of December 1998, S.S. Gupta, who loves to calculate, communicated the following results:

$$\pi^*_{2X^2-199}(10^7) = 2381779,$$
$$\pi^*_{X^2+X+41}(10^7) = 2208197.$$

I shall consider this question again in Chapter 6, Section IV, for polynomials $f(X) = X^2 + X + A$, in the light of a conjecture by Hardy and Littlewood.

At age 78, M.L. Greenwood found—without using a computer—the polynomials

$$h_1(X) = -4X^2 + 381X - 8524,$$
$$h_2(X) = -2X^2 + 185X - 31819,$$

which assume 50 even values and 48 distinct odd prime values for $k = 0, 1, \ldots, 99$. The amateur Greenwood, joined by the professional Boston, found in 1995, with the aid of computers, the polynomial $f(X) = 41X^2 - 4641X + 88007$. It assumes 90 distinct prime values $f(k)$ for $k = 0, 1, \ldots, 99$.

For polynomials of degree greater than 2, in the race for cubic polynomials, with $N = 500$ (comparable to the 500 Miles race of Indianapolis), masterminded by Goetgheluck (1989), the winner was

$$f(X) = 2X^3 - 489X^2 + 39847X - 1084553.$$

It assumes 267 prime values for $k \leq 500$. In this race, the polynomials needed to have a leading coefficient 1 or 2, plus other restrictions on the size of their coefficients.

Later, in Chapter 6, Section III, I shall also consider the contrary phenomenon of polynomials assuming composite values for all integers $n = 0, 1, 2, \ldots$ up to some large N.

The least prime factors race

For any non-zero integer m let $P_0[m]$ denote the smallest prime factor of m. If $f(X) = aX^2 + bX + c$ with integral coefficients, $a \geq 1$, $c \neq 0$, let

$$P_0[f(X)] = \min\{P_0[f(k)] \mid k = 0, 1, 2, \ldots\}.$$

Moreover, for $N \geq 1$ let

$$q_N = \min\{P_0[f(k)] \mid k = 0, 1, 2, \ldots, N\}.$$

Since $q_1 \geq q_2 \geq \ldots$, there exists N such that $q_N < N$. Then $P_0[f(X)] = q_N$, which facilitates the computation of $P_0[f(X)]$.

Indeed, if p is a prime, $p < q_N$, and $p \mid f(M)$ for some $M > N$, then let $M = dp + r$ with $0 \leq r < p < q_N < N$. From $f(M) \equiv f(r)$ (mod p) it follows $p \mid f(r)$, so $p \geq q_N$, which is a contradiction.

Now let $f_A(X) = X^2 + X + A$ for $A \geq 1$. It has been shown that for every prime q there is an $A < q\#$ such that $P_0[f_A(X)] = q$. The race is to find the largest value for $P_0[f_A(X)]$. Incidentally, $P_0[f_{41}(X)] = 41$.

Record

If A is assumed to be prime and is required to be the smallest value giving $P_0[f_A(X)] = q$, then

$$P_0[X^2 + X + 33239521957671707] = 257.$$

This was found by P. Carmody in 2001. Previous records were by L. Rodríguez Torres in 1996 and 1995, respectively:

$$P_0[X^2 + X + 67374467] = 107,$$
$$P_0[X^2 + X + 32188691] = 71.$$

If A is prime but not necessarily minimal, then the largest known value of $P_0[f_A(X)]$ was obtained by M.J. Jacobson and H.C. Williams in 2002, using a special electronic number sieve described in their paper (2003). For the 57-digit prime $A = 605069291083802407422281785816166476624287786946587507887$ they found $P_0[f_A(X)] = 373$.

If the primality condition on A is also dropped, then they reached $P_0[f_A(X)] = 401$ for the 68-digit integer $A = 47392132545934368303439248393872932657758235983472584357825592740917$ (which is the product of six primes). The previous record was established by Lukes, Patterson & Williams (1995), who found by extensive computation:

$$P_0[X^2 + X + 2457080965043150051] = 281.$$

IV   Functions Satisfying Condition (c)

We recall that condition (c) requires that the set of prime numbers coincides with the set of positive values of the function. Surprisingly, this is possible and was discovered as a by-product in the investigation of Hilbert's tenth problem. The ideas come from logic and the results are quite extraordinary—even if at the present they have not yet found immediate practical application.

In my presentation, I will not enter into technical details, which would take me too far from the prime numbers. Thus, I shall have to trade rigor for intuition, and I'm counting on the good will of the reader. Please do not interpret what I'm going to write in any undesirable way! The nice article of Davis (1973) is recommended for those intrigued with the results which will follow.

Hilbert's tenth problem asked about the solution in integers $(x_1, \ldots, x_n)$ of diophantine equations $P(X_1, \ldots, X_n) = 0$, where P is any polynomial with integral coefficients, and any number of indeterminates. More exactly: To give an algorithm that may be applied to any diophantine equation and will tell if it has solution in integers.

An algorithm should be understood as a definite procedure, which could be implemented as a computer program consisting of finitely many successive steps and leading to an answer "yes" or "no"—the kind of manipulations that mathematicians agree as legitimate.

To the study of sets S of n-tuples $(x_1, \ldots, x_n)$ of positive integers, the central concept is the following: S is called a *diophantine set* if there exists a polynomial P with integral coefficients, in indeterminates $X_1, \ldots, X_n, Y_1, \ldots, Y_m$ $(m \geq 0)$, such that $(x_1, \ldots, x_n) \in S$

if and only if there exist positive integers $y_1, \ldots, y_m$ satisfying

$$P(x_1, \ldots, x_n, y_1, \ldots, y_m) = 0.$$

First, the trivial examples. Every finite set S of n-tuples of positive integers is diophantine. Indeed, if S consists of the n-tuples $\left(a_1^{(i)}, \ldots, a_n^{(i)}\right)$ (where $i = 1, \ldots, k$ and $k \geq 1$), let $Y_j^{(i)}$ ($i = 1, \ldots, k$, $j = 1, \ldots, n$) be different indeterminates and let

$$P = \prod_{i=1}^{k} \left[(X_1 - Y_1^{(i)})^2 + \cdots + (X_n - Y_n^{(i)})^2 \right].$$

If we put $Y_j^{(i)}$ equal to $a_j^{(i)}$ (for all i, j), then $(x_1, \ldots, x_n) \in S$ if and only if $P\left(x_1, \ldots, x_n, a_1^{(1)}, \ldots, a_n^{(1)}, a_1^{(k)}, \ldots, a_n^{(k)}\right) = 0$.

Here is another example: the set S of all composite positive integers. Indeed, x is composite if and only if there exist positive integers y, z such that (x, y, z) is a solution of $X - (Y + 1)(Z + 1) = 0$.

The following fact, noted by Putnam in 1960, is not difficult to show:

A set S of positive integers is diophantine if and only if there exists a polynomial Q with integral coefficients (in $m \geq 1$ indeterminates) such that

$$S = \{Q(x_1, \ldots, x_m) \geq 1 \mid x_1 \geq 1, \ldots, x_m \geq 1\}.$$

The next step in this theory consists in establishing that the set of prime numbers is diophantine. For this purpose, it is necessary to examine the definition of prime numbers from the vantage of the theory of diophantine sets.

A positive integer x is a prime if and only if $x > 1$ and for any integers y, z such that $y \leq x$ and $z \leq x$, either $yz < x$, or $yz > x$, or $y = 1$, or $z = 1$. This definition of prime numbers contains bounded universally quantified occurrences of y, z, namely, $y \leq x$, $z \leq x$.

Another possible definition of prime numbers is the following. The positive integer x is a prime if and only if $x > 1$ and $\gcd((x-1)!, x) = 1$. The latter condition is rephrased as follows: There exist positive integers a, b such that $a(x - 1)! - bx = 1$; note that if a or b is negative, taking a sufficiently large integer k, then $a' = a + kx > 0$, $b' = b + k(x - 1)! > 0$ and $a'(x - 1)! - b'x = 1$.

Using one or the other characterization of prime numbers, and with the aid of the theory developed by Putnam, Davis, J. Robinson and Matijasevič, the following important theorem was produced:

The set of prime numbers is diophantine.

A combination of these results leads to the following astonishing result:

There exists a polynomial, with integral coefficients, such that the set of prime numbers coincides with the set of positive values assumed by this polynomial, as the variables range in the set of nonnegative integers.

It should be noted that this polynomial also takes on negative values, and that a prime number may appear repeatedly as a value of the polynomial.

In 1971, Matijasevič indicated a system of algebraic relations leading to such a polynomial (without writing it explicitly) with degree 37, in 24 indeterminates; in the English translation of his paper this was improved to degree 21, and 21 indeterminates.

An explicit polynomial with this property, of degree 25, in the 26 indeterminates $a, b, c, \ldots, z$, was given by Jones, Sato, Wada & Wiens in 1976:

$$
\begin{aligned}
(k+2)\{ & 1 - [wz + h + j - q]^2 - [(gk + 2g + k + 1)(h + j) + h - z]^2 \\
& - [2n + p + q + z - e]^2 - [16(k+1)^3(k+2)(n+1)^2 + 1 - f^2]^2 \\
& - [e^3(e+2)(a+1)^2 + 1 - o^2]^2 - [(a^2 - 1)y^2 + 1 - x^2]^2 \\
& - [16r^2y^4(a^2 - 1) + 1 - u^2]^2 - [((a + u^2(u^2 - a))^2 - 1)(n + 4dy)^2 \\
& + 1 - (x + cu)^2]^2 - [n + l + v - y]^2 \\
& - [(a^2 - 1)l^2 + 1 - m^2]^2 - [ai + k + 1 - l - i]^2 \\
& - [p + l(a - n - 1) + b(2an + 2a - n^2 - 2n - 2) - m]^2 \\
& - [q + y(a - p - 1) + s(2ap + 2a - p^2 - 2p - 2) - x]^2 \\
& - [z + pl(a - p) + t(2ap - p^2 - 1) - pm]^2 \}.
\end{aligned}
$$

One is obviously tempted to reduce the number of indeterminates, or the degree, or both. But there is a price to pay. If the number n of indeterminates is reduced, then the degree d increases, and vice versa, if the degree d is forced to be smaller, then n must increase.

This is illustrated in Table 12 which concerns prime-representing polynomials.

Table 12. Polynomials giving prime numbers

n = number of indeterminates	d = degree	Author	Year	Remarks
24	37	Matijasevič	1971	Not written explicitly
21	21	Same author	1971	
26	25	Jones, Sato, Wada & Wiens	1976	First explicit polynomial
42	5	Same authors	1976	Record low degree, not written explicitly
12	13697	Matijasevič	1976	
10	about 1.6×10^{45}	Same author	1977	Record low number of indeterminates, not written explicitly

It is not known which is the minimum possible number of variables (it cannot be 2). However, Jones showed that there is a prime-representing polynomial of degree at most 5.

The same methods used to treat the set of prime numbers apply also to other diophantine sets once their defining arithmetical properties are considered from the appropriate point of view.

This has been worked out by Jones. In a paper of 1975, Jones showed that the set of Fibonacci numbers is identical with the set of positive values at nonnegative integers, of the polynomial in 2 indeterminates and degree 5:

$$2xy^4 + x^2y^3 - 2x^3y^2 - y^5 - x^4y + 2y.$$

In 1979, Jones showed that each one of the sets of Mersenne primes, even perfect numbers, Fermat primes, corresponds in the same way to some polynomial in seven indeterminates, but with higher degree. He also wrote explicitly other polynomials with lower degree, and more indeterminates, representing the above sets.

Table 13. Polynomials giving various sets of numbers

Set	Number of indeterminates	Degree
Fibonacci numbers	2	5
Mersenne primes	13	26
	7	914
Even perfect numbers	13	27
	7	915
Fermat primes	14	25
	7	905

By a method of Skolem (see his book, 1938), for the three latter sets the degree may be reduced to 5; however, the number of variables increases to about 20.

For the set of Mersenne primes the polynomial in 13 indeterminates and degree 26 is the following:

$$n\{1 - [4b + 3 - n]^2 - b([2 + hn^2 - a]^2$$
$$+ [n^3d^3(nd + 2)(h + 1)^2 + 1 - m^2]^2$$
$$+ [db + d + chn^2 + g(4a - 5) - kn]^2$$
$$+ [(a^2 - 1)c^2 + 1 - k^2n^2]^2 + [4(a^2 - 1)i^2c^4 + 1 - f^2]^2$$
$$+ [(kn + lf)^2 - ((a + f^2(f^2 - a))^2 - 1)(b + 1 + 2jc)^2 - 1]^2)\}.$$

For the even perfect numbers, the polynomial in 13 indeterminates and degree 27 is

$$(2b + 2)n\{1 - [4b + 3 - n]^2 - b([2 + hn^2 - a]^2$$
$$+ [n^3d^3(nd + 2)(h + 1)^2 + 1 - m^2]^2$$
$$+ [db + d + chn^2 + g(4a - 5) - kn]^2$$
$$+ [(a^2 - 1)c^2 + 1 - k^2n^2]^2 + [4(a^2 - 1)i^2c^4 + 1 - f^2]^2$$
$$+ [(kn + lf)^2 - ((a + f^2(f^2 - a))^2 - 1)(b + 1 + 2jc)^2 - 1]^2)\}.$$

For the prime Fermat numbers, the polynomial in 14 indeterminates and degree 25 is

$$(6g + 5)\{1 - [bh + (a - 12)c + n(24a - 145) - d]^2$$
$$- [16b^3h^3(bh + 1)(a + 1)^2 + 1 - m^2]^2$$
$$- [3g + 2 - b]^2 - [2be + e - bh - 1]^2 - [k + b - c]^2$$
$$- [(a^2 - 1)c^2 + 1 - d^2]^2 - [4(a^2 - 1)i^2c^4 + 1 - f^2]^2$$
$$- [(d + lf)^2 - ((a + f^2(f^2 - a))^2 - 1)(b + 2jc)^2 - 1]^2\}.$$

4

How Are the Prime Numbers Distributed?

As I have already stressed, the various proofs of existence of infinitely many primes are not constructive and do not give an indication of how to determine the nth prime number. Equivalently, the proofs do not indicate how many primes are less than any given number N. By the same token, there is no reasonable formula or function representing primes.

It will, however, be possible to predict with rather good accuracy the number of primes smaller than N (especially when N is large); on the other hand, the distribution of primes in short intervals shows a kind of built-in randomness. This combination of "randomness" and "predictability" yields at the same time an orderly arrangement and an element of surprise in the distribution of primes. According to Schroeder (1984), in his intriguing book *Number Theory in Science and Communication*, these are basic ingredients of works of art. Many mathematicians will readily agree that this topic has a great aesthetic appeal.

Recall from Chapter 3 that for every real number $x > 0$, $\pi(x)$ denotes the number of primes p such that $p \leq x$; $\pi(x)$ is also called the *prime counting function*.

The matters to consider are the following:

(I) Properties of $\pi(x)$: the growth of $\pi(x)$, its order of magnitude, and comparison with other known functions.

(II) Results about the nth prime; the difference between consecutive primes, how small, how large, how irregular it may be. This includes the discussion of large gaps between consecutive primes, but leads also to several open problems, discussed below.

(III) Twin primes, their characterization and distribution.

(IV) Prime k-tuplets.

(V) Primes in arithmetic progressions.

(VI) Goldbach's famous conjecture.

(VII) The distribution of pseudoprimes and of Carmichael numbers.

Now I elaborate on these topics.

I The Function $\pi(x)$

The basic idea in the study of the function $\pi(x)$, or others related to the distribution of primes, is to compare with functions that are both classical and computable, and such that their values are as close as possible to the values of $\pi(x)$. Of course, this is not simple and, as one might expect, an error will always be present. So, for each approximating function, one should estimate the order of magnitude of the difference, that is, of the error. The following notions are therefore natural.

Let $f(x)$, $h(x)$ be positive real valued continuous functions, defined for $x \geq x_0 > 0$.

The notation $f(x) \sim h(x)$ means that $\lim_{x\to\infty}(f(x)/h(x)) = 1$; $f(x)$ and $h(x)$ are then said to be *asymptotically equal* as x tends to infinity. Note that their difference may actually tend to infinity.

If, under the above hypothesis, there exist constants C, C', $0 < C < C'$, and x_0, x_1, with $x_1 \geq x_0$, such that $C \leq f(x)/h(x) \leq C'$ for all $x \geq x_1$, then $f(x)$, $h(x)$ are said to have *the same order of magnitude*.

If $f(x)$, $g(x)$, $h(x)$ are real valued continuous functions defined for $x \geq x_0 > 0$, and $h(x) > 0$ for all $x \geq x_0$, the notation

$$f(x) = g(x) + O(h(x))$$

means that the functions $f(x)$ and $g(x)$ have a difference that is ultimately bounded (as x tends to infinity) by a constant multiple of $h(x)$; that is, there exists $C > 0$ and $x_1 \geq x_0$ such that for every $x \geq x_1$ the inequality $|f(x) - g(x)| \leq Ch(x)$ holds. This is a useful notation to express the size of the error when $f(x)$ is replaced by $g(x)$.

Similarly, the notation

$$f(x) = g(x) + o(h(x))$$

means that $\lim_{x \to \infty}[f(x) - g(x)]/h(x) = 0$, so, intuitively, the error is negligible in comparison to $h(x)$.

A   HISTORY UNFOLDING

It is appropriate to describe in historical order the various discoveries about the distribution of primes, culminating with the prime number theorem. This is how Landau proceeded in his famous treatise *Handbuch der Lehre von der Verteilung der Primzahlen*, which is the classical work on the subject. Another presentation, of historic interest and prior to Landau, may be found in a very long article by Torelli (1901), which is written in Italian.

Euler

First, I give a result of Euler which tells, not only that there are infinitely many primes, but also that "the primes are not so sparse as the squares." (This statement will be made clear shortly.)

Euler noted that for every real numer $\sigma > 1$ the series $\sum_{n=1}^{\infty}(1/n^\sigma)$ is convergent, and in fact, for every $\sigma_0 > 1$ it is uniformly convergent on the half-line $\sigma_0 \leq x < \infty$. Thus, it defines a function $\zeta(\sigma)$ (for $1 < \sigma < \infty$), which is continuous and differentiable. Moreover, $\lim_{\sigma \to \infty} \zeta(\sigma) = 1$ and $\lim_{\sigma \to 1+0}(\sigma - 1)\zeta(\sigma) = 1$. The function $\zeta(\sigma)$ is called the *zeta function*.

The link between the zeta function and the prime numbers is the following eulerian product, which expresses the unique factorization of integers as product of primes:

$$\sum_{n=1}^{\infty} \frac{1}{n^{\sigma}} = \prod_p \frac{1}{1 - \dfrac{1}{p^{\sigma}}} \quad \text{(for } \sigma > 1\text{)}.$$

In particular, this implies that $\zeta(\sigma) \neq 0$ for $\sigma > 1$.

With the same idea used in his proof of the existence of infinitely many primes (see Chapter 1), Euler proved in 1737:

The sum of the inverses of the prime numbers is divergent: $\sum_p (1/p) = \infty$.

Proof. Let N be an arbitrary natural number. Each integer $n \leq N$ is a product, in a unique way, of powers of primes p, $p \leq n \leq N$. Also for every prime p,

$$\sum_{k=0}^{\infty} \frac{1}{p^k} = \frac{1}{1 - \dfrac{1}{p}}.$$

Hence,

$$\sum_{n=1}^{N} \frac{1}{n} \leq \prod_{p \leq N} \left(\sum_{k=0}^{\infty} \frac{1}{p^k} \right) = \prod_{p \leq N} \frac{1}{1 - \dfrac{1}{p}}.$$

But

$$\log \prod_{p \leq N} \frac{1}{1 - \dfrac{1}{p}} = -\sum_{p \leq N} \log \left(1 - \frac{1}{p} \right),$$

and for each prime p,

$$-\log \left(1 - \frac{1}{p} \right) = \sum_{m=1}^{\infty} \frac{1}{m p^m} \leq \frac{1}{p} + \frac{1}{p^2} \left(\sum_{h=0}^{\infty} \frac{1}{p^h} \right)$$

$$= \frac{1}{p} + \frac{1}{p^2} \times \frac{1}{1 - \dfrac{1}{p}} = \frac{1}{p} + \frac{1}{p(p-1)}$$

$$< \frac{1}{p} + \frac{1}{(p-1)^2}.$$

Hence,

$$\log \sum_{n=1}^{N} \frac{1}{n} \le \log \prod_{p \le N} \frac{1}{1 - \frac{1}{p}} \le \sum_{p \le N} \frac{1}{p} + \sum_{p \le N} \frac{1}{(p-1)^2}$$

$$\le \sum_{p} \frac{1}{p} + \sum_{n=1}^{\infty} \frac{1}{n^2} \ .$$

But the series $\sum_{n=1}^{\infty}(1/n^2)$ is convergent. Since N is arbitrary and the harmonic series is divergent, then $\log \sum_{n=1}^{\infty}(1/n) = \infty$, and therefore the series $\sum_{p}(1/p)$ is divergent. □

As I have already mentioned, the series $\sum_{n=1}^{\infty}(1/n^2)$ is convergent. Thus, it may be said, somewhat vaguely, that the primes are not so sparsely distributed as the squares.

One of the beautiful discoveries of Euler was the sum of this series:

$$\sum_{n=1}^{\infty} \frac{1}{n^2} = \frac{\pi^2}{6} \ .$$

Euler also evaluated the sums $\sum_{n=1}^{\infty}(1/n^{2k})$ for every $k \ge 1$, thereby solving a rather elusive problem. For this purpose, he made use of the Bernoulli numbers, which are defined as follows:

$$B_0 = 1, \quad B_1 = -\frac{1}{2}, \quad B_2 = \frac{1}{6}, \quad \dots \ ,$$

B_k being recursively defined by the relation

$$\binom{k+1}{1} B_k + \binom{k+1}{2} B_{k-1} + \cdots + \binom{k+1}{k} B_1 + B_0 = 0.$$

These numbers are clearly rational, and it is easy to see that $B_{2k+1} = 0$ for every $k \ge 1$. They appear also as coefficients in the Taylor expansion:

$$\frac{x}{e^x - 1} = \sum_{k=0}^{\infty} \frac{B_k}{k!} x^k.$$

Using Stirling's formula,

$$n! \sim \frac{\sqrt{2\pi} n^{n+\frac{1}{2}}}{e^n} \quad \text{(as } n \to \infty),$$

it may also be shown that

$$|B_{2n}| \sim 4\sqrt{\pi n}\left(\frac{n}{\pi e}\right)^{2n};$$

hence the above series is convergent in the interval $|x| < 2\pi$.

Euler had already used the Bernoulli numbers to express the sums of equal powers of consecutive numbers:

$$\sum_{j=1}^{n} j^k = S_k(n) \quad (k \geq 1),$$

where

$$S_k(X) = \frac{1}{k+1}\left[X^{k+1} - \binom{k+1}{1}B_1 X^k + \binom{k+1}{2}B_2 X^{k-1}\right.$$
$$\left. + \cdots + \binom{k+1}{k}B_k X\right].$$

A similar expression was also obtained by Seki, in Japan, at about the same time.

Euler's formula giving the value of $\zeta(2k)$ is:

$$\zeta(2k) = \sum_{n=1}^{\infty}\frac{1}{n^{2k}} = (-1)^{k+1}\frac{(2\pi)^{2k}B_{2k}}{2(2k)!}.$$

In particular,

$$\zeta(2) = \sum_{n=1}^{\infty}\frac{1}{n^2} = \frac{\pi^2}{6} \quad \text{(already mentioned)},$$

$$\zeta(4) = \sum_{n=1}^{\infty}\frac{1}{n^4} = \frac{\pi^4}{90}, \quad \text{etc.}$$

Euler also considered the Bernoulli polynomials, defined by

$$B_k(X) = \sum_{i=0}^{k}\binom{k}{i}B_i X^{k-i} \quad (k \geq 0).$$

They may be used to rewrite the expression for $S_k(X)$, but more important is their application to a far-reaching generalization of

Abel's summation formula, namely, the well-known Euler-MacLaurin summation formulas:

If $f(x)$ is a continuous function, continuously differentiable as many times as required, if $a < b$ are integers, then, for every $k \geq 1$,

$$\sum_{n=a+1}^{b} f(n) = \int_a^b f(t)dt + \sum_{r=1}^{k}(-1)^r \frac{B_r}{r!}\{f^{(r-1)}(b) - f^{(r-1)}(a)\}$$
$$+ \frac{(-1)^{k-1}}{k!}\int_a^b B_k(t - [t])f^{(k)}(t)dt$$

(the notation $[t]$, explained earlier, denotes the integral part of t).

The reader is urged to consult the paper by Ayoub, *Euler and the zeta function* (1974), where there is a description of the many imaginative relations and findings of Euler concerning $\zeta(s)$—some fully justified, others only made plausible, but anticipating later works by Riemann.

Legendre

The first serious attempt to study the function $\pi(x)$ is due to Legendre (1808), who used the Eratosthenes sieve and proved that

$$\pi(N) = \pi(\sqrt{N}) - 1 + \sum \mu(d)\left[\frac{N}{d}\right].$$

The summation is over all divisors d of the product of all primes $p \leq \sqrt{N}$, and $\mu(n)$ denotes the Möbius function, which was already defined in Chapter 3, Section I.

As a consequence, Legendre showed that $\lim_{x\to\infty}(\pi(x)/x) = 0$, but this is a rather weak result.

Experimentally, Legendre conjectured in 1798 and again in 1808 that

$$\pi(x) = \frac{x}{\log x - A(x)},$$

where $\lim_{x\to\infty} A(x) = 1.08366\ldots$. It was shown forty years later by Tschebycheff (see below) that if $\lim_{x\to\infty} A(x)$ exists, it must be equal to 1. An easier proof was given by Pintz (1980).

Gauss

At age 15, in 1792, Gauss conjectured that $\pi(x)$ and the function logarithmic integral of x, defined by

$$\text{Li}(x) = \int_2^x \frac{dt}{\log t},$$

are asymptotically equal. Since $\mathrm{Li}(x) \sim x/\log x$, this implies that

$$\pi(x) \sim \frac{x}{\log x},$$

as was implicitly conjectured by Legendre. This conjecture was to be confirmed later, and is now known as the *prime number theorem*; I shall soon return to this matter.

The approximation of $\pi(x)$ by $x/\log x$ is only reasonably good, while it is much better using the logarithmic integral, as will be illustrated in Table 14.

Tschebycheff

Important progress for the determination of the order of magnitude of $\pi(x)$ was made by Tschebycheff, around 1850. He proved, using elementary methods, that for every $\varepsilon > 0$ there exists $x_0 > 0$ such that if $x > x_0$, then

$$(C' - \varepsilon)\frac{x}{\log x} < \pi(x) < (C + \varepsilon)\frac{x}{\log x},$$

where

$$C' = \log \frac{2^{1/2}\, 3^{1/3}\, 5^{1/5}}{30^{1/30}} = 0.92129\ldots, \qquad C = \frac{6}{5}C' = 1.10555\ldots .$$

Moreover, Tschebycheff showed that if the limit of

$$\frac{\pi(x)}{x/\log x}$$

exists (as $x \to \infty$), it must be equal to 1. He deduced also that Legendre's approximation of $\pi(x)$ cannot be true, unless 1.08366 is replaced by 1 (see Landau's book, page 17).

Tschebycheff also proved Bertrand's postulate that between any natural number $n \geq 2$ and its double there exists at least one prime. I shall discuss this proposition in more detail when I present the main properties of the function $\pi(x)$.

Tschebycheff worked with the function $\theta(x) = \sum_{p \leq x} \log p$, now called *Tschebycheff's function*, which yields basically the same information as $\pi(x)$, but is somewhat easier to handle.

Even though Tschebycheff came rather close, the proof of the fundamental prime number theorem, conjectured by Gauss, had to wait for about 50 more years, until the end of the century. During this time, important new ideas were contributed by Riemann.

Riemann

Riemann had the idea of defining the zeta function for complex numbers s having real parts greater than 1, namely,

$$\zeta(s) = \sum_{n=1}^{\infty} \frac{1}{n^s} .$$

The Euler product formula still holds, for every complex s with $\mathrm{Re}(s) > 1$.

Using the Euler-MacLaurin summation formula, $\zeta(s)$ is expressible as follows:

$$\zeta(s) = \frac{1}{s-1} + \frac{1}{2} + \sum_{r=2}^{k} \frac{B_r}{r!} s(s+1) \cdots (s+r-2)$$
$$- \frac{1}{k!} s(s+1) \cdots (s+k-1) \int_1^{\infty} B_k(x - [x]) \frac{dx}{x^{s+k}} .$$

Here k is any integer, $k \geq 1$, the numbers B_r are the Bernoulli numbers, which the reader should not confuse with $B_k(x - [x])$, the value of the kth Bernoulli polynomial $B_k(X)$ at $x - [x]$.

The integral converges for $\mathrm{Re}(s) > 1-k$, and since k is an arbitrary natural number, this formula provides the analytic continuation of $\zeta(s)$ to the whole plane. $\zeta(s)$ is everywhere holomorphic, except at $s = 1$, where it has a simple pole with residue 1, that is,

$$\lim_{s \to 1} (s-1)\zeta(s) = 1.$$

In 1859, Riemann established the functional equation for the zeta function. Since this equation involves the gamma function $\Gamma(s)$, I must first define $\Gamma(s)$. For $\mathrm{Re}(s) > 0$, a convenient definition is by means of the eulerian integral

$$\Gamma(s) = \int_0^{\infty} e^{-u} u^{s-1} du.$$

For arbitrary complex numbers s, it may be defined by

$$\Gamma(s) = \frac{1}{se^{\gamma s}} \prod_{n=1}^{\infty} \frac{e^{s/n}}{1 + \dfrac{s}{n}} ,$$

where γ is Euler's constant, equal to

$$\gamma = \lim_{n \to \infty} \left(1 + \frac{1}{2} + \cdots + \frac{1}{n} - \log n \right) = 0.577215665\ldots .$$

Euler's constant, also known with good reason as Mascheroni's constant by the Italians, is related to Euler's product by the following formula of Mertens:

$$e^\gamma = \lim_{n \to \infty} \frac{1}{\log p_n} \prod_{i=1}^{n} \frac{1}{1 - \dfrac{1}{p_i}} .$$

$\Gamma(s)$ is never equal to 0; it is holomorphic everywhere except at the points $0, -1, -2, -3, \ldots,$ where it has simple poles. For every positive integer n, $\Gamma(n) = (n-1)!$, so the gamma function is an extension of the factorial function. The gamma function satisfies many interesting relations, among which are the functional equations

$$\Gamma(s)\Gamma(1-s) = \frac{\pi}{\sin \pi s}, \qquad \Gamma(s+1) = s\Gamma(s),$$

and

$$\Gamma(s)\Gamma\left(s + \frac{1}{2}\right) = \frac{\sqrt{\pi}}{2^{2s-1}}\Gamma(2s).$$

Here is the functional equation for the Riemann zeta function:

$$\pi^{-s/2}\Gamma\left(\frac{s}{2}\right)\zeta(s) = \pi^{-(1-s)/2}\Gamma\left(\frac{1-s}{2}\right)\zeta(1-s).$$

For example, it follows from the functional equation that $\zeta(0) = -\frac{1}{2}$. The zeros of the zeta function are:

(a) Simple zeros at the points $-2, -4, -6, \ldots,$ which are called the trivial zeros.

(b) Zeros in the critical strip, consisting of the nonreal complex numbers s with $0 \le \mathrm{Re}(s) \le 1$.

Indeed, if $\mathrm{Re}(s) > 1$, then by the Euler product, $\zeta(s) \ne 0$. If $\mathrm{Re}(s) < 0$, then $\mathrm{Re}(1-s) > 1$, the right-hand side in the functional equation is not zero, so the zeros must be exactly at $s = -2, -4, -6, \ldots,$ which are the poles of $\Gamma(s/2)$.

The knowledge of the zeros in the critical strip has a profound influence on the understanding of the distribution of primes. A first thing to note is that the zeros in the critical strip are not real and they are symmetric about the real axis and the vertical line $\mathrm{Re}(x) = \frac{1}{2}$.

Riemann conjectured that all nontrivial zeros ρ of $\zeta(s)$ are on the critical line $\mathrm{Re}(s) = \frac{1}{2}$, that is, $\rho = \frac{1}{2} + i\gamma$. This is the famous *Riemann's hypothesis*, which has never been proved. It is undoubtedly ·one of the most difficult and important problems in number theory, and I am not wrong in saying, in the whole of mathematics. I shall return to it soon and narrate some modern developments.

Now I want to indicate very superficially how Riemann deduced a better approximation for $\pi(x)$. He counted also prime powers p^n, giving them the weight $1/n$. Thus he defined, for each real number $x > 0$, the expression

$$J(x) = \pi(x) + \frac{1}{2}\pi(x^{1/2}) + \frac{1}{3}\pi(x^{1/3}) + \frac{1}{4}\pi(x^{1/4}) + \cdots .$$

Note that the summands are 0 when $2^n > x$, so for each x the above expression is a finite sum. With the help of the Möbius function, by the Möbius inversion formula,

$$\pi(x) = \sum_{n=1}^{\infty} \frac{\mu(n)}{n} J(x^{1/n})$$

(once again a finite sum). The essential part of the work was the analytical formula for $J(x)$ in terms of the logarithmic integral with complex arguments.

Let $w = u + iv$, and let z be defined as πi, $-\pi i$, or 0, according to $v > 0$, $v < 0$, or $v = 0$. By definition,

$$\mathrm{Li}(e^w) = \int_C \frac{e^t}{t} dt,$$

where C is the horizontal line $C = \{s + iv \mid -\infty < s \leq u\}$. The fundamental analytic formula for the function $J(x)$, proved by Riemann, is the following: for all $x > 0$,

$$J(x) = \mathrm{Li}(x) - \sum_{\rho} \mathrm{Li}(x^\rho) - \log 2 + \int_x^{\infty} \frac{dt}{t(t^2 - 1)\log t} ;$$

here the sum is extended over all the nontrivial zeros ρ of the zeta function, in the upper half-plane. Substituting $J(x^{1/n})$ into the expression for $\pi(x)$ leads to the following expression of $\pi(x)$ in terms of the logarithmic integral:

$$\pi(x) = \sum_{n=1}^{\infty} \frac{\mu(n)}{n} \mathrm{Li}(x^{1/n}) + \text{other terms involving the roots.}$$

It goes without saying that the justification of the analytical manipulations is delicate and was challenging even for Riemann; moreover the estimation of the terms depending on the roots ρ is outright difficult. But, nevertheless, the function

$$R(x) = \sum_{n=1}^{\infty} \frac{\mu(n)}{n} \mathrm{Li}(x^{1/n}),$$

called the *Riemann function*, offers a rather excellent approximation to $\pi(x)$, confirmed by actual calculations, as it will be indicated in Table 14.

The Riemann function is computable by this quickly converging power series, given by Gram in 1893:

$$R(x) = 1 + \sum_{n=1}^{\infty} \frac{1}{n\zeta(n+1)} \times \frac{(\log x)^n}{n!}.$$

The work of Riemann on the distribution of primes is thoroughly studied in Edwards' book (1974), which I recommend without reservations. Other books on the Riemann zeta function are the classical treatise by Titchmarsh (1951) and the more recent volumes of Ivić (1985) and Patterson (1988).

de la Vallée Poussin and Hadamard

Riemann provided many of the tools for the proof of the fundamental *prime number theorem*:

$$\pi(x) \sim \frac{x}{\log x}.$$

Other tools came from the theory of complex analytic functions, which was experiencing a period of rapid growth.

The prime number theorem was raised to the status of a "most wanted theorem," and it was folklore to consider that he who would prove it would become immortal.

The theorem was established, not by one, but by two eminent analysts, independently, and in the same year (1896). No, they did not become immortals, as in some old Greek legend, but... almost! Hadamard lived to the age of 98, de la Vallée Poussin just slightly less, to the age of 96.

De la Vallée Poussin established the following fact: there exists $c > 0$ and $t_0 = t_0(c) > e^{2c}$, such that $\zeta(s) \neq 0$ for every $s = \sigma + it$ in the region:

$$\begin{cases} 1 - \dfrac{c}{\log t_0} \leq \sigma \leq 1, & \text{when } |t| \leq t_0, \\ 1 - \dfrac{c}{\log |t|} \leq \sigma \leq 1, & \text{when } t_0 \leq |t|. \end{cases}$$

Thus, in particular, $\zeta(1+it) \neq 0$ for every t, as shown by Hadamard.

The determination of a large zero-free region for $\zeta(s)$ was an important feature in the proof of the prime number theorem.

Not only did Hadamard and de la Vallée Poussin prove the prime number theorem. They have also estimated the error as being:

$$\pi(x) = \text{Li}(x) + O\big(xe^{-A\sqrt{\log x}}\big),$$

for some positive constant A. I shall soon tell how the error term was subsequently reduced by determining larger zero-free regions for the zeta function.

There have been many variants of proofs of the prime number theorem with analytical methods, and they appear in various books and papers; see, for example, Grosswald (1964). One which is particularly simple is due to Newman (1980).

There are other equivalent ways of formulating the prime number theorem. Using the Tschebycheff function, the theorem may be rephrased as follows:

$$\theta(x) \sim x.$$

Another formulation involves the summatory function of the *von Mangoldt function*. Let

$$\Lambda(n) = \begin{cases} \log p & \text{if } n = p^\nu \ (\nu \geq 1) \text{ and } p \text{ is a prime}, \\ 0 & \text{otherwise.} \end{cases}$$

This function, introduced by von Mangoldt, has the following interesting property relating to the logarithmic derivative of the zeta function:

$$-\frac{\zeta'(s)}{\zeta(s)} = \sum_{n=1}^{\infty} \frac{\Lambda(n)}{n^s}, \quad \text{for } \operatorname{Re}(s) > 1.$$

It is also related to the function $J(x)$ already encountered:

$$J(x) = \sum_{n \le x} \frac{\Lambda(n)}{\log n}.$$

The summatory function of $\Lambda(n)$ is defined to be

$$\psi(x) = \sum_{n \le x} \Lambda(n).$$

It is easily expressible in terms of Tschebycheff's function

$$\psi(x) = \theta(x) + \theta(x^{1/2}) + \theta(x^{1/3}) + \cdots.$$

The prime number theorem may also be formulated as:

$$\psi(x) \sim x.$$

Erdös and Selberg

It was believed for a long time that analytical methods could not be avoided in the proof of the prime number theorem. Thus, the mathematical community was surprised when both Erdös and Selberg showed, in 1949, how to prove the prime number theorem using essentially only elementary estimates of arithmetical functions.

Many such estimates of sums were already known, as, for example,

$$\sum_{n \le x} \frac{1}{n} = \log x + \gamma + O\left(\frac{1}{x}\right), \quad \text{where } \gamma \text{ is Euler's constant,}$$

$$\sum_{n \le x} \frac{1}{n^\sigma} = \frac{x^{1-\sigma}}{1-\sigma} + \zeta(\sigma) + O\left(\frac{1}{x^\sigma}\right), \quad \text{where } \sigma > 1,$$

$$\sum_{n \le x} \log n = x \log x - x + O(\log x),$$

$$\sum_{n \le x} \frac{\log n}{n} = \frac{1}{2}(\log x)^2 + C + O\left(\frac{\log x}{x}\right).$$

The above estimates are obtained using the Abel or Euler-Maclaurin summation formulas, and have really no arithmetical content. The following sums involving primes are more interesting:

$$\sum_{p \leq x} \frac{\log p}{p} = \log x + O(1),$$

$$\sum_{p \leq x} \frac{1}{p} = \log \log x + C + O\left(\frac{1}{\log x}\right), \quad \text{where } C = 0.2615\ldots,$$

$$\sum_{n \leq x} \frac{\Lambda(n)}{n} = \log x + O(1),$$

$$\sum_{n \leq x} \frac{\Lambda(n) \log n}{n} = \frac{1}{2}(\log x)^2 + O(\log x).$$

Selberg gave in 1949 the following estimate:

$$\sum_{p \leq x} (\log p)^2 + \sum_{pq \leq x} (\log p)(\log q) = 2x \log x + O(x),$$

where p, q are primes.

This estimate is, in fact, equivalent to each of the following:

$$\theta(x) \log x + \sum_{p \leq x} \theta\left(\frac{x}{p}\right) \log p = 2x \log x + O(x),$$

$$\sum_{n \leq x} \Lambda(n) \log n + \sum_{mn \leq x} \Lambda(m)\Lambda(n) = 2x \log x + O(x).$$

From his estimate, Selberg was able to give an elementary proof of the prime number theorem. At the same time, also using a variant of Selberg's estimate

$$\frac{\psi(x)}{x} + \frac{1}{\log x} \sum_{n \leq x} \frac{\psi(x/n)}{x/n} \frac{\Lambda(n)}{n} = 2 + O\left(\frac{1}{\log x}\right),$$

Erdös gave, with a different elementary method, his proof of the prime number theorem.

In 1970, Diamond & Steinig gave an elementary proof with an explicit error term. Diamond (1982) published a detailed and authoritative article about elementary methods in the distribution of prime numbers.

B   Sums Involving the Möbius Function

Even before Möbius had formally defined the function $\mu(n)$, Euler had already considered it. In 1748, based on experimental evidence, Euler conjectured that $\sum_{n=1}^{\infty} \mu(n)/n$ converges to 0. von Mangoldt proved this conjecture, as an application of the prime number theorem. Actually, the converse is also true: this fact implies the prime number theorem.

Also, for every s with $\mathrm{Re}(s) > 1$,

$$\sum_{n=1}^{\infty} \frac{\mu(n)}{n^s} = \frac{1}{\zeta(s)}.$$

In particular, with $s = 2$ it follows that for every $x > 1$,

$$\sum_{n \le x} \frac{\mu(n)}{n^2} = \frac{6}{\pi^2} + O\left(\frac{1}{x}\right).$$

The summatory function of the Möbius function is the *Mertens function*

$$M(x) = \sum_{n \le x} \mu(n).$$

It may be shown that the prime number theorem is also equivalent to the assertion that $\lim_{x \to \infty} M(x)/x = 0$. For details relative to the preceding statements, the reader may consult the books by Landau (1909), Ayoub (1963), or Apostol (1976).

In 1984, Daboussi gave an elementary proof that $\lim_{x \to \infty} M(x)/x = 0$, thus providing a new elementary proof of the prime number theorem, without appealing to Selberg's inequality.

Concerning the order of magnitude of $M(x)$, Mertens himself conjectured that $|M(x)| < \sqrt{x}$. This was a very important and difficult problem, investigated by classical number theorists like Stieltjes and Hadamard. Finally, in 1985, Odlyzko & te Riele proved that the conjecture is false, by showing that

$$\limsup_{x \to \infty} \frac{M(x)}{\sqrt{x}} > 1.06, \qquad \liminf_{x \to \infty} \frac{M(x)}{\sqrt{x}} < -1.009.$$

A paper by Pintz, published in 1987, gives an effective proof that the conjecture is false for some x_0 such that $\log x_0 < 3.21 \times 10^4$. See the paper of te Riele (1985) for details.

C   TABLES OF PRIMES

Now, I turn my attention to tables of prime numbers, and of factors of numbers (not divisible by 2, 3, or 5). The first somewhat extended tables are by Brancker in 1668 (table of least factor of numbers up to 100 000), Krüger in 1746 (primes up to 100 000), Lambert in 1770 (table of least factor of numbers up to 102 000), Felkel in 1776 (table of least factor of numbers up to 408 000), Vega in 1797 (primes up to 400 031), Chernac in 1811 (prime factors of numbers up to 1 020 000), and Burckhardt in 1816/7 (least factor of numbers up to 3 036 000).

Legendre and Gauss based their empirical observations on the available tables.

Little by little, the tables were extended. Thus, in 1856, Crelle presented to the Berlin Academy a table of primes up to 6 000 000, and this work was extended by Dase, before 1861, up to 9 000 000.

In this connection, the most amazing feat is Kulik's factor table of numbers to 100 330 220 (except for multiples of 2, 3, 5), entitled *Magnus Canon Divisorum pro omnibus numeris per 2, 3 et 5 non divisibilibus, et numerorum primorum interfacentium ad millies centena millia accuratius and* 100 300 201 *usque.* Kulik spent about 20 years preparing this table, and at his death in 1863, the eight manuscript volumes, with a total of 4212 pages, were deposited at the Academy of Sciences in Vienna (in February 1867).

In 1909, D. N. Lehmer published a table of factor numbers up to about 10 000 000, and in 1914 he published the list of primes up to that limit. This time, the volumes were widely distributed and easily accessible to mathematicians.

With the advent of computers, numerous tables became available on magnetic tapes. They may be easily generated in any interval of a reasonable size, by the Eratosthenes sieve.

To my readers, who have faithfully arrived up to this point, as a token of appreciation, and for their utmost convenience, I include a Table of Primes up to 10000 following the Bibliography. Enjoy yourself!

D THE EXACT VALUE OF $\pi(x)$ AND COMPARISON WITH $x/\log x$, $\mathrm{Li}(x)$, AND $R(x)$

Calculation of the exact value of $\pi(x)$

The exact value of $\pi(x)$ may be obtained by direct counting using tables, or by an ingenious method devised in 1871 by Meissel, which allowed him to go far beyond the range of the tables. In fact, to compute $\pi(x)$ the method requires the knowledge of the prime numbers $p \leq x^{1/2}$ as well as the values of $\pi(y)$ for $y \leq x^{2/3}$. It is based on the following formula:

$$\pi(x) = \varphi(x, m) + m(s + 1) + \frac{s(s-1)}{2} - 1 - \sum_{i=1}^{s} \pi\left(\frac{x}{p_{m+i}}\right),$$

where $m = \pi(x^{1/3})$, $n = \pi(x^{1/2})$, $s = n - m$, and $\varphi(x, m)$ denotes the number of integers a such that $a \leq x$ and a is not a multiple of $2, 3, \ldots, p_m$.

Even though the calculation of $\varphi(x, m)$ is long, when m is large, it offers no major difficulty. The calculation is based on the following simple facts:

Recurrence relation.

$$\varphi(x, m) = \varphi(x, m - 1) - \varphi\left(\left[\frac{x}{p_m}\right], m - 1\right).$$

Division property.   If $P_m = p_1 p_2 \cdots p_m$, if $a \geq 0$, $0 \leq r < P_m$, then

$$\varphi(a P_m + r, m) = a\varphi(P_m) + \varphi(r, m).$$

Symmetry property.   If $\frac{1}{2}P_m < r < P_m$, then

$$\varphi(r, m) = \varphi(P_m) - \varphi(P_m - r - 1, m).$$

Meissel determined, in 1885, the number $\pi(10^9)$ (however he found a value which is low by 56). A simple proof of Meissel's formula was given by Brauer in 1946. In 1959, Lehmer simplified and extended Meissel's method. In 1985, Lagarias, Miller & Odlyzko have further refined the method by incorporating new sieving techniques.

The exact value of $\pi(x)$ has been calculated for some x up to 4×10^{16} by Lagarias, Miller & Odlyzko (1985), and for some x up to

10^{18} by Deléglise & Rivat (1996). These computations were extended to $\pi(10^{20})$ by Deléglise (announced April 1996) and to $\pi(10^{21})$ by X. Gourdon (announced October 2000). The following table gives values of $\pi(x)$ up to that limit, which may be compared with the corresponding values of the functions $x/\log x$, $\mathrm{Li}(x)$, and $R(x)$.

Table 14.

Values of $\pi(x)$ and comparison with $x/\log x$, $\mathrm{Li}(x)$ and $R(x)$

x	$\pi(x)$	$(x/\log x) - \pi(x)$	$\mathrm{Li}(x) - \pi(x)$	$R(x) - \pi(x)$
10^8	5 761 455	$-332\,774$	754	97
10^9	50 847 534	$-2\,592\,592$	1 701	-79
10^{10}	455 052 511	$-20\,758\,030$	3 104	$-1\,828$
10^{11}	4 118 054 813	$-169\,923\,160$	11 588	$-2\,318$
10^{12}	37 607 912 018	$-1\,416\,705\,193$	38 263	$-1\,476$
10^{13}	346 065 536 839	$-11\,992\,858\,452$	108 971	$-5\,773$
10^{14}	3 204 941 750 802	$-102\,838\,308\,636$	314 890	$-19\,200$
10^{15}	29 844 570 422 669	$-891\,604\,962\,453$	1 052 619	73 218
10^{16}	279 238 341 033 925	$-7\,804\,289\,844\,393$	3 214 632	327 052
10^{17}	2 623 557 157 654 233	$-68\,883\,734\,693\,929$	7 956 589	$-598\,255$
10^{18}	24 739 954 287 740 860	$-612\,483\,070\,893\,537$	21 949 555	$-3\,501\,366$
10^{19}	234 057 667 276 344 607	$-5\,481\,624\,169\,369\,961$	99 877 775	23 884 333
10^{20}	2 220 819 602 560 918 840	$-49\,347\,193\,044\,659\,702$	222 744 643	$-4\,891\,825$
10^{21}	21 127 269 486 018 731 928	$-446\,579\,871\,578\,168\,707$	597 394 254	$-86\,432\,204$

It has also been determined that

$$\pi(4\,185\,296\,581\,467\,695\,669) = 10^{17}.$$

RECORD

The largest computed values of $\pi(x)$ were obtained through a distributed computation led by X. Gourdon and P. Demichel. The most recently computed values are:

x	$\pi(x)$	$\mathrm{Li}(x) - \pi(x)$	$R(x) - \pi(x)$
2×10^{21}	41 644 391 885 053 857 293	1 454 564 714	501 830 649
4×10^{21}	82 103 246 362 658 124 007	1 200 472 717	$-127\,211\,330$
10^{22}	201 467 286 689 315 906 290	1 932 355 207	$-127\,132\,665$
2×10^{22}	397 382 840 070 993 192 736	2 732 289 619	$-139\,131\,087$
4×10^{22}	783 964 159 847 056 303 858	5 101 648 384	1 097 388 163

Comparison of $\pi(x)$ with $x/\log x$

I have already mentioned Tschebycheff's inequalities for $\pi(x)$, obtained with elementary methods and prior to the prime number theorem. In 1892, Sylvester refined Tschebycheff's method, obtaining the explicit estimates

$$0.95695 \frac{x}{\log x} < \pi(x) < 1.04423 \frac{x}{\log x}$$

for every x sufficiently large (see also Langevin, 1977).

For teaching purposes, Erdös (1932) gave an elegant proof for the weaker inequalities

$$\log 2 \frac{x}{\log x} < \pi(x) < 2 \log 2 \frac{x}{\log x}.$$

In 1962, Rosser & Schoenfeld obtained, with a very delicate analysis, the estimates

$$\frac{x}{\log x} \left(1 + \frac{1}{2 \log x}\right) < \pi(x) \quad \text{for} \quad x \geq 59$$

and

$$\pi(x) < \frac{x}{\log x} \left(1 + \frac{3}{2 \log x}\right) \quad \text{for} \quad x > 1.$$

In his thesis, Dusart (1998) refined these and other results in the literature, showing

$$\frac{x}{\log x} \left(1 + \frac{1}{\log x}\right) \leq \pi(x)$$

for $x \geq 599$, and

$$\pi(x) \leq \frac{x}{\log x} \left(1 + \frac{1.2762}{\log x}\right)$$

for $x > 1$. Also noteworthy are the inequalities obtained by Dusart:

$$\pi(x) < \frac{x}{\log x - 1.1} \quad \text{for} \quad x > 60184,$$

$$\pi(x) > \frac{x}{\log x - 1} \quad \text{for} \quad x > 5393.$$

By the results of Rosser & Schoenfeld (1962), if $x \geq 11$, then $\pi(x) \geq x/\log x$.

Comparison of $\pi(x)$ with $\mathrm{Li}(x)$

Gauss and Riemann believed that $\mathrm{Li}(x) > \pi(x)$ for every sufficiently large x. Even though in the present range of tables this is true, it had been shown by Littlewood in 1914 that the difference $\mathrm{Li}(x) - \pi(x)$ changes sign infinitely often, say, at numbers $x_0 < x_1 < x_2 < \cdots$, where x_n tends to infinity.

Assuming Riemann's hypothesis, Skewes showed in 1933 that $x_0 < 10^{10^{10^{34}}}$. For a long time, this number was famous as being the largest number that appeared, in a somewhat natural way, in mathematics. One knows now, even without assuming Riemann's hypothesis, a much smaller upper bound for x_0.

RECORD

In a computation first reported in 1986 (published in 1987), te Riele showed that already between 6.62×10^{370} and 6.69×10^{370} there are more than 10^{180} successive integers x for which $\mathrm{Li}(x) < \pi(x)$.

E   THE NONTRIVIAL ZEROS OF $\zeta(s)$

I recall that the zeros of the Riemann zeta function are the trivial zeros $-2, -4, -6, \ldots$ and the nontrivial zeros $\sigma + it$, with $0 \leq \sigma \leq 1$, that is, zeros in the critical strip.

First, I shall discuss the zeros in the whole critical strip and then the zeros on the critical line $\mathrm{Re}(s) = \frac{1}{2}$.

Since $\zeta(\bar{s}) = \overline{\zeta(s)}$ (where the bar denotes the complex conjugate), then the zeros lie symmetrically with respect to the real axis; so, it suffices to consider the zeros in the upper half of the critical strip.

For each $t > 0$, the zeta function can have only finitely many zeros (if any) of the form $\sigma + it$ (for σ real number). Thus it is possible to enumerate the nontrivial zeros of the zeta function as $\rho_n = \sigma_n + it_n$, with $0 < t_1 \leq t_2 \leq t_3 \leq \ldots$.

For every $T > 0$, let $N(T)$ denote the number of zeros $\rho_n = \sigma_n + it_n$ in the critical strip, with $0 < t_n \leq T$. Similarly, let $N_0(T)$ denote the number of zeros $\frac{1}{2} + it$ of Riemann's zeta function, which lie on the critical line, such that $0 < t \leq T$.

Clearly, $N_0(T) \leq N(T)$ and Riemann's hypothesis is the statement that $N_0(T) = N(T)$ for every $T > 0$.

Here are the main results concerning $N(T)$. First of all, it was conjectured by Riemann, and proved by von Mangoldt:

$$N(T) = \frac{T}{2\pi} \left\{ \log\left(\frac{T}{2\pi}\right) - 1 \right\} + O(\log T).$$

It follows that there exist infinitely many zeros in the critical strip.

All the known nontrivial zeros of $\zeta(s)$ are simple and lie on the critical line. Montgomery showed in 1973, assuming Riemann's hypothesis, that at least two thirds of the nontrivial zeros are simple.

In 1974, Levinson proved that at least one third of the nontrivial zeros of Riemann's zeta function are on the critical line. More precisely, if T is sufficiently large, $L = \log(T/2\pi)$, and $U = T/L^{10}$, then

$$N_0(T + U) - N_0(T) > \frac{1}{3}\left(N(T + U) - N(T)\right).$$

Conrey improved this result in 1989, replacing $\frac{1}{3}$ by $\frac{2}{5}$.

Extensive computations of the zeros of $\zeta(s)$ have been made. These began with Gram in 1903, who computed the first 15 zeros (that is, ρ_n for $1 \leq n \leq 15$). Titchmarsh calculated in 1935 the zeros ρ_n for $n \leq 1041$. With the advent of computers, Lehmer brought this up to $n = 35\,337$. By 1969, Rosser, Yohe & Schoenfeld had computed the first $3\,500\,000$ zeros.

Just not to be shamefully absent from this book, here is a table with the smallest zeros $\rho_n = \frac{1}{2} + it_n$, $t_n > 0$:

Table 15.   Nontrivial zeros of the Riemann zeta function

n	t_n	n	t_n	n	t_n
1	14.134725	11	52.970321	21	79.337375
2	21.022040	12	56.446248	22	82.910381
3	25.010858	13	59.347044	23	84.735493
4	30.424876	14	60.831779	24	87.425275
5	32.935062	15	65.112544	25	88.809111
6	37.586178	16	67.079811	26	92.491899
7	40.918719	17	69.546402	27	94.651344
8	43.327073	18	72.067158	28	95.870634
9	48.005151	19	75.704691	29	98.831194
10	49.773832	20	77.144840	30	101.317851

In Edwards' book (1974), there is a detailed explanation of the method used by Gram, Backlund, Hutchinson, and Haselgrove to compute the smallest 300 zeros of $\zeta(s)$. In 1986, Wagon wrote a short account with the essential information.

Starting in 1977, with work by Brent, the computations have been largely extended. The latest published result is that of van de Lune, te Riele & Winter (1986) who determined that the first $1\,500\,000\,001$ nontrivial zeros of $\zeta(s)$ are all simple, lie on the critical line, and have imaginary part with $0 < t < 545\,439\,823.215$. This work has involved over 1000 hours on one of the most powerful computers in existence at that time.

RECORD

S. Wedeniwski has recently announced that he and a large team of collaborators have computed the first 100 billion nontrivial zeros of $\zeta(s)$, verifying Riemann's hypothesis for all t with $0 < t < 29\,538\,618\,432.236$.

In 1988, a fast method for the simultaneous computation of large sets of zeros of the zeta function was invented by Odlyzko & Schönhage. It has been used to determine 10 billion zeros near the zero numbered 10^{22}. As reported by Odlyzko (2001), the zeros were all on the critical line and provided evidence for additional conjectures that relate these zeros to eigenvalues of random matrices. According to Odlyzko, heuristics suggest that any counterexamples to the Riemann hypothesis (if they exist) would lie far beyond the ranges that can be reached with currently known algorithms.

One may wonder why just knowing that the zeros of $\zeta(s)$ are on that inviting critical line is so important. The fact is that mathematicians trying to prove Riemann's hypothesis did not succeed with a direct proof. So, the natural road became to assume that the hypothesis is true and to derive consequences. If one such corollary turns out to be false (when derived correctly from the hypothesis), this would mean that Riemann's hypothesis is invalid.

But no, it is just the opposite which is so tantalizing, as many long desired and marvelous results could be established under the assumption of the truth of Riemann's hypothesis. Of course, it cannot be excluded that the consequences will eventually be proved without appeal to Riemann's hypothesis. It is not useless to say that, among the specialists, many believe firmly in the Riemann hypothesis.

The exercise of finding consequences of Riemann's hypothesis—now we are so familiar, I shall write like everyone: the RH—has been extrapolated to other important arithmetic or geometric areas, and one introduces many kinds of ERH (extended Riemann hypothesis) for functions more general than the zeta function, mostly for the so-called *Dirichlet L-functions.*

A line of approach to prove the RH goes back to Hilbert—namely to find, in an appropriate Hilbert space, an operator whose eigenvalues coincide with the nontrivial zeros of the zeta function and then, from reasons of still undetected symmetry, to deduce that the eigenvalues are on the critical line. The great difficulty is to find the right space, operator, symmetry, and mostly how to incorporate the analytical fact into the picture. For the record, I call your attention to the work of Connes (1996).

On the other hand, just knowing—or even knowing—that all the 100 billion nontrivial zeros of the zeta function thus far calculated are on the critical line, gives no conceptual reason for the truth of RH. Deviations could occur further ahead for larger roots; a behaviour or phenomenon governed by a log log log function will remain undetected by calculations with our present capabilities.

F   Zero-Free Regions for $\zeta(s)$ and the Error Term in the Prime Number Theorem

The knowledge of larger zero-free regions for $\zeta(s)$ leads to better estimates of the various functions connected with the distribution of primes.

I have already indicated that de la Vallée Poussin determined a zero-free region, which he used in an essential way in his proof of the prime number theorem. There have been many extensions of his result. A very large zero-free region, which I do not describe explicitly here, was determined by Richert (and published in Walfisz's book, 1963). In a preprint of 2001, Kadiri found that the following region does not contain zeros of the Riemann zeta function:

$$\sigma \geq 1 - \frac{1}{5.70233 \log |t|} \qquad \text{for } |t| \geq 3.$$

It is to be remarked that up to now, no one has succeeded in showing that $\zeta(s)$ has a zero-free region of the form $\{\sigma + it \mid \sigma \geq \sigma_0\}$ with $\frac{1}{2} < \sigma_0 < 1$.

Using whatever is known about zero-free regions for $\zeta(s)$ it is possible to deduce an estimate for the error in the prime number theorem. Thus, Tschudakoff (1936) obtained

$$\pi(x) = \mathrm{Li}(x) + O\big(xe^{-C(\log x)^{\alpha}}\big);$$

with $\alpha < 4/7$, and $C > 0$.

In 1901, von Koch showed that Riemann's hypothesis is equivalent to the following form of the error:

$$\pi(x) = \mathrm{Li}(x) + O(x^{1/2} \log x).$$

The knowledge that many zeros of $\zeta(s)$ are on the critical line leads also to better estimates. Thus, Rosser & Schoenfeld proved in 1975 that

$$0.998684x < \theta(x) < 1.001102x$$

(the lower inequality for $x \geq 1319007$, the upper inequality for all x).

In 1999, Dusart used the knowledge of 1.5 billion zeros of $\zeta(s)$ to obtain sharper estimates, for example: for $x > 10544111$,

$$|\theta(x) - x| < 0.0066788 \frac{x}{\log x}.$$

There are similar estimates by Rosser & Schoenfeld and by Dusart for the function $\psi(x)$.

These estimates are often refinements of the work of numerous other authors who, like the record holders, for a time had the best results. It would be therefore unjust not to mention the work of Robin (1983) and of Massias & Robin (1996).

G   SOME PROPERTIES OF $\pi(x)$

In this respect, historically the first statement is Bertrand's experimental observation (in 1845):

Between $n \geq 2$ and $2n$ there is always a prime number.

Equivalently, this may be stated as

$$\pi(2n) - \pi(n) \geq 1 \quad \text{(for } n \geq 1\text{)},$$

or also as

$$p_{n+1} < 2p_n \quad \text{(for } n \geq 1\text{)}.$$

This statement has been known as "Bertrand's postulate". It was proved by Tschebycheff in 1852 as a by-product of his estimates already indicated for $\pi(x)$. The shortest and perhaps simplest proofs of Bertrand's postulate are the ones by Ramanujan (1919), by Erdös (1932), and notably also by Moser (1949). As a matter of fact, the following inequalities are more refined:

$$1 < \frac{1}{3}\frac{n}{\log n} < \pi(2n) - \pi(n) < \frac{7}{5}\frac{n}{\log n} \quad \text{(for } n \geq 5\text{)},$$

and clearly $\pi(4) - \pi(2) = 1$, $\pi(6) - \pi(3) = 1$, $\pi(8) - \pi(4) = 2$.

More generally, Erdös proved in 1949 that for every $\lambda > 1$ there exists $C = C(\lambda) > 0$, and $x_0 = x_0(\lambda) > 1$ such that

$$\pi(\lambda x) - \pi(x) > C\frac{x}{\log x}, \quad \text{for } x \geq x_0,$$

which is just a corollary of the prime number theorem.

Now, I focus on estimates for $\pi(xy)$ and $\pi(x+y)$ in terms of $\pi(x)$, $\pi(y)$. The following result of Ishikawa (1934) is also a consequence of Tschebycheff's theorem:

If $x \geq y \geq 2$, $x \geq 6$, then $\pi(xy) > \pi(x) + \pi(y)$.

The comparison of $\pi(x+y)$ with $\pi(x)$, $\pi(y)$ is very interesting. In 1923, Hardy & Littlewood made the following conjecture, based on heuristic considerations:

Conjecture. $\pi(x+y) \leq \pi(x) + \pi(y)$ for all $x \geq 2$, $y \geq 2$.

In a more concrete language, the conjecture says: for every $x > 0$, the number of primes in any interval $(y, y+x]$ (excluding y, including $y + x$), where y is arbitrary, is at most the number of primes in the interval $(0, x]$: $\pi(y + x) - \pi(y) \leq \pi(x)$.

The conjecture looks indeed very reasonable, at least it confirms the intuition that primes are rarifying, as one advances in the sequence of numbers.

I note that Rosser & Schoenfeld proved in 1975, as a consequence of their sharp estimates for the function $\theta(x)$, that $\pi(2x) < 2\pi(x)$ for $x \geq 5$. Landau proved the inequality for all sufficiently large x, in his treatise (1913). Using deep methods, Montgomery & Vaughan (1973) proved that

$$\pi(x+y) \leq \pi(x) + \frac{2y}{\log y}.$$

As it was mentioned in Part E, Rosser & Schoenfeld (1975) had shown that $y/(\log y) < \pi(y)$, hence $\pi(x+y) \leq \pi(x) + 2\pi(y)$.

It is a delicate problem to ascertain if the conjecture of Hardy & Littlewood is true. In the positive direction, besides the preceding results, Udrescu proved in 1975:

For every $\varepsilon > 0$, if $x, y \geq 17$ and $x + y \geq 1 + e^{4(1+1/\varepsilon)}$, then

$$\pi(x + y) < (1 + \varepsilon)(\pi(x) + \pi(y)).$$

In 2002, Dusart studied the set of pairs (x, y) for which the inequality of the conjecture is satisfied. He proved:

If $2 \leq x \leq y \leq (7/5)\, x \log x \log \log x$, then $\pi(x+y) \leq \pi(x) + \pi(y)$.

It follows that for every $t > e^{10}$, the ratio $A/t^2 < 5/(7 \log t \log \log t)$, where A represents the area of the set of all (x, y) for which $\pi(x+y) > \pi(x) + \pi(y)$.

In the negative direction, I shall discuss in Section IV the relation between the conjecture of Hardy & Littlewood and the "prime k-tuples conjecture". It will be explained how the two conjectures are incompatible.

Two statements which are still waiting to be proved, or disproved, are the following:

In 1882, Opperman stated that $\pi(n^2 + n) > \pi(n^2) > \pi(n^2 - n)$ for $n > 1$.

In 1904, Brocard asserted that $\pi(p_{n+1}^2) - \pi(p_n^2) \geq 4$ for $n \geq 2$; that is, between the squares of two successive primes greater than 2 there are at least four primes.

H   THE DISTRIBUTION OF VALUES OF EULER'S FUNCTION

I will gather here results concerning the distribution of values of Euler's function. They supplement the properties already stated in Chapter 2, Section II.

First, some indications about the growth of Euler's function. It is easy to show that

$$\varphi(n) \geq \log 2 \frac{n}{\log(2n)},$$

in particular, for every $\delta > 0$, $\varphi(n)$ grows ultimately faster than $n^{1-\delta}$. Even better, for every $\varepsilon > 0$ there exists $n_0 = n_0(\varepsilon)$ such that, if $n \geq n_0$, then

$$\varphi(n) \geq (1 - \varepsilon)e^{-\gamma} \frac{n}{\log \log n} \, .$$

On the other hand, it follows from the prime number theorem, that there exist infinitely many n such that

$$\varphi(n) < (1 + \varepsilon)e^{-\gamma} \frac{n}{\log \log n} \, .$$

So,

$$\liminf \frac{\varphi(n) \log \log n}{n} = e^{-\gamma}.$$

A proof of the above results may be found, for example, in the books by Landau (1909), or Apostol (1976).

And what is the average value of $\varphi(n)$?

From the relation $n = \sum_{d|n} \varphi(d)$, it is not difficult to show that

$$\frac{1}{x} \sum_{n \leq x} \varphi(n) = \frac{3x}{\pi^2} + O(\log x).$$

So, the mean value of $\varphi(n)$ is equal to $3n/\pi^2$.

As a consequence, if two integers m, $n \geq 1$ are randomly chosen, then the probability that m, n be relatively prime is $6/\pi^2$.

All these matters are well explained in the books of Hardy & Wright and Apostol (1976).

II   The nth Prime and Gaps Between Primes

The results in the preceding section concerned the function $\pi(x)$, its asymptotic behaviour, comparison with known functions, and a variety of other properties. But nothing was said about the behaviour of $\pi(x)$ in short intervals, nor about the size of the nth prime, the difference between consecutive primes, etc. These are questions putting in evidence the fine points in the distribution of primes, and much more irregularity is expected.

A   THE nTH PRIME

Now I shall consider specifically the nth prime.

The prime number theorem yields easily:

$$p_n \sim n \log n, \quad \text{that is,} \quad \lim_{n \to \infty} \frac{p_n}{n \log n} = 1.$$

In other words, for large indices n, the nth prime is about the size of $n \log n$. More precisely

$$p_n = n \log n + n(\log \log n - 1) + O\left(\frac{n \log \log n}{\log n}\right).$$

So, for large n, $p_n > n \log n$. But Rosser proved in 1938 that for every $n > 1$:

$$n \log n + n(\log \log n - 10) < p_n < n \log n + n(\log \log n + 8),$$

and also that for every $n \geq 1$, $p_n > n \log n$. In 1999, Dusart showed that $p_n < n(\log n + \log \log n - 0.9484)$ for $n > 39017$, and also that $p_n > n(\log n - \log \log n - 1)$ for all $n > 1$.

The following results by Ishikawa (1934) are consequences of Tschebycheff's theorems (see Trost's book, General References):

If $n \geq 2$, then $p_n + p_{n+1} > p_{n+2}$; if $m, n \geq 1$, then $p_m p_n > p_{m+n}$.

Dusart was able to show in 1998 that the conjecture of R. Mandl (see Rosser & Schoenfeld, 1975) is true:

$$\frac{p_1 + p_2 \cdots + p_n}{n} \leq \frac{1}{2} p_n \quad \text{for} \quad n \geq 9.$$

In a very interesting paper, Pomerance considered in 1979 the prime number graph, consisting of all the points (n, p_n) of the plane (with $n \geq 1$). He proved Selfridge's conjecture: there exist infinitely many n with $p_n^2 > p_{n-i} p_{n+i}$ for all positive $i < n$. Also, there are infinitely many n such that $2p_n < p_{n-i} + p_{n+i}$ for all positive $i < n$.

A new conjecture by F. Firoozbakht, dating from about 1992, was communicated to me by the author; as far as I know, it remains unpublished. The conjecture is that the sequence $(p_n^{1/n})_{n \geq 2}$ is strictly decreasing. One more conjecture for the seemingly unlimited collection of reasonable assertions about primes, all verified numerically up to high limits (otherwise they would be thrown in the "garbage of mathematics"). Yet we, smart mathematicians, cannot decide if—in so many cases—the statements are true or false.

B   GAPS BETWEEN PRIMES

The prime number theorem tells how the function $\pi(x)$ grows as x tends to infinity. To understand in more detail the distribution of primes, it is necessary to study the differences $d_n = p_{n+1} - p_n$ between consecutive primes.

The *gap* at the prime p is the number $g(p)$ of composite integers which follow p. Thus $p_{n+1} = p_n + g(p_n) + 1$. Note that $g(p)$ is odd for all $p > 2$. The number $g(p_n)$ is a *maximal gap*, if $g(p_n) > g(p_k)$ for all $p_k < p_n$.

Let $G = \{m \mid m = g(p) \text{ for some } p > 2\}$, the set of possible values of $g(p)$. For each $m \in G$ let $p[m]$ be the smallest prime p such that $g(p) = m$. In the literature, $p[m]$ is called the *first occurrence* of the gap m.

In the study of the gaps, or equivalently the difference between consecutive primes, the following topics will be discussed: the behaviour of d_n as n tends to infinity, the set G of possible gaps, the first occurrence of a gap, the rate of growth of d_n, and the iterated gaps.

The behaviour of d_n as n tends to infinity

It is easy to show that $\limsup d_n = \infty$. This means that for every $N > 1$ there exists a string of at least N consecutive composite integers; for example:

$$(N+1)! + 2, \ (N+1)! + 3, \ (N+1)! + 4, \ldots, (N+1)! + (N+1).$$

Actually, strings of N consecutive composite integers have been found experimentally between numbers much smaller than $(N+1)!$. One feels therefore that it would be more striking to locate big differences d_n when n is small—there is more "merit" in it, in the sense of a precise definition soon to be formulated.

In contrast to $\limsup d_n$, any assertion about $\liminf d_n$—apart its existence—is still undecided. I examine the conceivable possibilities:

• $\liminf d_n = \infty$. This means that for every k there are only finitely many primes p_n such that $d_n = 2k$. It is equivalent to saying that for every k there exists n_0 such that if $n > n_0$, then $d_n > 2k$. True or false?

• There exists $l \geq 1$ such that $\liminf d_n = l$. This means that there exist infinitely many primes p_n such that $d_n = l$ and l is the smallest integer with this property. For any l, it is not known if the statement is true or false.

These questions are intimately connected with the

Conjecture of Polignac (1849). *For every even natural number $2k$, $k \geq 1$, there exist infinitely many consecutive primes p_n, p_{n+1} such that $d_n = p_{n+1} - p_n = 2k$.*

For the special case $k = 1$, Polignac's conjecture includes the statement: There exist infinitely many primes p such that $p + 2$ is also a prime. This is the famous *twin prime conjecture*, which will be considered in Section III.

I hasten to stress that even the following statement has never been proven: For every $k \geq 1$ there exists *one* pair of primes p, q such that $q - p = 2k$ (without requiring them to be consecutive).

The set G of possible gaps

The following result is an easy application of the prime number theorem, and was proposed by Powell as a problem in the *American Mathematical Monthly* (1983; solution by Davies in 1984):

For every natural number M, there exists an even number $2k$, such that there are more than M pairs of successive primes with difference $2k$.

Proof. Let n be sufficiently large, consider the sequence of primes

$$3 = p_2 < p_3 < \cdots < p_n$$

and the $n - 2$ differences $p_{i+1} - p_i$ $(i = 2, \ldots, n - 1)$. If the number of distinct differences is less than

$$\left[\frac{n - 2}{M} \right],$$

then one of these differences, say $2k$, would appear more than M times. In the alternative case,

$$p_n - p_2 \geq 2 + 4 + \cdots + 2 \left[\frac{n - 2}{M} \right].$$

But the right-hand side is asymptotically equal to n^2/M^2, while the left-hand side is asymptotically equal to $n \log n$, by the prime number theorem. This is impossible. $\qquad \square$

The above result can also be expressed as follows: For every natural number M there exists an odd number $m \in G$ such that $g(p) = m$ for more than M primes p.

It is not known if every even positive number is the difference of two consecutive primes—this was already mentioned in connection with Polignac's conjecture. Thus, it is not known if G is the set of all odd positive numbers.

Efforts have been made to identify gaps (between reasonably "small" primes) of every given size, up to a certain limit, and also to find exceptionally large gaps between consecutive primes. Dubner has developed an algorithm which allowed him to determine a prime p followed by exactly m composite integers, for each odd $m < 10180$. In this range the established upper bound for $p[m]$ is a prime which may have up to 398 digits.

RECORD

The largest gap between consecutive primes that has been determined explicitly, consists of a string of 112193 composite integers which follow a prime number with 3474 digits (J.L. Gómez Pardo, March 2002). Primality of the limiting primes was verified by M. Ladra and M. Seijas, respectively, using M. Martin's excellent ECPP implementation.

The first occurrence and the merit of a gap

Various tables of $p[m]$ have been produced, and from these tables one may identify the primes giving maximal gaps. In order of publication: Lander & Parkin (1967), Brent (1973, 1980), Young & Potler (1989), and Nicely (1999). These computations were extended by Nicely, B. Nyman, and T. Oliveira e Silva, who jointly examined all primes up to $N = 6 \times 10^{16}$.

RECORD

For $p < N$, the largest maximal gap, found by Oliveira e Silva in September 2002, is $m = 1197$, with $p[1197] = 55350776431903243$. The previous records were 1183 (in 2002) and 1131 (in 1999), both by Nyman. Other largest maximal gaps had been 803, by Young & Potler (1989), and 651, by Brent (1973).

Of course, if m is not a value of $g(p)$ in the range of the tables, nothing may be said without further specific investigation. A particular gap whose first occurrence remained uncertain for some years was $m = 999$, but recently (May 2001) Nyman found that $p[999] = 22439962446379651$. The smallest gap whose first occurrence is presently uncertain is $m = 1047$. The best known upper bound $88089672331629091 \geq p[1047]$ was given by Oliveira e Silva.

Concerning the asymptotic behaviour of $p[m]$, in 1964 Shanks conjectured that $\log p[m] \sim \sqrt{m}$ (when m tends to infinity). Based on his own extensive calculations, Weintraub suggested in 1991 that

$$\log p[m] \sim \sqrt{1.165746m}.$$

When a gap $g(p)$ following a prime p is found (not necessarily a first occurrence gap), it is natural to ask to which extent this is an "unusually large" gap between primes of that magnitude. As a consequence of the prime number theorem, the average gap between primes near that point is approximately $\log p$. The *merit* of the gap $g(p)$ is defined to be $g(p)/(\log p)$: the greater the merit, the more unusual the gap.

The gap with the largest known merit is at $p = 1693182318746371$, with $g(p) = 1131$ (a maximal gap) and was discovered by Nyman; its merit is 32.25. Next, with merit 31.05, is Oliveira e Silva's above record gap of 1197. For comparison, the largest computed gap of 112193 has merit 14.03.

The rate of growth of d_n

The guiding idea in this investigation is simple: to find functions $f(p_n)$ with real positive values, as simple and easily computable as possible, which may be compared to d_n, and to compare $f(p_n)$ to d_n. Usually $f(p_n)$ involves powers or logarithms, and the comparison amounts to answering questions like:

$$\text{Is } d_n = O(f(p_n)) ? \quad \text{Is } d_n = o(f(p_n)) ? \quad \text{Is } d_n \sim f(p_n) ?$$

To begin, the old result of Tschebycheff on Bertrand's postulate tells that $d_n < p_n$ for every $n \geq 1$.

By the prime number theorem,

$$\lim_{n \to \infty} \frac{d_n}{p_n} = 0.$$

Clearly, since $d_n < p_n$ for every $n \geq 1$, then $d_n = O(p_n)$. This leads to the question of finding the best function $f(p_n)$ such that $d_n = O(f(p_n))$.

Mathematicians are hoping to prove, without assuming Riemann's hypothesis, that $d_n = O(p_n^{(1/2)+\varepsilon})$ for every $\varepsilon > 0$. The race to reach this bound has been tight, from the first paper of Hoheisel (1930), with $d_n = O(p_n^\theta)$ and θ just below 1, through papers by Ingham (1937), Montgomery (1969), Huxley (1972), Iwaniec & Jutila (1979), Heath-Brown & Iwaniec (1979), Iwaniec & Pintz (1984), until the more recent records.

RECORD

The present record is $\theta = 0.525$. It is shared by Baker, Harman & Pintz (2001). In 1986, Mozzochi reached $\theta = 1051/1920 \approx 0.5474$, while Lou & Yao could go slightly lower, $\theta = 6/11 \approx 0.5454$ (in 1993).

The preceding results are assertions about the growth of d_n as n increases. In contrast, explicit computations by Ramaré & Saouter (2003) give the comforting knowledge of effective short intervals containing at least one prime. Let $x_0 = 10\,726\,905\,041$ and let $\Delta = 28\,314\,000$. If n is such that $p_n > x_0$, then $d_n < p_{n-1}/\Delta$.

It follows from the prime number theorem that the order of magnitude of d_n is $\log p_n$, that is $(d_n/\log p_n) \sim 1$. But gaps do deviate from the expected value.

Assuming the Riemann hypothesis, Cramér showed in 1937 that $d_n = O(p_n^{1/2} \log p_n)$. Based on probability arguments, Cramér conjectured that $d_n = O((\log p_n)^2)$.

Concerning smaller than expected gaps, the following results have been established. First, Erdös proved in 1940 that there exists a constant C, $0 < C < 1$, such that

$$\liminf \frac{d_n}{\log p_n} < C.$$

The value of C has been estimated by various authors. In 1966, Bombieri & Davenport showed that C may be taken to be 0.467. This was somewhat improved by Huxley (1977), but the best result is now by Maier (1985), where the constant is replaced by 0.248.

Regarding unusually large gaps, Westzynthius proved in 1931 that $\limsup(d_n/\log p_n) = \infty$, which means that for every $t > 0$, there exist infinitely many $n > 0$ such that $p_{n+1} > p_n + t\log p_n$.

Starting in 1938, Rankin continued the previous work of Erdös (1935) and proved in 1963 the sharper result: there exist infinitely many n such that

$$\frac{d_n}{\log p_n} \geq l_n e^\gamma = l_n \times 1.78107\ldots,$$

where γ is Euler's constant and

$$l_n = \frac{(\log_2 p_n)(\log_4 p_n)}{(\log_3 p_n)^2} > 1.$$

Erdös had offered a prize of US$ 10,000 to anyone who would prove that e^γ may be replaced by ∞ in the Erdös-Rankin inequality. Unfortunately, since 1996 the prize has become more difficult to collect.

Here is another open problem: to show that

$$\lim_{n\to\infty} \left(\sqrt{p_{n+1}} - \sqrt{p_n}\right) = 0.$$

If true, this would establish (for n sufficiently large) the conjecture of D. Andrica that $\sqrt{p_{n+1}} - \sqrt{p_n} < 1$ for all $n \geq 1$. In turn, from this inequality, if true, it would follow that between the squares of any two consecutive integers, there is always a prime. This seems indeed true, but has yet to be proved. Note that this is weaker than Opperman's conjecture.

It is not difficult to check Andrica's conjecture up to high limits, and this has been done up to $2^{42} \approx 4.39 \times 10^{12}$. But it has also been noted that the above inequality is equivalent to $g(p_n) < 2\sqrt{p_n}$, where $g(p_n)$ is the gap between p_n and p_{n+1}. Therefore, as can easily be seen, it suffices to verify the latter inequality only for the maximal gaps existing below a given limit N, in order to establish Andrica's inequality for all $p_n < N$. This is readily done for $N = 6 \times 10^{16}$ using Nicely's list of first occurrence gaps.

The iterated gaps

In 1878, Proth studied iterated gaps between primes. Let $p_1 = 2$, $p_2 = 3$, $\ldots$, p_n, p_{n+1}, $\ldots$ be the sequence of primes. Let $\delta_1(n) = |p_{n+1} - p_n| = d_n$ for $n \geq 1$. More generally, for $k \geq 1$ let $\delta_{k+1}(n) = |\delta_k(n+1) - \delta_k(n)|$. Thus, one has the following sequences:

2	3	5	7	11	13	17	19	23	29	...
1	2	2	4	2	4	2	4	6		
1	0	2	2	2	2	2	2			
1	2	0	0	0	0	0				
1	2	0	0	0	0					
1	2	0	0	0						
1	2	0	0							
1	2	0								etc.

For each k in this table, $1 \le k \le 7$, $\delta_k(1) = 1$. Proth claimed to have shown that $\delta_k(1) = 1$ for every $k \ge 1$; however, his proof was incorrect. N.L. Gilbreath, unaware of Proth's work, stated the same fact as a conjecture in the decade of 1950 (unpublished). Killgrove & Ralston ckecked in 1959 the veracity of the conjecture for $k \le 63419$. In 1993, Odlyzko verified that $\delta_k(1) = 1$ for all $k \le 3.46 \times 10^{11}$, in other words, for the sequence of primes up to 10^{13}.

Gaps and the marriage theorem

What has marriage to do with gaps between primes? You will find the answer in the Appendix 1, dealing with "Marriage and prime numbers". This appendix is a gift from M. Ram Murty, a specialist on half of this topic, to me—who have been practicing the other half for over fifty years. An example to be followed, like in other arts, a poem as a gift to a friend.

III Twin Primes

If p and $p + 2$ are primes, they are called *twin primes*.

The smallest pairs of twin primes are $(3, 5)$, $(5, 7)$, $(11, 13)$ and $(17, 19)$. Twin primes have been characterized by Clement in 1949 as follows.

Let $n \ge 2$. The integers n, $n + 2$ form a pair of twin primes if and only if

$$4\big[(n - 1)! + 1\big] + n \equiv 0 \pmod{n(n + 2)}.$$

Proof. If the congruence is satisfied, then $n \ne 2, 4$, and

$$(n - 1)! + 1 \equiv 0 \pmod{n},$$

so by Wilson's theorem n is a prime. Also

$$4(n-1)! + 2 \equiv 0 \pmod{n+2};$$

hence, multiplying by $n(n+1)$,

$$4[(n+1)! + 1] + 2n^2 + 2n - 4 \equiv 0 \pmod{n+2}$$

and then

$$4[(n+1)! + 1] + (n+2)(2n-2) \equiv 0 \pmod{n+2};$$

from Wilson's theorem $n+2$ is also a prime.

Conversely, if n, $n+2$ are primes, then $n \neq 2$ and

$$(n-1)! + 1 \equiv 0 \pmod{n},$$
$$(n+1)! + 1 \equiv 0 \pmod{n+2}.$$

But $n(n+1) = (n+2)(n-1) + 2$, so $2(n-1)! + 1 = k(n+2)$, where k is an integer. From $(n-1)! \equiv -1 \pmod{n}$, then $2k+1 \equiv 0 \pmod{n}$ and substituting $4(n-1)! + 2 \equiv -(n+2) \pmod{n(n+2)}$, therefore $4[(n-1)! + 1] + n \equiv 0 \pmod{n(n+2)}$.   □

However, this characterization has no practical value in the determination of twin primes.

The main problem is to ascertain whether there exist infinitely many twin primes.

In 1919, Brun proved the famous result that the sum

$$\left(\frac{1}{3} + \frac{1}{5}\right) + \left(\frac{1}{5} + \frac{1}{7}\right) + \left(\frac{1}{11} + \frac{1}{13}\right) + \cdots + \left(\frac{1}{p} + \frac{1}{p+2}\right) + \cdots,$$

extended to all primes p such that $p+2$ is also a prime, converges to a number denoted by B, now called *Brun's constant*.

This result does not exclude that there are infinitely many pairs of twin primes, but tells that these pairs are found farther and farther apart, so the sum of their reciprocals stays finite.

Based on heuristic considerations about the distribution of twin primes, B has been calculated, for example, by Shanks & Wrench (1974), by Brent (1976), and more recently by Nicely (2001), who obtained:

$$B = 1.9021605823\ldots.$$

Brun also proved that for every $m \geq 1$ there exist m successive primes which are not twin primes.

For every $x > 1$, let $\pi_2(x)$ denote the number of primes p such that $p + 2$ is also prime and $p + 2 \leq x$.

Brun announced in 1919 that there exists an effectively computable integer x_0 such that, if $x \geq x_0$, then

$$\pi_2(x) < \frac{100x}{(\log x)^2}.$$

The proof appeared in 1920.

The upper bound of $\pi_2(x)$ has been reduced by a determination of the constant and of the size of the error. This was done, among others, by Bombieri & Davenport, in 1966. It is an application of sieve methods and its proof may be found, for example, in the book of Halberstam & Richert.

Here is the result:

$$\pi_2(x) \leq 2C \prod_{p>2} \left(1 - \frac{1}{(p-1)^2}\right) \frac{x}{(\log x)^2}.$$

The best values obtained thus far for C have been: $C = 3.5$ by Bombieri, Friedlander & Iwaniec (1986) and $C = 3.13$ by S. Lou (private communication).

Earlier, on heuristic grounds Hardy & Littlewood (1923) had conjectured that

$$\pi_2(x) \sim C_2 \frac{x}{(\log x)^2},$$

where the infinite product

$$C_2 = \prod_{p>2} \left(1 - \frac{1}{(p-1)^2}\right)$$

(see Chapter 6, Section IV) is called the *twin-prime constant*; its value is $0.66016\ldots$. It had been calculated by Wrench in 1961.

To figure out the constant C_2, Hardy and Littlewood's heuristic reasoning was later explained by other authors, and more simply by Golomb (1960). It is not without interest to describe the reasoning that leads to the constant C_2, even though the argument cannot yet be made rigorous.

From the prime number theorem $\pi(x)/x \sim 1/(\log x)$, hence the probabilty that the positive integer n is a prime is $1/(\log n)$. The probability that $n + 2$ is also a prime is essentially the same, so, if the two events were independent, then the probability that n and $n + 2$ be simultaneously primes would be $1/(\log n)^2$. But the events are not independent. If n is prime, $n > 2$, then n is odd, so is $n + 2$ and the probability for $n + 2$ to be a prime has to be corrected by a factor 2, to take into account that $n + 2$ belongs to the subset of odd numbers.

For each odd prime $p \neq n$ the number n belongs to one of $(p-1)/p$ residue classes; if $n + 2$ is also a prime and p does not divide $n + 2$, then it belongs to one of $(p-2)/(p-1)$ among the preceding classes. So for each prime $p > 2$ a factor

$$\frac{p-2}{p-1} \bigg/ \frac{p-1}{p} = \frac{p(p-2)}{(p-1)^2} = 1 - \frac{1}{(p-1)^2}$$

must be taken in the calculation of the probability, which is, on these heuristic grounds, $C_2/(\log n)^2$.

To give a feeling for the growth of $\pi_2(x)$, I reproduce in Table 16 part of the calculations of Brent (1975, 1976) and Nicely (1995, 2001).

Table 16. Number of twin prime pairs

x	$\pi_2(x)$
10^3	35
10^4	205
10^5	1 224
10^6	8 169
10^7	58 980
10^8	440 312
10^9	3 424 506
10^{10}	27 412 679
10^{11}	224 376 048
10^{12}	1 870 585 220
10^{13}	15 834 664 872
10^{14}	135 780 321 665
10^{15}	1 177 209 242 304

RECORD

The largest exact value for the number of twin primes below a given limit has been determined by Nicely in February 2003:

$$\pi_2(4.35 \times 10^{15}) = 4\,698\,614\,557\,533.$$

RECORDS

The largest known twin prime pairs are presented in the following table.

Table 17. The largest known twin prime pairs

Prime pair	Digits	Discoverer	Year
$33218925 \times 2^{169690} \pm 1$	51090	D. Papp, P. Jobling, G. Woltman and Y. Gallot	2002
$60194061 \times 2^{114689} \pm 1$	34533	D. Underbakke, G. Woltman and Y. Gallot	2002
$1765199373 \times 2^{107520} \pm 1$	32376	J. McElhatton and Y. Gallot	2002
$318032361 \times 2^{107001} \pm 1$	32220	D. Underbakke, P. Carmody, C. Nash et al.	2001
$1807318575 \times 2^{98305} \pm 1$	29603	D. Underbakke, P. Carmody and Y. Gallot	2001
$665551035 \times 2^{80025} \pm 1$	24099	D. Underbakke, P. Carmody and Y. Gallot	2000
$781134345 \times 2^{66445} \pm 1$	20011	D. Underbakke, P. Carmody, C. Nash et al.	2001
$1693965 \times 2^{66443} \pm 1$	20008	G. La Barbera, P. Jobling and Y. Gallot	2000
$83475759 \times 2^{64955} \pm 1$	19562	D. Underbakke, P. Jobling and Y. Gallot	2000
$291889803 \times 2^{60090} \pm 1$	18098	D. Boivin and Y. Gallot	2001

A set of $2k$ consecutive primes consisting of k pairs of twin primes is called a *cluster of twin primes of order* k. The existence of clusters of twin primes of arbitrary order has never been proved. Yet, it follows from the unproven conjecture of Dickson; see Chapter 6, Section I, (D_3). In 1996, N.B. Backhouse communicated to me the smallest clusters of twin primes of order 1 to 7. Their initial primes are 3, 5, 5, 9419, 909 287, 325 267 931, and 678 771 479, respectively.

RECORDS

The smallest cluster of twin primes of order 8 has initial prime 1 107 819 732 821 and was found by P. Carmody in January 2001. The smallest cluster of order 9 has initial prime 170 669 145 704 411 and was discovered by D. DeVries and P. Sebah in March 2002.

Sieve theory has been used in attempts to prove that there exist infinitely many twin primes. Many authors have worked with this method.

To begin, in his famous paper of 1920, Brun showed that 2 may be written, in infinitely many ways, in the form $2 = m - n$, where m, n are products of at most 9 primes (not necessarily distinct).

The best result to date, with sieve methods, is due to Chen (announced in 1966, published in 1973, 1978); he proved that 2 may be written in infinitely many ways in the form $2 = m - p$, with p prime, and m a product of at most two primes (not necessarily distinct).

The sieve methods used for the study of twin primes are also appropriate for the investigation of Goldbach's conjecture (see Section VI).

The general Polignac conjecture (see the previous section) can be, in part, treated like the twin prime conjecture.

For every $k \geq 1$ and $x > 1$, let $\pi_{2k}(x)$ denote the number of integers $n > 1$ such that $p_{n+1} \leq x$ and $p_{n+1} - p_n = 2k$.

With Brun's method, it may be shown that there exists a constant $C_k > 0$ such that

$$\pi_{2k}(x) < C_k \frac{x}{(\log x)^2} \, .$$

IV   Prime k-Tuplets

I have considered above pairs of twin-primes $(p, p+2)$. They consist of two consecutive prime numbers with the smallest possible difference, which is 2.

Similarly, I define *triplets* of primes (p_0, p_1, p_2), where $p_0 < p_1 < p_2$ are consecutive primes with the smallest possible difference $p_2 - p_0$. There are two kinds of triplets of primes: $(p, p+2, p+6)$, as for example $(11, 13, 17)$, and $(p, p+4, p+6)$, like $(7, 11, 13)$. On the other hand, it is clear that if p, $p+2$, $p+4$ are primes, then $p = 3$, because one of the three numbers must be divisible by 3.

The quadruple (p_0, p_1, p_2, p_3) is called a *prime quadruplet* if $p_0 < p_1 < p_2 < p_3$ are consecutive primes and $p_3 - p_0$ is smallest possible. Since p, $p+2$, $p+4$, $p+6$ cannot be simultaneously prime, the smallest difference is not 6. But 11, 13, 17, 19 are primes, so the minimal difference is 8 and $(11, 13, 17, 19)$ is a prime quadruplet. Like any other prime quadruplet, it is of the form $(p, p+2, p+6, p+8)$, for some prime p.

More generally, let $k \geq 2$,

(i) let $b_1 < b_2 < \cdots < b_{k-1}$,

(ii) let $p, p + b_1, \ldots, p + b_{k-1}$ be consecutive primes,

(iii) assume there is no sequence of primes $q_0 < q_1 < \cdots < q_{k-1}$ with $q_{k-1} - q_0 < b_{k-1}$.

Then $(p, p + b_1, \ldots, p + b_{k-1})$ is called a *prime k-tuplet*, and in this case $(b_1, b_2, \ldots, b_{k-1})$ is called the *type* of the prime k-tuplet.

Now I introduce the following notation: for every real $x > 0$, let

$$\pi_{2,6}(x) = \#\{(p, p+2, p+6) \mid (p, p+2, p+6) \text{ is a}$$
$$\text{triplet of primes and } p \leq x\},$$
$$\pi_{4,6}(x) = \#\{(p, p+4, p+6) \mid (p, p+4, p+6) \text{ is a}$$
$$\text{triplet of primes and } p \leq x\}.$$

Similarly,

$$\pi_{2,6,8}(x) = \#\{(p, p+2, p+6, p+8) \mid (p, p+2, p+6, p+8) \text{ is a}$$
$$\text{quadruplet of primes and } p \leq x\}.$$

Also, let

$$B_{2,6} = \sum \left(\frac{1}{p} + \frac{1}{p+2} + \frac{1}{p+6}\right),$$
$$B_{4,6} = \sum \left(\frac{1}{p} + \frac{1}{p+4} + \frac{1}{p+6}\right),$$
$$B_{2,6,8} = \sum \left(\frac{1}{p} + \frac{1}{p+2} + \frac{1}{p+6} + \frac{1}{p+8}\right),$$

where the summation is extended over all triplets (resp. quadruplets) of the form indicated. Just like Brun's result for the sum of reciprocals of pairs $(p, p+2)$ of twin primes, the sums indicated above are convergent. Of course, this is not contrary to the possibility that the corresponding sets of prime k-tuplets be infinite.

RECORDS

The following data were communicated to me by T.R. Nicely in 1999. Let $N = 1.5 \times 10^{15}$. Then

$$\pi_{2,6}(N) = 110261940034,$$
$$\pi_{4,6}(N) = 110262203391,$$
$$\pi_{2,6,8}(N) = 4737286827.$$

RECORDS

The explicit examples below of largest known k-tuplets (with $k \geq 3$) are taken from a list maintained by T. Forbes.

(1) Triplet $(p, p + 2, p + 6)$, with
      $p = (90159302514 \times d(d + 1) + 210)(d - 1)/35 + 5,$
      where $d = 4436 \times 3251\#$
      (4135 digits, D. Broadhurst et al. in 2002).

(2) Triplet $(p, p + 4, p + 6)$, with
      $p = (108748629354 \times d(d + 1) + 210)(d - 1)/35 + 7,$
      where $d = 4436 \times 3251\#$
      (4135 digits, D. Broadhurst et al. in 2002).

(3) Quadruplet $(p, p + 2, p + 6, p + 8)$, with
      $p = 10271674954 \times 2999\# + 3461$
      (1284 digits, M. Bell et al. in 2002).

(4) Quintuplet $(p, p + 2, p + 6, p + 8, p + 12)$, with
      $p = 31969211688 \times 2399\# + 16061$
      (1034 digits, N. Luhn et al. in 2002).

As an illustration of the calculations that have been carried out, the largest prime 17-tuplet is

$$(p, p + 6, p + 8, p + 12, p + 18, p + 20, p + 26, p + 32, p + 36,$$
$$p + 38, p + 42, p + 48, p + 50, p + 56, p + 60, p + 62, p + 66),$$
$$\text{with } p = 2845372542509911868266811$$

(25 digits, J. Waldvogel and P. Leikauf in 2000).

As it was indicated, when k is small there are many known examples of prime k-tuplets. But when k is very large, it is a totally different matter even to find *one* prime k-tuplet. What could you do when $k = 10^{10^{10}}$? How is it possible to know that there is at least one prime k-tuplet? This is a very interesting question, which I shall now discuss in some detail.

Let $k \geq 2$. A $(k-1)$-tuple $(b_1, b_2, \ldots, b_{k-1})$ of integers is said to be *admissible* when

(i) $b_1 < b_2 < \cdots < b_{k-1}$,

(ii) for every prime $q \leq k$ the set of residue classes $\{0 \bmod q,$ $b_1 \bmod q, b_2 \bmod q, \ldots, b_{k-1} \bmod q\}$ is properly contained in the set of all residue classes modulo q.

If $(b_1, b_2, \ldots, b_{k-1})$ is admissible, taking $q = 2$ it follows that each b_i is even. Among the admissible $(k-1)$-tuples those with minimal b_{k-1} are said to be *tight*. Thus for $k \leq 4$ the tight admissible $(k-1)$-tuples are (2), $(2, 6)$, $(4, 6)$ and $(2, 6, 8)$.

Following Hensley & Richards (1974), and with their notation, let $\rho^*(x) = k$ if there exists an admissible $(k-1)$-tuple $(b_1, b_2, \ldots, b_{k-1})$ with $b_{k-1} < x$, but none such with more than $k-1$ components. The calculation of $\rho^*(x)$ may be done, for each x, in finitely many steps, but when x is large, this is an intricate combinatorial problem.

Similar functions have been considered in the literature by Hardy & Littlewood (1923), by Schinzel & Sierpiński (1958), and other authors. In particular,

$$\rho(x) = \limsup_{y \to \infty} \big(\pi(x + y) - \pi(y)\big).$$

I note that $\rho(x) \leq \rho^*(x)$.

Proof. Indeed, let $\rho(x) = k$, so there exist $y \geq k$ and k primes $p + b_i$ (with $b_0 = 0$) such that $y < p < p + b_1 < \cdots < p + b_{k-1} \leq y + x$. Then $b_{k-1} < x$. If there exists a prime $q \leq k$ such that $\{b_i \bmod q \mid i = 0, 1, \ldots, k-1\}$ is the set of all congruence classes modulo q, then there exists i such that $-p \equiv b_i \pmod{q}$, hence q divides $p + b_i$, so $q \leq k \leq y \leq p + b_i = q$, which is absurd. Thus $(b_1, b_2, \ldots, b_{k-1})$ is admissible, hence $k \leq \rho^*(x)$. □

It is of great interest to compare $\rho^*(x)$ with $\pi(x)$. Numerical calculations were started by Schinzel (1961); Selfridge (unpublished) showed that $\rho^*(x) \leq \pi(x)$ for all $x \leq 500$. The following theorem of Hensley & Richards (1974) shows that this situation is ultimately reversed:

For every ε, $0 < \varepsilon < \log 2$, there exists $x_0 > 1$ such that if $x \geq x_0$, then

$$\rho^*(x) - \pi(x) > (\log 2 - \varepsilon)\frac{x}{(\log x)^2}.$$

In particular, $\lim_{x \to \infty} \big(\rho^*(x) - \pi(x)\big) = \infty$.

With a well-conceived computer program by W. Stenberg, it was shown in 1974 that $\rho^*(20000) > \pi(20000)$. A difficult problem is to determine the order of magnitude of $\rho^*(x)$. Is it true that $\rho^*(x) \sim \pi(x)$?

The following conjecture implies in particular the existence of prime k-tuplets.

Prime k-Tuples Conjecture. *If $k \geq 2$ and $(b_1, b_2, \ldots, b_{k-1})$ is an admissible $(k-1)$-tuple of positive integers, there exist infinitely many primes p such that $p, p + b_1, \ldots, p + b_{k-1}$ are simultaneously prime.*

In particular, if $(b_1, b_2, \ldots, b_{k-1})$ is admissible and tight, there exist infinitely many prime k-tuplets of type $(b_1, b_2, \ldots, b_{k-1})$.

In Chapter 6, Section I, I shall consider a conjecture by Dickson, thoroughly studied by Schinzel & Sierpiński (1958), which asserts that, under certain conditions, linear polynomials have simultaneous prime values. The prime k-tuples conjecture states that if $(b_1, b_2, \ldots, b_{k-1})$ is admissible, then the polynomials $X, X + b_1, \ldots, X + b_{k-1}$ assume, infinitely often, simultaneous prime values.

The proof of the prime k-tuples conjecture cannot be achieved before the proof of the twin prime conjecture. Yet, there are many devoted believers in the truth of this conjecture, and notably Erdös was among them. At this stage of ignorance, to believe or not to believe in the conjecture is purely emotional.

Arguments of believers. There is general belief that the twin prime conjecture is true. And then, why not the prime k-tuples conjecture for $k > 2$? One feels that these problems are of the same order of difficulty for every $k \geq 2$, and the methods which will prove the twin prime conjecture will also serve for the wider k-tuples conjecture. The "house of primes" is very welcoming—in its vast expanses it can accomodate pairs of twins, triplets, quadruplets, . . . and all sorts of legitimate families of primes.

Arguments of non-believers. If you cannot see or touch, then it does not exist. Indeed, nobody will ever face an admissible $10^{10^{10}}$-tuple of primes. It is mind-boggling, and the prime k-tuples conjecture has absolutely no supporting evidence. In a more scientific vein, I present now the beautiful result of Hensley & Richards:

> *The conjecture of Hardy & Littlewood* (Section I, H) *and the prime k-tuples conjecture cannot be simultaneously true.*

Proof. I assume that the prime k-tuples conjecture is true and I show that $\rho^*(x) \leq \rho(x)$, hence $\rho^*(x) = \rho(x)$, for all $x > 1$.

Let $\rho^*(x) = k$, so there exists an admissible $(k-1)$-tuple $(b_1, b_2, \ldots, b_{k-1})$ with $b_{k-1} < x$. By the prime k-tuples conjecture, there exist infinitely many primes p such that $p, p + b_1, \ldots, p + b_{k-1}$ are primes; note that they satisfy $p - 1 < p < p + b_1 < \cdots < p + b_{k-1} \leq (p-1) + x$. Hence, by definition of $\rho(x)$, $\rho(x) \geq k = \rho^*(x)$. From the theorem already mentioned, there exists x_0 such that $\rho^*(x) > \pi(x)$ for all $x \geq x_0$. So for each x, from $\rho^*(x) \leq \rho(x)$ it follows that there exist infinitely many y such that $\pi(x) < \rho(x) = \pi(x + y) - \pi(y)$, and this shows that the conjecture of Hardy & Littlewood cannot be true if the prime k-tuples conjecture is assumed to be true.   □

A conjecture of the family of the prime k-tuples conjecture, concerning translates of sequences, was enunciated by Golomb (1992), as an open problem:

Golomb's Conjecture. *There exists an increasing sequence of positive integers $1 \leq a_1 < a_2 < \ldots$ and an integer $B \geq 1$ such that for every $n \in \mathbb{Z}$, the number of primes in the sequence $a_1 + n < a_2 + n < \ldots$ is at most equal to B.*

If no bound B is imposed, here is an interesting example: Note that for every $n \in \mathbb{Z}$ there are only finitely many primes of the form $((2k)!)^3 + n$. Indeed, this is trivial if $n = 0$ or $|n| \geq 2$. Finally $((2k)!)^3 - 1 = \left[(2k)! - 1\right]\left[(2k)!)^2 + (2k)! + 1\right]$ and $((2k)!)^3 + 1 = \left[(2k)! + 1\right]\left[(2k)!)^2 - (2k)! + 1\right]$. But of course for this sequence there is no bound B as stated in the conjecture.

In 1995, Ford gave a neat proof of the following theorem:

The prime k-tuples conjecture and Golomb's conjecture cannot be simultaneously true.

Proof. Assume that the sequence $A = (a_i)_{i \geq 1}$ and $B \geq 1$ satisfy the condition indicated in Golomb's conjecture. As it is known, there exists a constant $c > 0$ such that for every $l \geq 2$,

$$\prod_{p \leq l} \left(1 - \frac{1}{p}\right) > \frac{c}{\log l}.$$

Let l be such that $cl/(\log l) > B$. Now let $A_l = \{a_1, a_2, \ldots, a_l\}$ and $E_2 = A_l \setminus (A_l \cap C)$, where C is a congruence class of $\mathbb{Z}$ modulo 2, such that $\#(A_l \cap C)$ is minimal. Then $\#(A_l \cap C) \leq l/2$, hence $\#(E_2) \geq l(1 - 1/2)$. By definition, no element of E_2 belongs to the congruence class C modulo 2.

Let $E_3 = E_2 \setminus (E_2 \cap C')$, where C' is a congruence class of $\mathbb{Z}$ modulo 3, such that $\#(E_2 \cap C')$ is minimal; hence $\#(E_2 \cap C') \leq \#(E_2)/3$ and

$$\#(E_3) \geq \#(E_2) \left(1 - \frac{1}{3}\right) \geq l \left(1 - \frac{1}{2}\right)\left(1 - \frac{1}{3}\right).$$

Note that no element of E_3 belongs to the congruence class C' modulo 3.

This argument is repeated for all primes $p \leq l$, leading to a set E^* such that

$$\#(E^*) \geq l \prod_{p \leq l} \left(1 - \frac{1}{p}\right) > \frac{cl}{\log l} > B.$$

By definition $\#(E^*) < l$ and for every prime $q \leq \#(E^*) < l$, the set of congruence classes $\{b \bmod q \mid b \in E^*\}$ is properly contained in the set of all congruence classes modulo q. This E^* is an admissible set. By the prime k-tuples conjecture, there exist infinitely many primes p such that p and all $p + b$ (with $b \in E^*$) are primes. For any such prime p, the set $\{a_i + p \mid i \geq 1\}$ contains more than B primes—which is a contradiction. $\square$

The following consequence of the prime k-tuples conjecture was communicated by A. Granville in a letter of 1989, for the polynomials $X^2 + X + c$ (see Mollin, 1997). Let $f(X) = aX^2 + bX + c$ with $a \geq b$. Assume that $a + b$ is even and if q is an odd prime dividing a then q divides b. Assuming the truth of the prime k-tuples conjecture, for every $M > 1$ there exist infinitely many $n \geq 1$ such that $f(0) + n$, $f(1) + n, \ldots, f(M) + n$ are primes. With the terminology of Chapter 3, Section III, C, the translated polynomials $g_n(X) = f(X) + n$ have prime-producing length greater than M.

Proof. Let $S = \{f(0), f(1), \ldots, f(M)\}$. I show that S is an admissible set. Indeed $f(0) \equiv f(1) \equiv \cdots \equiv f(M) \equiv c \pmod{2}$. If q is an odd prime and $q \mid a$, then also $q \mid b$ and again $f(0) \equiv f(1) \equiv \cdots \equiv f(M) \equiv c \pmod{q}$. Now let $q \neq 2$ and q not a factor of a. Let u, a' be such that $2u \equiv 1 \pmod{q}$ and $aa' \equiv 1 \pmod{q}$. For every s,

$$as^2 + bs + c \equiv a(s^2 + a'bs) + c \equiv a(s + ua'b)^2 + (c - a'u^2b^2).$$

The set of residue classes $at^2 \pmod{q}$ contains $(q+1)/2$ elements. So there exists y such that $y \not\equiv at^2 \pmod{q}$ for all t. Let $z = y + (c - a'u^2b^2)$. If there exists s, $0 \leq s \leq M$, such that $z \equiv f(s) \pmod{q}$, then

$$y + (c - a'u^2b^2) \equiv a(s + ua'b)^2 + (c - a'u^2b^2).$$

This implies that $y \equiv a(s + ua'b)^2 \pmod{q}$, which is a contradiction. So the set S is admissible, hence by the prime k-tuples conjecture there exist infinitely many n such that $g_n(X) = f(X) + n$ has prime values at $s = 0, 1, \ldots, M$. $\qquad\square$

V   Primes in Arithmetic Progression

A   THERE ARE INFINITELY MANY!

A classical and most important theorem was proved by Dirichlet in 1837. It states:

   If $d \geq 2$ and $a \neq 0$ are integers that are relatively prime, then the arithmetic progression

$$a, \quad a + d, \quad a + 2d, \quad a + 3d, \quad \ldots$$

contains infinitely many primes.

   Many special cases of this theorem were already known, including of course Euclid's theorem (when $a = 1$, $d = 2$). Indeed, if $d = 4$ or $d = 6$ and $a = -1$, the proof is very similar to Euclid's proof.

Using simple properties of quadratic residues, it is also easy to show that each of these arithmetic progressions contain infinitely many primes:

$d = 4$, $a = 1$;
$d = 6$, $a = 1$;
$d = 3$, $a = 1$;
$d = 8$, $a = 3$, or $a = 5$, or $a = 7$
(this includes the progressions with $d = 4$);
$d = 12$, $a = 5$, or $a = 7$, or $a = 11$
(this includes also the progressions with $d = 6$).

For $d = 8, 16, \ldots$, or more generally $d = 2^r$, and $a = 1$, the ingredients of a simple proof are: to consider $f(N)$ where

$$f(X) = X^{2^{r-1}} + 1, \quad N = 2p_1 p_2 \cdots p_n,$$

each p_i prime, $p_i \equiv 1 \pmod{2^r}$, and then to use Fermat's little theorem. These are hints for a reader wanting to find this proof by himself.

The proof when d is arbitrary and $a = 1$ or $a = -1$ is also elementary, though not so simple, and requires the cyclotomic polynomials and some of their elementary properties.

A detailed discussion of Dirichlet's theorem, with several variants of proofs, is in Hasse's book *Vorlesungen über Zahlentheorie* (now available in English translation, from Springer-Verlag).

In 1949, Selberg gave an elementary proof of Dirichlet's theorem, similar to his proof of the prime number theorem.

Concerning Dirichlet's theorem, de la Vallée Poussin established the following additional density result. For a, d as before, and $x \geq 1$ let

$$\pi_{d,a}(x) = \#\{p \text{ prime} \mid p \leq x, \ p \equiv a \pmod{d}\}.$$

Then

$$\pi_{d,a}(x) \sim \frac{1}{\varphi(d)} \cdot \frac{x}{\log x}.$$

Note that the right-hand side is the same, for any a, such that $\gcd(a, d) = 1$. It follows that

$$\lim_{x \to \infty} \frac{\pi_{d,a}(x)}{\pi(x)} = \frac{1}{\varphi(d)},$$

and this may be stated by saying that the set of primes in the arithmetic progression $\{a + kd \mid k \geq 1\}$ has natural density $1/\varphi(d)$ (with respect to the set of all primes).

Despite the fact that the asymptotic behavior of $\pi_{d,a}(x)$ is the same, for every a, $1 \leq a < d$, with $\gcd(a, d) = 1$, Tschebycheff had noted, already in 1853, that $\pi_{3,1}(x) < \pi_{3,2}(x)$ and $\pi_{4,1}(x) < \pi_{4,3}(x)$ for small values of x; in other words, there are more primes of the form $3k + 2$ than of the form $3k + 1$ (resp. more primes $4k + 3$ than primes $4k + 1$) up to x (for x not too large). Are these inequalities true for every x? The situation is somewhat similar to that of the inequality $\pi(x) < \text{Li}(x)$. Once again, as in Littlewood's theorem, it may be shown that these inequalities are reversed infinitely often. Thus, Leech computed in 1957 that $x_1 = 26861$ is the smallest prime for which $\pi_{4,1}(x) > \pi_{4,3}(x)$; see also Bays & Hudson (1978), who found that $x_1 = 608\,981\,813\,029$ is the smallest prime for which $\pi_{3,1}(x) > \pi_{3,2}(x)$.

In 1977, Hudson derived a formula, similar to Meissel's formula for $\pi(x)$, in order to compute $\pi_{d,a}(x)$, which represents the exact number of primes $p < x$ in the arithmetic progression $\{a + kd \mid k \geq 0\}$. In the same year, Hudson & Brauer studied in more detail the particular arithmetic progressions $4k \pm 1$, $6k \pm 1$.

A recent theorem concerns sequences of consecutive primes in arithmetic progression. In the discussion below, $(p_n)_{n \geq 1}$ is the increasing sequence of all prime numbers. In his Ph.D. Thesis, Shiu (1996) proved, using sophisticated analytical methods, the following theorem (see also his paper, 2000).

Let a and d be coprime natural numbers such that $1 \leq a < d$. Then there exist positive real numbers x_0 and C (which depend on a, d) with the following property: for every real number $x > x_0$ there exist $n \geq 1$ and

$$k \geq C \left[\frac{\log_2 x \, \log_4 x}{(\log_3 x)^2} \right]^{1/\varphi(d)}$$

such that $p_{n+k} \leq x$ and $p_{n+1} \equiv p_{n+2} \equiv \cdots \equiv p_{n+k} \equiv a \pmod{d}$.

Therefore, each allowable arithmetic progression contains arbitrarily long sequences of consecutive primes. The reason is that when x tends to infinity, so does k. It is however not asserted that these consecutive primes are themselves in arithmetic progression.

B THE SMALLEST PRIME IN AN ARITHMETIC PROGRESSION

With $d \geq 2$ and $a \geq 1$, relatively prime, let $p(d, a)$ be the smallest prime in the arithmetic progression $\{a + kd \mid k \geq 0\}$. Can one find an upper bound, depending only on a, d, for $p(d, a)$?

Let $p(d) = \max\{p(d, a) \mid 1 \leq a < d,\ \gcd(a, d) = 1\}$. Again, can one find an upper bound for $p(d)$ depending only on d? And how about lower bounds?

Linnik's theorem of 1944, which is one of the deepest results in analytic number theory asserts:

There exists $d_0 \geq 2$ and $L > 1$ such that $p(d) < d^L$ for every $d \geq d_0$.

Note that the absolute constant L, called *Linnik's constant*, is effectively computable.

It is clearly important to compute the value of L. Pan (Cheng-Dong) was the first to evaluate Linnik's constant, giving $L \leq 5448$ in 1957. Subsequently, a number of papers have appeared, where the estimate of the constants was improved.

RECORD

Heath-Brown (1992) has shown that $L \leq 5.5$, superseding the estimate $L \leq 13.5$ of Chen & Liu (1989). Previous records were due to Chen (1965), Jutila (1977), and Graham (1981).

Schinzel & Sierpiński (1958) and Kanold (1963) have conjectured that $L = 2$, that is, $p(d) < d^2$ for every sufficiently large $d \geq 2$. Explicitly, this means that if $1 \leq a < d$, $\gcd(a, d) = 1$, there is a prime number among the numbers $a, a + d, a + 2d, \ldots, a + (d-1)d$.

Heath-Brown has advanced in 1978 the conjecture that $p(d) \leq Cd(\log d)^2$ and Wagstaff sustained in 1979 that $p(d) \sim \varphi(d)(\log d)^2$ on heuristic grounds.

Concerning the lower bounds for $p(d)$, I first mention the following result of Schatunowsky (1893), obtained independently by Wolfskehl[1] in 1901.

---

[1] Paul Wolfskehl is usually remembered as the rich mathematician who endowed a substantial prize for the discovery of a proof of Fermat's last theorem. I tell the story

d = 30 is the largest integer with the following property: if $1 \leq a <$ d and $\gcd(a, d) = 1$, then $a = 1$ or a is prime.

The proof is elementary, and may be found in Landau's book *Primzahlen* (1909), page 229. It follows at once that if $d > 30$ then $p(d) > d + 1$.

From the prime number theorem, it follows already that for every $\varepsilon > 0$ and for all sufficiently large d:

$$p(d) > (1 - \varepsilon)\varphi(d) \log d.$$

This implies that

$$\liminf \frac{p(d)}{\varphi(d) \log d} \geq 1.$$

In 1980, Pomerance proved the sharper result that

$$\liminf \frac{p(d)}{\varphi(d) \log d} \geq e^{\gamma} = 1.78107 \ldots$$

(where γ is Euler's constant). On the other hand, let Q be the set of all integers $d \geq 2$ with more than $\exp(\log_2 d / \log_3 d)$ distinct prime factors. Then, for every d outside the set Q,

$$\liminf \frac{p(d)}{\varphi(d) \log d} \times t_d \geq e^{\gamma},$$

where

$$t_d = \frac{(\log_3 d)^2}{(\log_2 d)(\log_4 d)} ;$$

note that $\lim_{d \to \infty} t_d = 0$. In particular, the set of numbers $p(d)/(\varphi(d) \log d)$ is unbounded.

In this connection, I note that Prachar & Schinzel had obtained in 1961 and 1962 some results about this question.

It has to be said that the set Q has density equal to zero, since the number of distinct prime factors of d is, on average, equal to $\log_2 d$.

In 1990, Granville & Pomerance conjectured that

$$p(d) \geq C\varphi(d)(\log d)^2$$

for $d \geq 2$ and some constant $C > 0$.

---

in my book *13 Lectures on Fermat's Last Theorem*. Here, I wish to salute his memory, on behalf of all the young assistants who, in the last 90 years, have been thrilled finding mistakes in an endless and continuous flow of "proofs" of Fermat's theorem. Eventually, the theorem was proved and the prize awarded to Andrew Wiles in 1997.

C STRINGS OF PRIMES IN ARITHMETIC PROGRESSION

Now I consider the question of existence of sequences of k primes $p_1 < p_2 < p_3 < \cdots < p_k$ with difference $p_2 - p_1 = p_3 - p_2 = \cdots = p_k - p_{k-1}$, so these primes are in arithmetic progression.

In 1939, van der Corput proved that there exist infinitely many sequences of three primes in arithmetic progression (see Section VI); this was proved again by Chowla in 1944, and again by Heath-Brown in 1985, as a corollary of a more general theorem.

Even though it is easy to give examples of four (and sometimes more) primes in arithmetic progression, the following question remains open: Is it true that for every $k \geq 4$, there exist infinitely many arithmetic progressions consisting of k primes?

In examining the first case, $k = 4$, one already discovers how very difficult this question is. In fact, to date the best result was obtained by Heath-Brown, who showed in 1981 that there exist infinitely many arithmetic progressions consisting of four numbers, of which three are primes and the other is the product of two, not necessarily distinct, prime factors.

It is conjectured that for every $k > 3$ there exists at least one arithmetic progression consisting of k prime numbers.

There have been extensive computer searches for long strings of primes in arithmetic progression.

RECORD

The longest known string of primes in arithmetic progression contains 22 primes, of which the smallest is $p = 11\,410\,337\,850\,553$ and the difference is $d = 4\,609\,098\,694\,200$. This sequence was discovered on March 17, 1993 (published in 1995). More than 60 computers were used in this search, which was coordinated by P. Pritchard (Griffith University in Queensland, Australia). This was a truly international project: the 22-term sequence was discovered in Bergen, Norway. Many sequences of 21 primes were also found in the search.

The previous records were by Young & Fry (20 terms, 1987), Pritchard (19 terms in 1985, 18 terms in 1982) and Weintraub (17 terms in 1977). Much computation is required in the search of long strings of primes in arithmetic progression.

In this connection the following statement, due to M. Cantor (1861), and quoted in Dickson's *History of the Theory of Numbers*, Vol. I, p. 425, is easy to prove:

Let $d \geq 2$, let a, $a + d, \dots, a + (n-1)d$ be n prime numbers in arithmetic progression. Let q be the largest prime such that $q \leq n$. Then either $\prod_{p \leq q} p$ divides d, or $a = q$ and $\prod_{p < q} p$ divides d.

Proof. First an easy remark. If p is a prime, not dividing d, and if a, $a + d, \dots, a + (p-1)a$ are primes, then these numbers are pairwise incongruent modulo p and p divides exactly one of these numbers. Now assume that $\prod_{p \leq q} p$ does not divide d, so that there exists a prime $p \leq n$ such that p does not divide d. Choose the smallest such prime p. By the remark, there exists j, $0 \leq j \leq p-1$, such that p divides $a + jd$, so $p = a + jd$, since $a + jd$ is a prime number. But a is a prime; if $a \neq a + jd$, then a divides d (by the choice of p), so a divides p, that is $a = p + jd$. This proves that $p = a$. If $p < q$, then $p \leq n - 1$, so p divides $a + pd$, hence $p = a + pd = p(1 + d)$, which is an absurdity. I have therefore established that if $\prod_{p \leq q} p$ does not divide d, then $q = a$ and $\prod_{p < q} p$ divides d. □

A particular case of this proposition had been established by Lagrange.

At this point, I recall that in Chapter 3, Section II, I have discussed the even more difficult question of finding strings of p primes in arithmetic progression, of which the smallest number is equal to p itself.

A related and even more difficult problem is the following. Are there arbitrarily long arithmetic progressions of *consecutive* primes?

RECORD

The longest known string of consecutive primes in arithmetic progression contains ten terms. The initial prime is

$$p = 100996972469714247637786655587969840329509324689190041803603417758904341703348882159067229719,$$

and the difference of the arithmetic progression is 210.

This sequence of consecutive primes was discovered on March 2, 1998 by M. Toplic, one from over 100 international participants in a project directed by H. Dubner, T. Forbes, N. Lygeros, M. Mizony and P. Zimmermann. Previously, on January 24, 1998, it was once again the "winning lottery ticket" for M. Toplic, who found 9 consecutive primes in arithmetic progression: the difference is 210 und the smallest prime is

$$p = 99679432066701086484490653695853556163898236408-$$
$$099161839577404858552907147546111479967769465.$$

VI   Goldbach's Famous Conjecture

In a letter of 1742 to Euler, Goldbach expressed the belief that:

(G) *Every integer $n > 5$ is the sum of three primes.*

Euler replied that this is easily seen to be equivalent to the following statement:

(G') *Every even integer $2n \geq 4$ is the sum of two primes.*

Indeed, if (G') is assumed to be true and if $2n \geq 6$, then $2n - 2 = p + p'$ so $2n = 2 + p + p'$, where p, p' are primes. Also $2n + 1 = 3 + p + p'$, which proves (G).

Conversely, if (G) is assumed to be true, and if $2n \geq 4$, then $2n + 2 = p + p' + p''$, with p, p', p'' primes; then necessarily $p'' = 2$ (say) and $2n = p + p'$.

Note that it is trivial that (G') is true for infinitely many even integers: $2p = p + p$ (for every prime).

A related, but weaker, problem is this: Is it true that every odd integer greater than 5 is the sum of three primes? This is called the *odd Goldbach statement*. It would imply, in turn, that every integer greater than 6 is the sum of at most four primes.

Very little progress was made in the study of these conjectures before the development of refined analytical methods and sieve theory. And despite all the attempts, the problems are still unsolved.

There have been three main lines of attack, reflected, perhaps inadequately, by the key words "asymptotic," "almost primes," "basis."

(**A**) An asymptotic statement is one which is true for all sufficiently large integers.

The first important result is due to Hardy & Littlewood in 1923—it is an asymptotic theorem. Using the circle method and a modified form of the Riemann hypothesis, they proved that there exists n_0 such that every odd number $n \geq n_0$ is the sum of three primes.

Later, in 1937, Vinogradov gave a proof of Hardy & Littlewood's theorem, without any appeal to the Riemann hypothesis. In 1985, Heath-Brown gave a different proof of this theorem, with no effective value for n_0.

By examining Vinogradov's proof, Borodzkin showed in 1956 that one may take $n_0 = 3^{3^{15}} \approx 10^{7000000}$. Further improvement is due to Chen & Wang, who established a bound of $n_0 = 10^{43000}$ in 1989, and of $n_0 = 10^{7194}$ in 1996—but this is still too large to allow computer verification for all smaller odd numbers.

In 1997, Deshouillers, Effinger, te Riele & Zinoviev solved the problem of three primes for every odd integer greater than 5; nevertheless, they required a conjecture similar to Riemann's hypothesis.

(B) Let $k \geq 1$. A natural number of the form $p_1 p_2 \cdots p_r$, where $p_1, p_2, \ldots, p_r$ are primes (not necessarily distinct) and $r \leq k$, is called a k-*almost-prime*. The set of k-almost-primes is denoted by P_k.

The approach via almost-primes consists in showing that there exist h, $k \geq 1$ such that every sufficiently large even integer is the sum of a k-almost-prime and a h-almost-prime. What is intended is, of course, to show that h, k can be taken to be 1.

In this direction, the first result is due to Brun (1919, *C.R. Acad. Sci. Paris*): Every sufficiently large even number is the sum of two 9-almost-primes.

Much progress has been achieved, using more involved types of sieve. In 1950, Selberg showed that every sufficiently large even integer is in the set $P_2 + P_3$ of sums of integers of P_2 and of P_3.

While these results involved summands which were both composite, Rényi proved in 1947 that there exists an integer $k \geq 1$ such that every sufficiently large even integer is in $P_1 + P_k$. Subsequent work provided explicit values of k.

The best result to date—and the closest one has come to establishing Goldbach's conjecture—is by Chen (announcement of results in 1966; proofs in detail in 1973, 1978). In his famous paper, Chen proved:

Every sufficiently large even integer may be written as $2n = p+m$, where p is a prime and $m \in P_2$.

At the same time, Chen proved the "conjugate" result that there are infinitely many primes p such that $p + 2 \in P_2$; this is very close to showing that there are infinitely many twin primes.

The same method is good to show that for every even integer $2k \geq 2$, there are infinitely many primes p such that $p + 2k \in P_2$; so $2k$ is the difference $m - p$ ($m \in P_2$, p prime) in infinitely many ways.

A proof of Chen's theorem is given in the book of Halberstam & Richert. See also the simpler proof given by Ross (1975).

(C) The "basis" approach began with the celebrated theorem of Schnirelmann (1930), proved, for example, in Landau's book (1937) and in Gelfond & Linnik's book (translated in 1965):

There exists a positive integer S, such that every sufficiently large integer is the sum of at most S primes.

It follows that there exists a positive integer $S_0 \geq S$ such that every integer (greater than 1) is a sum of at most S_0 primes. S_0 is called the *Schnirelmann constant*. Goldbach's conjecture is expressed by saying that $S_0 = 3$.

In his small and neat book (1947), Khinchin wrote an interesting and accessible chapter on Schnirelmann's ideas of bases and density of sequences of numbers.

Schnirelmann's constant S_0 has been effectively estimated by numerous computations.

RECORD

The best estimate for Schnirelmann's constant up to now is $S_0 \leq 6$, obtained by Ramaré in 1995. Previously, the best estimate was $S_0 \leq 19$, due to Riesel & Vaughan (1983).

In 1949, Richert proved the following analogue of Schnirelmann's theorem: Every integer $n > 6$ is the sum of distinct primes.

Here I note that Schinzel showed in 1959 that Goldbach's conjecture implies (and so, it is equivalent to) the statement: Every integer $n > 17$ is the sum of exactly three distinct primes. Thus, Richert's

result will be a corollary of Goldbach's conjecture (if and when it
will be shown true).

(D) The number of representations

Now I shall deal with the number $r_2(2n)$ of representations of $2n \geq 4$
as sums of two primes. *A priori*, $r_2(2n)$ might be zero (until Goldbach's conjecture is established).

Hardy & Littlewood gave, in 1923, the asymptotic formula below,
which at first relied on a modified Riemann's hypothesis; later work
of Vinogradov removed this dependence:

$$r_2(2n) \leq C \frac{2n}{(\log 2n)^2} \log \log 2n.$$

For $n > 2$, let $\pi^*(n)$ be the number of primes p such that $n/2 \leq
p \leq n - 2$. Clearly, $r_2(n) \leq \pi^*(n)$. In 1993, Deshouillers, Granville,
Narkiewicz & Pomerance proved that $n = 210$ is the largest integer
for which $r_2(n) = \pi^*(n)$.

In 1985, Powell proposed to give an elementary proof of the following fact (problem in *Mathematics Magazine*): for every $k > 0$
there exist infinitely many even integers $2n$, such that $r_2(2n) > k$.
A solution by Finn & Frohliger was published in 1986.

Here is my own proof, which requires only the knowledge that
there are at least $x/(2 \log x)$ primes $p \leq x$, thus a weak version of
Tschebycheff's inequality.

Proof. Let x be such that $x/(2 \log x) > \sqrt{2kx} + 1$. Let P be a set of
odd primes $p \leq x$ with at least $x/(2 \log x)$ elements. Let P_2 be the
set of pairs (p, q) with $p < q$, and $p, q \in P$. The set P_2 has at least

$$\frac{1}{2} \cdot \frac{x}{2 \log x} \left(\frac{x}{2 \log x} - 1 \right)$$

elements. Now let $f(p, q) = p + q$, so the image of f is contained in
the set of even integers at most equal to $2x - 2$; hence the image has
at most $x - 4$ elements. So there exists $n \leq 2x - 2$ such that the set
of pairs $(p, q) \in P_2$ with $p + q = n$ has at least

$$\frac{P_2}{x - 4} > \frac{1}{2x} \left(\frac{x}{2 \log x} - 1 \right)^2 > k$$

elements.   □

(E) The exceptional set.

For every $x \geq 4$ let

$$G'(x) = \#\{2n \mid 2n \leq x, 2n \text{ is not a sum of two primes}\}.$$

Van der Corput (1937), Estermann (1938), and Tschudakoff (1938) proved independently that $\lim G'(x)/x = 0$, and in fact, $G'(x) = O(x/(\log x)^\alpha)$, for every $\alpha > 0$. Another proof was given by Heath-Brown in 1985.

The best result in this direction is the object of a deep paper by Montgomery & Vaughan (1975), and it asserts that there exists an effectively computable constant α, $0 < \alpha < 1$, such that for every sufficiently large x, $G'(x) < x^{1-\alpha}$. In 1980, Chen & Pan showed that $\alpha = 1/100$ is a possible choice. In a second paper (1983), Chen suceeded in taking $\alpha = 1/25$ (also done independently by Pan).

Concerning numerical calculations about Goldbach's conjecture, I give now the records.

RECORD

A. Consider first the problem of three primes. In 1998, Saouter verified that every odd integer below 10^{20} is the sum of at most three prime numbers.

B. Now the Goldbach problem. In 1998, Deshouilliers, te Riele & Saouter verified Goldbach's conjecture up to 10^{14}. Richstein (2001) extended these calculations to 4×10^{14}. More recently, T. Oliveira e Silva confirmed the conjecture up to 8×10^{15}, and is currently pursuing his computations.

Previously, the conjecture was verified by Sinisalo (1993) up to 4×10^{11}, and by Granville, van de Lune & te Riele (1989) up to 2×10^{10}.

A sort of Goldbach's problem

This time the question is to express any odd integer as a sum of a prime and (of course, not another prime but) a power of 2. So the problem bears resemblances to both the Goldbach problem and the twin primes problem. The question was raised by Prince A. de Polignac, who asserted in 1849 that every odd natural number is the sum of a prime and a power of 2. But soon he recognized his error, as 959 less any power of 2 is not a prime. See Dickson's *History of the Theory of Numbers*, Vol. I, p. 424.

Yet the problem remains to study the set $A = \{p + 2^k \mid p \text{ is an}$ odd prime and $k \geq 1\}$. I review just a few of the results. Romanoff proved in 1934 that A has positive density, that is, there exists $C > 0$ such that $\#\{m \in A \mid m \leq x\}/x > C$ (for all $x \geq 1$).

Erdös studied the problem in 1950. First of all, by the prime number theorem $\#\{m \in A \mid m \leq x\} = O(x)$. He also showed that there exists an arithmetic progression consisting of odd integers, but containing no integer of the form $p + 2^k \in A$.

The integers $n = 7, 15, 21, 45, 75, 105$ are such that $n - 2^k$ is a prime for all k such that $2^k < n$. Erdös conjectured that these are the only such examples. Let $R(n) = \#\{(p, k) \mid p \text{ is an odd prime}$, $k \geq 1$, and $p + 2^k = n\}$. Erdös proved that there exists $C > 0$ such that, for infinitely many n, $R(n) > C \log \log n$.

VII   The Distribution of Pseudoprimes and of Carmichael Numbers

Now I shall indicate results on the distribution of pseudoprimes and of Carmichael numbers.

A   DISTRIBUTION OF PSEUDOPRIMES

Let $P\pi(x)$ denote the number of pseudoprimes (to the base 2) less than or equal to x, and let $(\text{psp})_1 < (\text{psp})_2 < \cdots < (\text{psp})_n < \cdots$ be the increasing sequence of pseudoprimes.

In 1949 and 1950, Erdös gave the following estimates:

$$C \log x < P\pi(x) < \frac{x}{e^{\frac{1}{3}(\log x)^{1/4}}}$$

(for x sufficiently large and $C > 0$). Using the method of Lehmer given in Chapter 2, Section VIII, to generate infinitely many pseudoprimes in base 2, it is easy to show, more explicitly, that if $x \geq 341$, then $0.171 \log x \leq P\pi(x)$. These estimates were later much improved, as I shall soon indicate.

From these estimates, it is easy to deduce that

$$\sum_{n=1}^{\infty} \frac{1}{(\text{psp})_n}$$

is convergent (first proved by Szymiczek in 1967), while

$$\sum_{n=1}^{\infty} \frac{1}{\log{(\mathrm{psp})_n}}$$

is divergent (first shown by Mąkowski in 1974).

It is convenient to introduce the following notation for the counting functions of pseudoprimes, Euler and strong pseudoprimes in arbitrary bases $a \geq 2$:

$$P\pi_a(x) = \#\{n \mid 1 \leq n \leq x, \, n \text{ is psp}(a)\}, \qquad P\pi(x) = P\pi_2(x),$$
$$EP\pi_a(x) = \#\{n \mid 1 \leq n \leq x, \, n \text{ is epsp}(a)\}, \quad EP\pi(x) = EP\pi_2(x),$$
$$SP\pi_a(x) = \#\{n \mid 1 \leq n \leq x, \, n \text{ is spsp}(a)\}, \quad SP\pi(x) = SP\pi_2(x).$$

Clearly, $SP\pi_a(x) \leq EP\pi_a(x) \leq P\pi_a(x)$.

Now we consider the estimates of upper and lower bounds for these functions.

For the upper bound of $P\pi(x)$, improving on a previous result of Erdös (1956), Pomerance showed in 1981, that for all large x,

$$P\pi(x) \leq \frac{x}{l(x)^{1/2}},$$

with

$$l(x) = e^{\log x \, \log \log \log x / \log \log x}.$$

The same bound is also good for $P\pi_a(x)$, with arbitrary basis $a \geq 2$.

Concerning lower bounds, the best result to date is by Pomerance, in 1982 (see Remark 3 of his paper):

$$e^{(\log x)^\alpha} \leq SP\pi_a(x),$$

where $\alpha = 5/14$. Using the method of Lehmer to generate pseudoprimes to the base 2, which is elementary (see Chapter 2, Section VIII, A), it is possible to obtain the explicit, but weaker lower bound $0.171 \log x \leq P\pi(x)$.

The tables of pseudoprimes suggest that for every $x \geq 170$ there exists a pseudoprime between x and $2x$. However, this has not yet been proved. In this direction, I note the following results of Rotkiewicz (1965):

If n is an integer, $n > 19$, there exists a pseudoprime between n and n^2. Also, for every $\varepsilon > 0$ there exists $x_0 = x_0(\varepsilon) > 0$ such that, if $x > x_0$, then there exists a pseudoprime between x and $x^{1+\varepsilon}$.

Concerning pseudoprimes in arithmetic progression, Rotkiewicz proved in 1963 and 1967:

If $a \geq 1$, $d \geq 1$ and $\gcd(a, d) = 1$, there exist infinitely many pseudoprimes in the arithmetic progression $\{a + kd \mid k \geq 1\}$.

Let $\mathrm{psp}(d, a)$ denote the smallest pseudoprime in this arithmetic progression. Rotkiewicz showed in 1972:

For every $\varepsilon > 0$ and for every sufficiently large d, $\log \mathrm{psp}(d, a) < d^{4L^2 + L + \varepsilon}$, where L is Linnik's constant (seen in Section IV).

The above results have been extended by van der Poorten & Rotkiewicz in 1980: if $a, d \geq 1$, $\gcd(a, d) = 1$, then the arithmetic progression $\{a + kd \mid k \geq 1\}$ contains infinitely many odd strong pseudoprimes for each base $b \geq 2$.

B   DISTRIBUTION OF CARMICHAEL NUMBERS

I now turn my attention to the distribution of Carmichael numbers. Let $CN(x)$ denote the number of Carmichael numbers n such that $n \leq x$.

First, I discuss the upper bounds for $CN(x)$. In 1956, Erdös showed that there exists a constant $\alpha > \frac{1}{2}$ such that, for every sufficiently large x,

$$CN(x) \leq \frac{x}{l(x)^\alpha},$$

where $l(x)$ was defined above.

Pomerance, Selfridge & Wagstaff improved this estimate in 1980: for every $\varepsilon > 0$ there exists $x_0(\varepsilon) > 0$ such that, if $x \geq x_0(\varepsilon)$, then

$$CN(x) \leq \frac{x}{l(x)^{1-\varepsilon}}.$$

The problem of finding a lower bound for $CN(x)$ is difficult. In the proof by Alford, Granville & Pomerance (1994), of the existence of infinitely many Carmichael numbers, it was also established that $CN(x) \geq x^{2/7}$ for all sufficiently large x.

There is reason to believe that $CN(x) \geq x/l(x)^{1-\varepsilon}$ (for x sufficiently large), as was conjectured by Pomerance, Selfridge & Wagstaff. An easy to read presentation of these results was given by Granville in 1992.

I refer now to tables of pseudoprimes and of Carmichael numbers. In 1938 Poulet determined all the (odd) pseudoprimes (in base 2) up to 10^8. Carmichael numbers were starred in Poulet's table. In 1975, Swift compiled a table of Carmichael numbers up to 10^9, while in 1979 Yorinaga went up to 10^{10}.

The table of Pomerance, Selfridge & Wagstaff (1980) comprises pseudoprimes, Euler pseudoprimes, strong pseudoprimes (in base 2) and Carmichael numbers and goes to 25×10^9. It was extended by Pinch to 10^{12} in 1992 and to 10^{13} in 2000.

Table 18. $P\pi(x), EP\pi(x), SP\pi(x)$ and $CN(x)$

x	$P\pi(x)$	$EP\pi(x)$	$SP\pi(x)$	$CN(x)$
10^3	3	1	0	1
10^4	22	12	5	7
10^5	78	36	16	16
10^6	245	114	46	43
10^7	750	375	162	105
10^8	2057	1071	488	255
10^9	5597	2939	1282	646
10^{10}	14884	7706	3291	1547
25×10^9	21853	11347	4842	2163
10^{11}	38975	20417	8607	3605
10^{12}	101629	53332	22412	8241
10^{13}	264239	124882	58897	19279

In 1990, Jaeschke extended the table of Carmichael numbers up to 10^{12}. It was further extended (and slightly amended) by Pinch to 10^{15} in 1993 and to 10^{16} in 1998. Table 19 gives the number of Carmichael numbers up to 10^M, $3 \leq M \leq 16$, having k distinct prime factors.

Pinch has also published other tables about Carmichael numbers:

(1) List of the smallest Carmichael number having k prime factors, for $3 \leq k \leq 20$.

(2) Number of Carmichael numbers in the residue classes modulo 5, 7, 11, 12, up to 25×10^9 and also up to 10^M for $11 \leq M \leq 16$.

Table 19. Number of prime factors of Carmichael numbers

				k					
M	3	4	5	6	7	8	9	10	Total
3	1	0	0	0	0	0	0	0	1
4	7	0	0	0	0	0	0	0	7
5	12	4	0	0	0	0	0	0	16
6	23	19	1	0	0	0	0	0	43
7	47	55	3	0	0	0	0	0	105
8	84	144	27	0	0	0	0	0	255
9	172	314	146	14	0	0	0	0	646
10	335	619	492	99	2	0	0	0	1547
11	590	1179	1336	459	41	0	0	0	3605
12	1000	2102	3156	1714	262	7	0	0	8241
13	1858	3639	7082	5270	1340	89	1	0	19279
14	3284	6042	14938	14401	5359	655	27	0	44706
15	6083	9938	29282	36907	19210	3622	170	0	105212
16	10816	16202	55012	86696	60150	16348	1436	23	246683

(3) Frequency of each prime $p \leq 97$ as a factor (resp. as the smallest factor) of a Carmichael number, up to the bounds just indicated.

C DISTRIBUTION OF LUCAS PSEUDOPRIMES

The Lucas pseudoprimes were studied in Chapter 2, Section X. Recall that if P, Q are nonzero integers, $D = P^2 - 4Q$, and the Lucas sequence is defined by

$$U_0 = 0, \quad U_1 = 1, \quad U_n = PU_{n-1} - QU_{n-2} \quad \text{(for } n \geq 2),$$

then the composite integer n, relatively prime to D, is a Lucas pseudoprime (with parameters (P,Q)) when n divides $U_{n-(D|n)}$.

Since the concept of Lucas pseudoprime is quite recent, much less is known about the distribution of such numbers. My source is the paper of Baillie & Wagstaff (1980) quoted in Chapter 2. Here are the main results:

If x is sufficiently large, the number $L\pi(x)$ of Lucas pseudoprimes (with parameters (P,Q)) less or equal x, is bounded as follows:

$$L\pi(x) < \frac{x}{e^{Cs(x)}}$$

where $C > 0$ is a constant and $s(x) = (\log x \log\log x)^{1/2}$.

It follows (as in Szymiczek's result for pseudoprimes) for any given parameters (P, Q) that $\sum(1/U_n)$ is convergent (sum for all Lucas pseudoprimes with these parameters).

On the other hand Erdös, Kiss & Sárközy (1988) showed that there is a constant $C > 0$ such that for any nondegenerate Lucas sequence, and x sufficiently large, $L\pi(x) > \exp\{(\log x)^C\}$.

Similarly, there is the following lower bound for the number $SL\pi(x)$ of strong Lucas pseudoprimes (with parameters (P, Q)) less or equal x (see the definition of Chapter 2, Section X): $SL\pi(x) > C' \log x$ (valid for all x sufficiently large) where $C' > 0$ is a constant.

5
Which Special Kinds of Primes Have Been Considered?

We have already encountered several special kinds of primes, for example, those which are Fermat numbers, or Mersenne numbers (see Chapter 2). Now I shall discuss other families of primes, among them the regular primes, the Sophie Germain primes, the Wieferich primes, the Wilson primes, the prime repunits, the primes in second-order linear recurring sequences.

Regular primes, Sophie Germain primes, and Wieferich primes have directly sprung from attempts to prove Fermat's last theorem. The interested reader may wish to consult my book *13 Lectures on Fermat's Last Theorem*, where these matters are considered in more detail. In particular, there is an ample bibliography including many classical papers, which will not be listed in the bibliography of this book.

I   Regular Primes

Regular primes appeared in the work of Kummer in relation with Fermat's last theorem. In a letter to Liouville, in 1847, Kummer stated that he had succeeded in proving Fermat's last theorem for all primes p satisfying two conditions. Indeed, he showed that if p satisfies these conditions, then there do not exist integers x, y, $z \neq 0$

such that $x^p + y^p = z^p$. He went on saying that "it remains only to find out whether these properties are shared by all prime numbers."

To describe these properties I need to explain some concepts that were first introduced by Kummer.

Let p be an odd prime, let

$$\zeta = \zeta_p = \cos \frac{2\pi}{p} + i \sin \frac{2\pi}{p}$$

be a primitive pth root of 1. Note that $\zeta^{p-1} + \zeta^{p-2} + \cdots + \zeta + 1 = 0$, because $X^p - 1 = (X - 1)(X^{p-1} + X^{p-2} + \cdots + X + 1)$ and $\zeta^p = 1$, $\zeta \neq 1$. Thus, ζ^{p-1} is expressible in terms of the lower powers of ζ. Let K be the set of all numbers $a_0 + a_1\zeta + \cdots + a_{p-2}\zeta^{p-2}$, with $a_0, a_1, \ldots, a_{p-2}$ rational numbers. Let A be the subset of K consisting of those numbers for which $a_0, a_1, \ldots, a_{p-2}$ are integers. Then K is a field, called the field of p-cyclotomic numbers. A is a ring, called the ring of p-cyclotomic integers. The units of A are the numbers $\alpha \in A$ which divide 1, that is $\alpha\beta = 1$ for some $\beta \in A$. The element $\alpha \in A$ is called a cyclotomic prime if α cannot be written in the form $\alpha = \beta\gamma$ with $\beta, \gamma \in A$, unless β or γ is a unit.

I shall say that the arithmetic of p-cyclotomic integers is ordinary if every cyclotomic integer is equal, in a unique way, up to units, to the product of cyclotomic primes.

Kummer discovered as early as 1847, that if $p \leq 19$ then the arithmetic of the p-cyclotomic integers is ordinary; however it is not so for $p = 23$.

To find a way to deal with nonuniqueness of factorization, Kummer introduced ideal numbers. Later, Dedekind considered certain sets of cyclotomic integers, which he called ideals. I am refraining from defining the concept of ideal, assuming that it is well known to the reader. Dedekind ideals provided a concrete description of Kummer ideal numbers, so it is convenient to state Kummer's results in terms of Dedekind ideals. A prime ideal P is an ideal which is not equal to 0 nor to the ring A, and which cannot be equal to the product of two ideals, $P = IJ$, unless I or J is equal to P. Kummer showed that for every prime $p > 2$, every ideal (different from 0 and A) of the ring of p-cyclotomic integers is equal, in a unique way, to the product of prime ideals.

In this context, it was natural to consider two nonzero ideals I, J to be equivalent if there exist nonzero cyclotomic integers $\alpha, \beta \in A$ such that $A\alpha.I = A\beta.J$. The set of equivalence classes of ideals forms a commutative semigroup, in which the cancellation property holds. Kummer showed that this set is finite, hence it is a group, called the ideal class group. The number of its elements is called the class number, and denoted by $h = h(p)$. It is a very important arithmetic invariant.

The concepts of fractional ideal, classes of ideals and the finiteness of the number of classes are of central importance in the theory of algebraic number fields. Besides the present situation of cyclotomic fields, we had considered earlier (Chapter 3, Section III, B) the case of quadratic fields.

The class number $h(p)$ is equal to 1 exactly when every ideal of A is principal, that is, of the form $A\alpha$, for some $\alpha \in A$. Thus $h(p) = 1$ exactly when the arithmetic of the p-cyclotomic integers is ordinary. So, the size of $h(p)$ is one measure of the deviation from the ordinary arithmetic.

Let it be said here that Kummer developed a very deep theory, obtained an explicit formula for $h(p)$ and was able to calculate $h(p)$ for small values of p.

One of the properties of p needed by Kummer in connection with Fermat's last theorem was the following: p does not divide the class number $h(p)$. Today, a prime with this property is called a *regular prime*.

The second property mentioned by Kummer concerned units, and he showed later that it is satisfied by all regular primes. This was another beautiful result of Kummer, today called Kummer's lemma on units.

In his regularity criterion, Kummer established that the prime p is regular if and only if p does not divide the numerators of the Bernoulli numbers $B_2, B_4, B_6, \ldots, B_{p-3}$ (already defined in Chapter 4, Section I, A).

Kummer was soon able to determine all irregular primes less than 163, namely, 37, 59, 67, 101, 103, 131, 149, 157. He maintained the hope that there exist infinitely many regular primes. This is a truly difficult problem to settle, even though the answer should be positive, as it is supported by excellent numerical evidence.

In 1964, Siegel proved, under heuristic assumptions about the residues of Bernoulli numbers modulo primes, that the density of regular primes among all the primes is $1/\sqrt{e} \cong 61\%$.

On the other hand, it was somewhat of a surprise when Jensen proved, in 1915, that there exist infinitely many irregular primes. The proof was actually rather easy, involving some arithmetical properties of the Bernoulli numbers.

Let $\pi_{\mathrm{reg}}(x)$ be the number of regular primes p such that $2 \leq p \leq x$, and

$$\pi_{\mathrm{ir}}(x) = \pi(x) - \pi_{\mathrm{reg}}(x).$$

For each irregular prime p, the pair $(p, 2k)$ is called an *irregular pair* if $2 \leq 2k \leq p - 3$ and p divides the numerator of B_{2k}. The *irregularity index* of p, denoted $\mathrm{ii}(p)$, is the number of irregular pairs $(p, 2k)$.

For $s \geq 1$ let $\pi_{\mathrm{ii}s}(x)$ be the number of primes $p \leq x$ such that $\mathrm{ii}(p) = s$.

Record

The most important calculations about regular primes were done successively by Johnson (1975), Wagstaff (1978), Tanner & Wagstaff (1989), Buhler, Crandall & Sompolski (1992), Buhler, Crandall, Ernvall & Metsänkylä (1993), and just recently by Buhler, Crandall, Ernvall, Metsänkylä & Shokrollahi (2001). All the irregular primes up to $N = 12 \times 10^6$, together with their irregularity index, were determined. Here are the results (the prime 2 is neither counted as regular nor as an irregular prime):

$$\pi(N) = 788060$$
$$\pi_{\mathrm{reg}}(N) = 477616$$
$$\pi_{\mathrm{ir}}(N) = 310443$$
$$\pi_{\mathrm{ii}1}(N) = 239483 \qquad \text{(the smallest is } 37)$$
$$\pi_{\mathrm{ii}2}(N) = 59710 \qquad \text{(the smallest is } 157)$$
$$\pi_{\mathrm{ii}3}(N) = 9824 \qquad \text{(the smallest is } 491)$$
$$\pi_{\mathrm{ii}4}(N) = 1282 \qquad \text{(the smallest is } 12613)$$
$$\pi_{\mathrm{ii}5}(N) = 127 \qquad \text{(the smallest is } 78233)$$
$$\pi_{\mathrm{ii}6}(N) = 13 \qquad \text{(the smallest is } 527377)$$
$$\pi_{\mathrm{ii}7}(N) = 4 \qquad \text{(the smallest is } 3238481)$$
$$\pi_{\mathrm{ii}s}(N) = 0, \text{ for } s \geq 8.$$

This is our present knowledge: The largest regular prime is $p = 11999989$. The longest sequence of consecutive regular primes has 27 primes and begins with 17881. The longest sequence of consecutive irregular primes has 14 primes and begins with 670619.

The only irregular pairs $(p, 2k)$, $(p, 2k + 2)$ which are "successive" are $p = 491$, $2k = 336$ or $p = 587$, $2k = 90$. There are no known triples $(p, 2k)$, $(p, 2k + 2)$, $(p, 2k + 4)$ of irregular pairs.

For all primes $p \geq 11$, p is a Wolstenholme prime (see Chapter 2, Section II, C) if and only if p divides the numerator of the Bernoulli number B_{p-3}, or, in other words, if $(p, p - 3)$ is an irregular pair.

It is conjectured, but it has never been proved, that there are primes with arbitrarily high irregularity index.

It follows from the theorem of Kummer, from a criterion of Vandiver, and the calculations reported above, that Fermat's last theorem is true for every prime exponent $p < 12 \times 10^6$.

The regularity of a prime is relevant in many questions of number theory. In connection with Fermat's last theorem, the role of regular primes is now mainly of historical interest, since the complete proof of the theorem. This extraordinary mathematical feat was the conjunction of work by G. Frey, K.A. Ribet, J.P. Serre, A. Wiles and R. Taylor.

II   Sophie Germain Primes

I have already encountered the Sophie Germain primes, in Chapter 2, in connection with a criterion of Euler about divisors of Mersenne numbers.

I recall that p is a *Sophie Germain prime* when $2p + 1$ is also a prime. They were first considered by Sophie Germain and she proved the beautiful theorem:

If p is a Sophie Germain prime, then there are no integers x, y, z, different from 0 and not multiples of p, such that $x^p + y^p = z^p$.

In other words, for Sophie Germain's primes, the "first case of Fermat's last theorem" is true. For a detailed discussion, see my own book (1979), or my more recent book (1999).

It is presumed that there are infinitely many Sophie Germain primes. However, the proof would be of the same order of difficulty as that of the existence of infinitely many twin primes.

Now I wish to explain in more detail the relations between the first case of Fermat's last theorem and primes like Sophie Germain primes.

Sophie Germain's theorem was extended by Legendre, and also by Dénes (1951), and more recently by Fee & Granville (1991).

Now I indicate estimates for the number of Sophie Germain primes less than any number $x \geq 1$. More generally, let a, $d \geq 1$, with ad even and $\gcd(a, d) = 1$. For every $x \geq 1$, let

$$S_{d,a}(x) = \#\{p \text{ prime} \mid p \leq x, a + pd \text{ is a prime}\}.$$

If $a = 1$, $d = 2$, then $S_{2,1}(x)$ counts the Sophie Germain primes $p \leq x$.

The same sieve methods of Brun, used to estimate the number $\pi_2(x)$ of twin primes less than x, yield here the similar bound

$$S_{d,a}(x) < \frac{Cx}{(\log x)^2}.$$

By the prime number theorem,

$$\lim_{x \to \infty} \frac{S_{d,a}(x)}{\pi(x)} = 0.$$

It is then reasonable to say that the set of primes p such that $a + pd$ is prime has density 0. In particular, the set of Sophie Germain primes, and by the same token, the set of twin primes, have density 0.

In 1980, Powell gave a proof of the above facts, avoiding the use sieve methods.

Table 20. Number $S_{2,1}(x)$ of
Sophie Germain primes below x

x	$S_{2,1}(x)$
10^3	37
10^4	190
10^5	1 171
10^6	7 746
10^7	56 032
10^8	423 140
10^9	3 308 859
10^{10}	26 569 515

The last two frequencies were calculated by W. Keller and independently by C.F. Kerchner.

Many large Sophie Germain primes have now been found.

RECORDS

Table 21. The largest known Sophie Germain primes

Sophie Germain prime	Digits	Discoverer	Year
$2540041185 \times 2^{114729} - 1$	34547	D. Underbakke, G. Woltman and Y. Gallot	2003
$18912879 \times 2^{98395} - 1$	29628	M.J. Angel, D. Augustin, P. Jobling and Y. Gallot	2002
$1213822389 \times 2^{81131} - 1$	24432	M.J. Angel, D. Augustin, P. Jobling and Y. Gallot	2002
$109433307 \times 2^{66452} - 1$	20013	D. Underbakke, P. Jobling and Y. Gallot	2001
$984798015 \times 2^{66444} - 1$	20011	D. Underbakke, P. Jobling and Y. Gallot	2001
$3714089895285 \times 2^{60000} - 1$	18075	K.-H. Indlekofer, A. Járai and H.-G. Wassing	2000
$37561665 \times 2^{34090} - 1$	10270	C. Abraham	2003
$831264873 \times 2^{33539} - 1$	10106	K. Schoenberger and Y. Gallot	2003
$168851511 \times 2^{33250} - 1$	10018	D.O. Kremelberg and Y. Gallot	2003
$918522549 \times 2^{33216} - 1$	10008	J.A. Rouse and Y. Gallot	2003

A topic closely related to Sophie Germain primes is the following: An increasing sequence of primes $q_1 < q_2 < \cdots < q_k$ is called a *Cunningham chain of first kind* (resp. *of second kind*) *of length* k if $q_{i+1} = 2q_i + 1$ (resp. $q_{i+1} = 2q_i - 1$) for $i = 1, 2, \ldots, k-1$. Thus, the numbers in a Cunningham chain of the first kind are Sophie Germain primes.

It is not known if for every $k > 2$ there exists a Cunningham chain (of either kind) having length at least equal to k.

RECORD

The longest known Cunningham chain of first kind has length 14, its smallest prime being 14374829242422552838039. The longest known

Cunningham chain of second kind has length 16 and its smallest prime is 3203000719597029781. These chains were discovered by T. Forbes in 1997.

The previous record chains were obtained by G. Löh in 1989, with length 12 (first kind) and length 13 (second kind).

III   Wieferich Primes

A prime p satisfying the congruence

$$2^{p-1} \equiv 1 \pmod{p^2}$$

ought to be called a *Wieferich prime*. Indeed, it was Wieferich who proved in 1909 the difficult theorem:

If the first case of Fermat's last theorem is false for the exponent p, then p satisfies the congruence indicated above.

It should be noted that, contrary to the congruence $2^{p-1} \equiv 1$ (mod p) which is satisfied by every odd prime, the Wieferich congruence is very rarely satisfied.

Before the computer age, Meissner discovered in 1913, and Beeger in 1922, that the primes $p = 1093$ and $p = 3511$ satisfy Wieferich's congruence. If you have not been a passive reader, you must have already calculated that $2^{1092} \equiv 1$ (mod 1093^2), in Chapter 2, Section III. It is just as easy to show that 3511 has the same property.

Record

Lehmer has shown in 1981 that, with the exceptions of 1093 and 3511, there are no primes $p < 6 \times 10^9$, satisfying Wieferich's congruence. These computations were extended by Crandall, Dilcher & Pomerance (1997) up to 4×10^{12}, and subsequently by R. McIntosh to 8×10^{12}, by R. Brown to 49×10^{12}, and by J.K. Crump (with collaborators) to 2×10^{14}. Most recently, an Internet based search conducted by J. Knauer and J. Richstein attained the limit of 1.25×10^{15} (for a description, see their note of 2003). No other Wieferich prime was found.

According to results already quoted in Chapter 2, Sections III and IV, the above computations say that the only possible factors p^2 (where p is a prime less than 1.25×10^{15}) of any pseudoprime, must have $p = 1093$ or $p = 3511$. This is confirmed by the numerical calculations of Pinch (2000). Below the limit of 10^{13}, he has detected 54 pseudoprimes with a repeated factor.

In 1910, Mirimanoff proved the following theorem, analogous to Wieferich's theorem:

If the first case of Fermat's last theorem is false for the prime exponent p, then $3^{p-1} \equiv 1 \pmod{p^2}$.

It may be verified that 1093 and 3511 do not satisfy Mirimanoff's congruence.

These two results were the origin of a new line to attack the first case of Fermat's theorem. Thanks to the work of Vandiver, Frobenius, Pollaczek, Morishima, Rosser and, more recently, Granville & Monagan (1988) and Suzuki (1994), it was possible to extend considerably the validity of the first case of Fermat's last theorem. In this respect, it was very important the conjunction of several criteria by a combinatorial method of Gunderson. This is described in my own books already quoted, where one may find complete references to the relevant papers.

With the complete proof of Fermat's last theorem, these developments are now relegated to the history of Fermat's theorem. But the congruences mentioned above have kept their importance in other questions of number theory.

More generally, for any base $a \geq 2$ (where a may be prime or composite) one may consider the primes p such that p does not divide a and $a^{p-1} \equiv 1 \pmod{p^2}$. In fact, it was Abel who first asked for such examples (in 1828); these were provided by Jacobi, who indicated the following congruences, with $p \leq 37$:

$$3^{10} \equiv 1 \pmod{11^2}$$
$$9^{10} \equiv 1 \pmod{11^2}$$
$$14^{28} \equiv 1 \pmod{29^2}$$
$$18^{36} \equiv 1 \pmod{37^2}$$

The quotient

$$q_p(a) = \frac{a^{p-1} - 1}{p}$$

has been called the *Fermat quotient of p, with base a*. The residue modulo p of the Fermat quotient behaves somewhat like a logarithm (this was observed already by Eisenstein in 1850): if p does not divide ab, then

$$q_p(ab) \equiv q_p(a) + q_p(b) \pmod{p}.$$

Also

$$q_p(p-1) \equiv 1 \pmod{p}, \qquad q_p(p+1) \equiv -1 \pmod{p}.$$

In my article *1093* (1983), I indicated many truly interesting properties of the Fermat quotient. As an illustration, I quote the following congruence, which is due to Eisenstein (1850):

$$q_p(2) \equiv \frac{1}{p}\left(1 - \frac{1}{2} + \frac{1}{3} - \frac{1}{4} + \cdots - \frac{1}{p-1}\right) \pmod{p}.$$

The following problems are open:

(1) Given $a \geq 2$, do there exist infinitely many primes p such that $a^{p-1} \equiv 1 \pmod{p^2}$?

(2) Given $a \geq 2$, do there exist infinitely many primes p such that $a^{p-1} \not\equiv 1 \pmod{p^2}$?

The answer to (1) should be positive, why not? But I am stating it with no base whatsoever, since the question is no doubt very difficult. The next question concerns a fixed prime and a variable base:

(3) If p is an odd prime, does there exist one or more bases a, $2 \leq a < p$, such that $a^{p-1} \equiv 1 \pmod{p^2}$?

Few results are known. Kruyswijk showed in 1966 that there exists a constant C such that for every odd prime p:

$$\#\{a \mid 2 \leq a < p, \ a^{p-1} \equiv 1 \pmod{p^2}\} < p^{\frac{1}{2} + \frac{C}{\log\log p}}.$$

So, not too many bases are good for each prime p.

In 1987, Granville proved

$$\#\{q \text{ prime} \mid 2 \leq q < p, \ q^{p-1} \equiv 1 \pmod{p^2}\} < p^{1/2}$$

and, more generally, if $u \geq 1$ and p is a prime such that $p \geq u^{2u}$, then

$$\#\{q \text{ prime} \mid 2 \leq q \leq u^{1/u}, \ q^{p-1} \equiv 1 \pmod{p^2}\} \geq up^{u/2}.$$

Also,

$$\#\{q \text{ prime} \mid 2 \leq q < p, \ q^{p-1} \not\equiv 1 \pmod{p^2}\} \geq \pi(p) - p^{1/2}.$$

RECORD

In 2001, Keller & Richstein determined that for $p = 6692367337$ there are 16 bases a, $2 \leq a < p$, for which $a^{p-1} \equiv 1 \pmod{p^2}$, namely, $a = 5^k$ for $k = 1, 2, \ldots, 14$, along with $a = 4961139411$ and $a = 6462265338$. For $p = 188748146801$, the same number of solutions $a < p$ is observed, which in this case are $a = 5^k$ for $k = 1, 2, \ldots, 16$.

The previous record was by Ernvall & Metsänkylä (1997), with $p = 1645333507$ and 14 bases $a < p$. Note that all three values of p satisfy the congruence $5^{p-1} \equiv 1 \pmod{p^2}$; compare the following table.

Table 22. Fermat quotients which are divisible by p

Base	Primes p such that $a^{p-1} \equiv 1 \pmod{p^2}$
2	1093 3511
3	11 1006003
5	20771 40487 53471161 1645333507^{M}
	6692367337^{K} 188748146801^{K}
7	5 491531
11	71
13	863 1747591
17	3 46021 48947
19	3 7 13 43 137 63061489
23	13 2481757 13703077 15546404183^{R}
29	None
31	7 79 6451 2806861^{K}
37	3 77867 76407520781^{R}
41	29 1025273 138200401^{K}
43	5 103
47	None
53	3 47 59 97
59	2777
61	None
67	7 47 268573
71	3 47 331
73	3
79	7 263 3037 1012573^{K} 60312841^{K}
83	4871 13691 315746063^{C}
89	3 13
97	7 2914393^{K} 76704103313^{R}

Powell showed that if $p \not\equiv 7 \pmod 8$, then there is at least one prime $q < \sqrt{p}$ such that $q^{p-1} \not\equiv 1 \pmod{p^2}$ (problem posed in 1982 to the *American Mathematical Monthly*, solution published in 1986, by Tzanakis).

With stronger methods, it can be proved that for every prime $p \geq 11$, there exists a prime q, $2 \leq q < (\log p)^2$ such that $q^{p-1} \not\equiv 1 \pmod{p^2}$.

Inspired by the calculations of Lehmer for the Fermat quotient with base 2, Riesel (1964), Kloss (1965), and Brillhart, Tonascia & Weinberger (1971) published tables for bases up to 100, and larger and larger exponents.

These results were further extended by Aaltonen & Inkeri (1991), Montgomery (1993), and Keller & Richstein (2001). The latest table concerns prime bases $a < 1000$ and prime exponents $p < 10^{11}$; when $a = 3$ ou 5, then p is taken up to 10^{13}.

Table 22 is an excerpt restricted to prime bases $a < 100$. A solution marked with C was found by D. Clark, a solution with K was discovered by W. Keller, a solution with M was discovered by P.L. Montgomery, while solutions with R were found by J. Richstein.

IV   Wilson Primes

This is a very short section—almost nothing is known.

Wilson's theorem states that if p is a prime, then $(p-1)! \equiv -1 \pmod p$, thus the so-called *Wilson quotient*

$$W(p) = \frac{(p-1)! + 1}{p}$$

is an integer.

The number p is called a *Wilson prime* when $W(p) \equiv 0 \pmod p$, or equivalently $(p-1)! \equiv -1 \pmod{p^2}$. For example, $p = 5$, 13 are Wilson primes. It is not known whether there are infinitely many Wilson primes. In this respect, Vandiver wrote:

> This question seems to be of such a character that if I should come to life any time after my death and some mathematician were to tell me it had been definitely settled, I think I would immediately drop dead again.

RECORD

There is only one other known Wilson prime, besides 5, 13. It is 563, which was discovered by Goldberg in 1953 (one of the first successful computer searches).

The search for Wilson primes was continued by E.H. Pearson, K.E. Kloss, W. Keller, H. Dubner and also by Gonter & Kundert, in 1988, up to 10^7. In 1997, the search was extended by Crandall, Dilcher & Pomerance, up to 5×10^8; no new Wilson prime has been found.

V Repunits

There has been a great curiosity about numbers all of whose digits (in the base 10) are equal to 1: 1, 11, 111, 1111, ... They are called *repunits*. When are such numbers prime?

The notation Rn is commonly used for the number

$$111\ldots1 = \frac{10^n - 1}{9},$$

with n digits equal to 1. If Rn is a prime, then n must be a prime, because if $n, m > 1$ then

$$\frac{10^{nm} - 1}{9} = \frac{10^{nm} - 1}{10^m - 1} \times \frac{10^m - 1}{9}$$

and both factors are greater than 1.

RECORD

The only known prime repunits are $R2$, $R19$, $R23$ and, in the age of computers, $R317$ (discovered by Williams in 1978) and $R1031$ (discovered by Williams & Dubner in 1986).

On the other hand, Dubner had determined by 1992 that for every other prime $p < 20000$, the repunit Rp is composite. The computations were continued by J. Young, T. Granlund and H. Dubner up to $p < 60000$. In September 1999 (published in 2002), Dubner discovered that $R49081$ is a probable prime number, and even more recently, in October 2000, L. Baxter et al. found $R86453$ to be probably prime. At present, there is virtually no hope of proving true primality for numbers of that size.

The complete factorization of repunit numbers Rp is now known for every $p \leq 211$.

Open problem: are there infinitely many prime repunits?

For more information about repunits see, for example, the book by Yates (1982).

It is very easy to see that a repunit (different from 1) cannot be a square. It is more delicate to show that it cannot be a cube (see Rotkiewicz, 1987). A repunit cannot be also a fifth power (kindly communicated to me by R. Bond and also by K. Inkeri, both in 1989). It is not known whether a repunit may be a kth power, where k is not a multiple of 2, 3 or 5 (see Obláth, 1956).

In 1979, Williams & Seah also considered numbers of the form $(a^n - 1)/(a - 1)$, where $a \neq 2$, 10 (for $a = 2$, one gets the numbers $2^n - 1$, for $a = 10$, the repunits). These numbers are now called *generalized repunits* to the base a. As for "ordinary" repunits, they can only be prime for prime exponents n. It is usually difficult to establish the primality of large numbers of this type.

Table 23. Primes of the form $(a^n - 1)/(a - 1)$

a	n								
3	3	7	13	71	103	541	1091	1367	1627
	4177^{DB}	9011*	9551*	36913*	[42700]				
5	3	7	11	13	47	127	149	181	619	929
	3407*	10949*	13241*	13873*	16519*				
	[31400]								
6	2	3	7	29	71	127	271	509	1049
	6389*	6883^S	10613*	19889*	[29800]				
7	5	13	131	149	1699^{DB}	14221*	[28200]		
11	17	19	73	139	907	1907*	2029*	4801^B	
	5153*	10867*	20161*	[24000]					
12	2	3	5	19	97	109	317	353	701	9739*
	14951*	[26300]							

In this table, the limits on n are given in brackets. In his paper published in 1993, Dubner had covered $n \leq 12000$ for $a = 3$, 5 or 6, $n \leq 10700$ for $a = 7$, $n \leq 11000$ for $a = 11$, and $n \leq 10400$ for

$a = 12$. His much more extensive table included all bases $a \leq 99$. The extension of the above table to the current limits is due to A. Steward.

Probable prime numbers are marked with an asterisk. The largest primes in the list that were rigorously verified are highlighted by the initials of the provers: DB for H. Dubner and R.P. Brent (in 1996), B for D. Broadhurst (in 2000), S for A. Steward (in 2001).

VI   Numbers $k \times b^n \pm 1$

As I stated in Chapter 2, the factors of Fermat numbers are of the form $k \times 2^n + 1$. This property brought these numbers into focus, so it became natural to investigate their primality.

Besides the Mersenne numbers (with $k = 1$), other numbers of the form $k \times 2^n - 1$ have also been tested for primality.

From Dirichlet's theorem on primes in arithmetic progressions, given $n \geq 1$, there exist infinitely many integers $k \geq 1$, such that $k \times 2^n + 1$ is a prime, and also infinitely many integers $k' \geq 1$, such that $k' \times 2^n - 1$ is a prime.

A very interesting question is the following, in which the multiplier k is fixed: Given $k \geq 1$, does there exist some integer $n \geq 1$, such that $k \times 2^n + 1$ (resp. $k \times 2^n - 1$) is prime? This question was asked by Bateman and answered by Erdös & Odlyzko (in 1979). I quote special cases of their results.

For any real number $x \geq 1$, let $N(x)$ denote the number of odd integers k, $1 \leq k \leq x$, such that there exists $n \geq 1$ for which $k \times 2^n + 1$ (resp. $k \times 2^n - 1$) is a prime. Then there exists $C_1 > 0$, C_1 being effectively computable, such that $N(x) \geq C_1 x$ (for every $x \geq 1$). The method developed also serves to study other similar sequences.

Even though a positive proportion of integers $k \geq 1$ has the property that $k \times 2^n + 1$ (or $k \times 2^n - 1$) is a prime for some n, Riesel found in 1956 that if $k = 509203$, then $k \times 2^n - 1$ is composite for all $n \geq 1$. His paper, written in Swedish, was certainly not known to Sierpiński, who in 1960 proved the following interesting theorem:

There exist infinitely many odd integers k such that $k \times 2^n + 1$ is composite for every $n \geq 1$.

The numbers k with the above property are called *Sierpiński numbers*. It is natural to call the odd integer k a *Riesel number* if $k \times 2^n - 1$ is composite for every $n \geq 1$.

It follows from Dirichlet's theorem on primes in arithmetic progressions and Sierpiński's result that there exist infinitely many Sierpiński numbers which are primes. Similarly, there exist infinitely many Riesel numbers which are primes.

RECORD

The smallest known Sierpiński number is $k = 78557$, determined by Selfridge in 1963. The smallest known Riesel number is Riesel's own $k = 509203$.

During many years, Keller has made great efforts in attempting to prove the conjecture that no smaller Sierpiński number k exists. He showed (1991) that $k \geq 4847$, and that only 35 odd integers k in the interval $4847 \leq k < 78557$ are "possible" Sierpiński numbers. Of these, 14 were eliminated by J. Young in 1997.

Another four were eliminated with the aid of Gallot's program: two by M. Thibeault (in 1999), one by L. Baxter in April 2001, and one by J. Szmidt in November 2001. Finally, a distributed computing project conducted by L. Helm and D. Norris suceeded in eliminating five more candidates within a period of only a few weeks, near the end of 2002. Thus, only the following 12 multipliers remain to be examined further:

$$k = 4847, 5359, 10223, 19249, 21181, 22699,$$
$$24737, 27653, 28433, 33661, 55459, 67607.$$

Regarding possible Riesel numbers k with $k < 509203$, Keller has shown that $k \geq 659$. As the result of a coordinated search involving many individual participants, there remain 101 candidates which have not yet been excluded.

The computations aimed at eliminating candidates have led to the discovery of very large primes. Thus, $k = 54767$ was eliminated as a possible Sierpiński number by the discovery of a 402569 digit prime (see Table 24). Also, $k = 204223$ was excluded as a possible Riesel number when O. Haeberlé (in March 2003) determined the prime $204223 \times 2^{696891} - 1$ (with 209791 digits), the numbers $204223 \times 2^n - 1$ being composite for all $n < 696891$.

RECORDS

The largest known prime of the form $k \times 2^n + 1$ is $3 \times 2^{2145353} + 1$ (645817 digits), shown in Table 24, and which is also the largest known divisor of a Fermat number (see Chapter 2, Section VI). The largest known prime of the form $k \times 2^n - 1$ is $138847 \times 2^{1283793} - 1$ (386466 digits), found by J. Penné and P. Jobling in 2003.

While the four largest primes currently known are Mersenne numbers, the following list gives the largest known primes which are not Mersenne numbers. Note that the top two primes in the list did surpass one of the Mersenne primes discovered by GIMPS.

RECORDS

Table 24. The largest known non-Mersenne primes

Prime number	Digits	Discoverer	Year
$3 \times 2^{2145353} + 1$	645817	J. Cosgrave, P. Jobling, G. Woltman and Y. Gallot	2003
$62722^{2^{17}} + 1$	628808	M. Angel, P. Carmody and Y. Gallot	2003
$1483076^{2^{16}} + 1$	404434	D. Heuer, J. Fougeron and Y. Gallot	2003
$1478036^{2^{16}} + 1$	404337	D. Heuer, J. Fougeron and Y. Gallot	2002
$54767 \times 2^{1337287} + 1$	402569	P. Coels, L. Helm, D. Norris, G. Woltman and Y. Gallot	2002
$1361846^{2^{16}} + 1$	402007	A.J. Penrose, J. Fougeron and Y. Gallot	2002
$1266062^{2^{16}} + 1$	399931	D. Underbakke and Y. Gallot	2002
$5 \times 2^{1320487} + 1$	397507	M. Toplic and Y. Gallot	2002

Generalized Fermat numbers

Strikingly, most of the above primes are of the form $b^{2^m} + 1$, a special case of $k \times b^n + 1$ with $k = 1$, $n = 2^m$, b even. Numbers $b^{2^m} + 1$, with $b \geq 2$, $m \geq 1$, are called *generalized Fermat numbers*. In 1985, Dubner for the first time listed large primes of this form (briefly known as *generalized Fermat primes*), where the largest one was $150^{2^{11}} + 1$, with 4457 digits.

Around 1998, Y. Gallot noticed that generalized Fermat numbers could be tested for primality at a similar speed as Mersenne numbers of the same size. He then developed and successively optimized a computer program which showed that this was a practical reality. In fact, the test is clearly faster than for any number of the form $k \times 2^n \pm 1$ with $k > 1$. The arithmetic involved uses the discrete weighted transform (DWT) devised by Crandall & Fagin (1994), and was also used to discover the five largest known Mersenne primes.

Historically, the largest known prime has almost always been a Mersenne prime. Only in August 1989 the prime $391581 \times 2^{216193} - 1$, discovered by the six devoted numerologists J. Brown, L.C. Noll, B. Parady, G. Smith, J. Smith, and S. Zarantonello, had dethroned the Mersenne prime M_{216091}. Poor Mersenne, who for a while turned around in his tomb, full of worries and sadness; he could rest in peace again, with the performance of his Mersenne primes. But how long will his peace last?

As Dubner & Gallot (2002) explain in their recent paper, many more generalized Fermat primes of comparable magnitude are expected to exist and therefore, in their words, a properly organized search could soon change the status of the largest known primes. About 120 primes of this form with more than 100000 digits have already been determined.

Other records of interest involving numbers of the form $k \times b^n + 1$ with $b > 2$ are the following.

RECORDS

A. The prime $62722^{2^{17}} + 1$, given in Table 24, is the largest known prime of the form $N^2 + 1$. Recall that it is not yet known whether there are infinitely many primes of that form.

B. The largest known prime of the form $k \times b^n + 1$ with $k > 1$ and b odd is $82960 \times 31^{82960} + 1$, with 123729 digits, discovered by G. Löh and Y. Gallot in 2002.

Cullen numbers

The numbers of the form $Cn = n \times 2^n + 1$ are known as *Cullen numbers*. In 1958, Robinson showed that $C141$ is a prime number, and that Cn is composite for all other n, $1 < n \le 1000$. For 25 years, this was the only known Cullen prime, apart from $C1 = 3$, of course.

In 1987 (published in 1995), Keller determined all the Cullen primes Cn with $n \leq 30000$. These computations were extended by J. Young (1997) up to $n \leq 100000$, and more primes were found thanks to the impetus given by Y. Gallot's program. The following list of primes is believed to be complete up to $n \leq 633000$.

Table 25. Cullen primes Cn

n	Discoverer	Year
481899	M. Morii and Y. Gallot	1998
361275	D. Smith and Y. Gallot	1998
262419	D. Smith and Y. Gallot	1998
90825	J. Young	1997
59656	J. Young	1997
32469	M. Morii	1997
32292	M. Morii	1997
18496	W. Keller	1984
6611	W. Keller	1984
5795	W. Keller	1984
4713	W. Keller	1984
141	R.M. Robinson	1958
1	–	–

In his book (1976), Hooley has indicated that almost all Cullen numbers Cn are composite, that is,

$$\lim_{x \to \infty} \frac{C\pi(x)}{x} = 0$$

where $C\pi(x)$ denotes the number of Cullen numbers $Cn \leq x$ which are prime. It is not yet known, however, if there exist infinitely many Cullen primes Cn.

The numbers $Wn = n \times 2^n - 1$ are called *Woodall numbers*, or also Cullen numbers of the second kind.

If $n \leq 20000$, Wn is prime only for $n = 2, 3, 6, 30, 75, 81$ (Riesel, 1969), 115, 123, 249, 362, 384, 462, 512, 751, 822, 5312, 7755, 9531, 12379, 15822, and 18885 (Keller, 1987). The computations were continued by J. Young up to $n \leq 100000$, and by Y. Gallot and collaborators up to $n \leq 416000$. The following table gives the Woodall primes known for $n > 20000$.

Table 26. The largest known Woodall primes Wn

n	Discoverer	Year
667071	M. Toplic and Y. Gallot	2000
151023	K. O'Hare and Y. Gallot	1998
143018	R. Ballinger and Y. Gallot	1998
98726	J. Young	1997
23005	J. Young	1997
22971	J. Young	1997

W. Keller & W. Niebuhr (1995) have also obtained the complete factorization of numbers Cn and Wn for all $n \leq 300$. These complete factorizations were extended by P. Leyland, who had covered all $n \leq 400$ by November 1998, and all $n \leq 450$ by August 2000. With the assistance of other contributors the factor tables were recently completed up to $n \leq 500$ (January 2002).

Cullen numbers (of both kinds) may be generalized to the forms $n \times b^n + 1$ and $n \times b^n - 1$, where $b > 2$. *Generalized Cullen numbers* $n \times b^n + 1$ were introduced by Dubner in 1989. He studied their possible primality status and remarked that for prime bases $b > 3$ there seemed to be almost an absence of primes. In fact, for $b = 13$, 17, 19, 23, 29, 31, 41, 47, 53, 71, 73 not a single prime was found. It seemed unlikely, however, that for any of these bases non-existence of primes could be proved.

By extensive computation, it was subsequently shown that for a given base a first prime might occur for quite a large exponent n. With the help of Gallot's program, which also examines numbers of this kind, "smallest" primes were finally found for $b = 19$, 23 (Keller and Gallot, 1998) and for $b = 17$, 71 (Löh and Gallot, 2000). Most recently, for $b = 31$, Löh discovered the prime $82960 \times 31^{82960} + 1$, already mentioned in the preceding Section VI because of its remarkable size.

VII   Primes and Second-Order Linear Recurrence Sequences

In this section I shall consider sequences $T = (T_n)_{n \geq 0}$, defined by second-order linear recurrences.

General second-order linear recurrence sequences

Let P, Q be given nonzero integers such that $D = P^2 - 4Q \neq 0$. These integers P, Q are the parameters of the sequence T, to be defined now. Let T_0, T_1 be integers (not both equal to 0), and for every $n \geq 2$, let

$$T_n = PT_{n-1} - QT_{n-2}.$$

The characteristic polynomial of the sequence T is $X^2 - PX + Q$; its roots are

$$\alpha = \frac{P + \sqrt{D}}{2}, \qquad \beta = \frac{P - \sqrt{D}}{2}.$$

So $\alpha + \beta = P$, $\alpha\beta = Q$, $\alpha - \beta = \sqrt{D}$.

The sequences $(U_n)_{n \geq 0}$, $(V_n)_{n \geq 0}$, with parameters (P, Q) and $U_0 = 0$, $U_1 = 1$ (resp. $V_0 = 2$, $V_1 = P$), are precisely the Lucas sequences, already considered in Chapter 2, Section IV.

Let $\gamma = T_1 - T_0\beta$, $\delta = T_1 - T_0\alpha$, then it is easy to show that

$$T_n = \frac{\gamma\alpha^n - \delta\beta^n}{\alpha - \beta} = T_1\frac{\alpha^n - \beta^n}{\alpha - \beta} - QT_0\frac{\alpha^{n-1} - \beta^{n-1}}{\alpha - \beta},$$

for every $n \geq 0$.

If $U = (U_n)_{n \geq 0}$ is the Lucas sequence with the same parameters, then $T_n = T_1U_n - QT_0U_{n-1}$ (for $n \geq 2$).

It is also possible to define the companion sequence $W = (W_n)_{n \geq 0}$. Let

$$W_0 = 2T_1 - PT_0, \qquad W_1 = T_1P - 2QT_0$$

and

$$W_n = PW_{n-1} - QW_{n-2}, \quad \text{for } n \geq 2.$$

Again, $W_n = \gamma\alpha^n + \delta\beta^n = T_1V_n - QT_0V_{n-1}$, where $V = (V_n)_{n \geq 0}$ is the companion Lucas sequence with parameters (P, Q).

I could now proceed establishing many algebraic relations and divisibility properties of the sequences, just like for the Lucas sequences, in Chapter 2, Section IV. However, my aim is just to discuss properties related with prime numbers.

The prime divisors of a sequence T

Consider the set

$$\mathcal{P}(T) = \{p \text{ prime} \mid \text{there exists } n \text{ such that } T_n \neq 0 \text{ and } p \mid T_n\}.$$

The sequence T is called *degenerate* if $\alpha/\beta = \eta$ is a root of unity. Then $\beta/\alpha = \eta^{-1}$ is also a root of unity; thus $|\eta + \eta^{-1}| \leq 2$. But

$$\eta + \eta^{-1} = \frac{\alpha^2 + \beta^2}{\alpha\beta} = \frac{P^2 - 2Q}{Q},$$

so, if T is degenerate, then $P^2 - 2Q = 0, \pm Q, \pm 2Q$.

It is not difficult to show that if T is degenerate, then $\mathcal{P}(T)$ is finite.

In 1954, Ward proved that the converse is true:

If T is any nondegenerate sequence, then $\mathcal{P}(T)$ is infinite.

A natural problem is to ask whether $\mathcal{P}(T)$ has necessarily a positive density (in the set of all primes) and, if possible, to compute it.

The pioneering work was done by Hasse (1966). He wanted to study the set of primes p such that the order of 2 modulo p is even. This means that there exists $n \geq 1$ such that p divides $2^{2n} - 1$, but p does not divide $2^m - 1$, for $1 \leq m < 2n$. Hence $2^n \equiv -1 \pmod{p}$, so p divides $2^n + 1$, and conversely.

The sequence $H = (H_n)_{n \geq 0}$, with $H_n = 2^n + 1$, is the companion Lucas sequence with parameters $P = 3$, $Q = 2$. Let

$$\pi_H(x) = \#\{p \in \mathcal{P}(H) \mid p \leq x\}, \quad \text{for every } x \geq 1.$$

Hasse showed that

$$\lim_{x \to \infty} \frac{\pi_H(x)}{\pi(x)} = \frac{17}{24}.$$

The number $17/24$ represents the density of primes p dividing the sequence H, that is, for which an n exists such that $p \mid H_n$.

In 1985, Lagarias reworked Hasse's method and showed, among other results, that for the sequence $V = (V_n)_{n \geq 0}$ of Lucas numbers, $\mathcal{P}(V)$ has density $2/3$.

The prevailing conjecture is that for every nondegenerate sequence T, the set $\mathcal{P}(T)$ has a positive density.

Prime terms in sequences T

Now, I turn to another very interesting and difficult problem.

Let $T = (T_n)_{n \geq 0}$ be a second-order linear recurring sequence. For example, the Fibonacci or Lucas sequences. These sequences do contain prime numbers, but it is not known, and certainly difficult to decide, whether they contain infinitely many prime numbers.

It follows from Chapter 2, Section IV, relations (IV.15) and (IV.16):

If U_m is a prime, then $m = 4$ or m is a prime.

If V_m is a prime, then m is a power of 2 or m is a prime.

Of course, the converse need not be true.

The search for Fibonacci and Lucas primes, as well as the factorization of these numbers (see Chapter 2, Section XI, D), has been the object of much computational work. Since the numbers in these sequences are growing rapidly, one is faced with difficult problems in factoring and primality proving.

From the published work I mention Jarden's book of 1958, its third edition, revised and enlarged by Brillhart (1973), and the papers by Brillhart, Montgomery & Silverman (1988), and by Dubner & Keller (1999). The current state of our knowledge is as follows.

For $n < 360000$, the Fibonacci number U_n is prime (or probably prime) when

$$n = 3, 4, 5, 7, 11, 13, 17, 23, 29, 43, 47, 83, 131, 137, 359, 431, 433,$$
$$449, 509, 569, 571, 2971^{\mathrm{W}}, 4723^{\mathrm{WM}}, 5387^{\mathrm{WM}}, 9311^{\mathrm{DK}}, 9677^{\mathrm{deW}},$$
$$14431^{\mathrm{BdeW}}, 25561^{\mathrm{BdeW}}, 30757^{\mathrm{BdeW}}, 35999^{\mathrm{BdeW}}, 37511^*, 50833^*,$$
$$81839^{\mathrm{BdeW}}, 104911^*, 130021^*, 148091^*, 201107^*.$$

The letters attached to some of these numbers indicate that the primality of the corresponding Fibonacci number was determined by H.C. Williams (W), by H.C. Williams and F. Morain (WM), by H. Dubner and W. Keller (DK), by B. de Water (deW), or by D. Broadhurst and B. de Water (BdeW).

An asterisk marking an index n tells us that U_n is only known as a probable prime. This means that U_n was subjected to a number of tests which did not reveal it as composite. The probable primes U_{37511} and U_{50833} were found by Dubner, U_{104911} was discovered by de Water, U_{130021} by D. Fox, U_{148091} by T.D. Noe, and U_{201107} by H. Lifchitz. Note that the proven prime U_{81839} has 17103 digits, the proof being a remarkable achievement.

For $n < 260000$, the Lucas number V_n is prime (or probably prime) when

$$
\begin{aligned}
n = \ & 2, 4, 5, 7, 8, 11, 13, 16, 17, 19, 31, 37, 41, 47, 53, 61, 71, 79, \\
& 113, 313, 353, 503^{\mathrm{W}}, 613^{\mathrm{W}}, 617^{\mathrm{W}}, 863^{\mathrm{W}}, 1097^{\mathrm{DK}}, 1361^{\mathrm{DK}}, \\
& 4787^{\mathrm{DK}}, 4793^{\mathrm{DK}}, 5851^{\mathrm{DK}}, 7741^{\mathrm{DK}}, 8467^{\mathrm{deW}}, 10691^{\mathrm{DK}}, \\
& 12251^{\mathrm{BdeW}}, 13963^{\mathrm{Oak}}, 14449^{\mathrm{DK}}, 19469^{\mathrm{BdeW}}, 35449^{\mathrm{deW}}, \\
& 36779^*, 44507^*, 51169^{\mathrm{BdeW}}, 56003^*, 81671^*, 89849^*, 94823^*, \\
& 140057^*, 148091^*, 159521^*, 183089^*, 193201^*, 202667^*.
\end{aligned}
$$

Here the same notational conventions are applied as above, the prime-prover M. Oakes (Oak) being added. The probable primes V_{36779}, V_{44507}, V_{81671}, V_{89849} are due to Dubner, V_{140057} and V_{148091} were found by de Water, and H. Lifchitz discovered V_{94823}, V_{159521} V_{183089} V_{193201} and V_{202667}.

If you look again at the above lists, you shall notice that for the primes $n = 5$, 7, 11, 13, 17, 47, both U_n and V_n are primes, and if you continue advancing in the list, nothing more happens until—surprise—for $n = 148091$, U_n and V_n are both probable primes. When (if) it will be shown that these numbers are indeed primes, we will acquire the feeling that, perhaps, there exist (also) infinitely many primes n for which U_n and V_n are both primes. This problem will be a "hard nut to crack". Don't lose your nights of sleep, nuts are hard to digest.

Sequences T consisting only of composite integers

It should be noted that, if T is neither a Lucas sequence nor a companion Lucas sequence, then T need not to contain any prime number. In 1964, Graham discovered an example with $P = 1$, $Q = -1$; however, an error was found in the computation of T_0 and T_1. Then, in 1990, Knuth gave the correct values,

$$
\begin{aligned}
T_0 &= 331635635998274737472200656430763, \\
T_1 &= 1510028911088401971189590305498785,
\end{aligned}
$$

and also another example with smaller values of T_0, T_1:

$$T_0 = 62638280004239857,$$
$$T_1 = 49463435743205655.$$

Now let $P = 3$, $Q = 2$. The Lucas sequence and companion Lucas sequence with these parameters are U, V with $U_n = 2^n - 1$, $V_n = 2^n + 1$. These sequences contain prime numbers. If $T_0 = k + 1$, $T_1 = 2k + 1$, and the parameters are $P = 3$, $Q = 2$, one obtains the sequence with $T_n = k \times 2^n + 1$. For $T_0' = k - 1$, $T_1' = 2k - 1$, one gets $T_n' = k \times 2^n - 1$.

These sequences were discussed in the preceding section, when it was stated that there are infinitely many odd integers k (the Sierpiński numbers) such that each T_n is composite; similarly, there are infinitely many integers k (the Riesel numbers) such that T_n' is composite.

In 2002, Izotov described infinitely many pairs (P, Q) of parameters, and in each case infinitely many pairs of initial terms, such that the corresponding sequence consists only of composite numbers.

The NSW numbers

NSW does not mean North-South-West, nor New South Wales, but stands for Newman, Shanks & Williams (1980), and I had the privilege of getting acquainted with their paper, while still in preprint form. This was just after Dan Shanks' visit to Queen's University, memorable for more than one reason.

The NSW numbers, as they appear in the paper, are defined for odd indices: $S_1 = 1$, $S_3 = 7$, $S_5 = 41$, $S_7 = 239$, $S_9 = 1393$, ... These numbers arise when addressing the question of the existence of finite simple groups whose order is a square.

The numbers $W_n = S_{2n+1}$, $n \geq 0$, are the terms of the second-order linear recurring sequence with parameters $P = 6$, $Q = 1$ and initial terms $W_0 = 1$, $W_1 = 7$. Therefore, for every $n \geq 2$,

$$W_n = \frac{(1 + \sqrt{2})^{2n+1} + (1 - \sqrt{2})^{2n+1}}{2}.$$

It is not known if there exist infinitely many prime NSW numbers. On the other hand, it was proved by Sellers & Williams (2002) that the sequence $(W_n)_{n \geq 0}$ (and many other similar sequences) contain infinitely many composite numbers.

In the original notation, if S_{2n+1} is a prime, then $2n+1$ must also be a prime. The following values of $p < 2000$ yield prime NSW numbers S_p:

$$p = 3, 5, 7, 19, 29, 47, 59, 163, 257, 421, 937, 947, 1493, 1901.$$

F. Morain has shown in 1989 that the last two values in fact yielded primes. In 1999, H. Dubner determined that in the interval $2000 < p < 80000$, S_p is a probable prime (PRP) for

$$p = 6689, 8087, 9679, 28953, 79043,$$

and for no other prime p in that interval. Also in 1999, true primality of S_{6689} was established by Dubner and Keller.

Since these results remained unpublished, the first three PRPs were rediscovered independently by W. Roonguthai in 2000, and the fourth by A. Walker in 2001. The primality of S_{8087}, S_{9679} (and of S_{6689}, again) was established in July 2001 by D. Broadhurst. In the case of S_{8087}, this was a very difficult proof.

6
Heuristic and Probabilistic Results About Prime Numbers

The word "heuristic" means: based on, or involving, trial and error. Heuristic results are formulated following the observation of numerical data from tables or from extended calculations. Sometimes these results express the conclusions of some statistical analysis.

There are also probabilistic methods. The idea is quite well explained in Cramér's paper (1937) already quoted in Chapter 4:

> In investigations concerning the asymptotic properties of arithmetic functions, it is often possible to make an interesting use of probability arguments. If, e.g., we are interested in the distribution of a given sequence S of integers, we then consider S as a member of an infinite class C of sequences, which may be concretely interpreted as the possible realizations of some game of chance. It is then in many cases possible to prove that, *with a probability equal to* 1, a certain relation R holds in C, i.e., that in a definite mathematical sense, "almost all" sequences of C satisfy R. Of course we cannot, in general, conclude that R holds for the particular sequence S, but results suggested in this way may sometimes afterwards be rigorously proved by other methods.

Heuristic and probabilistic methods, if not properly handled with caution and intelligence, may give rise to "dream mathematics," far removed from the reality. Hasty conjectures, misinterpretation of numerical evidence have to be avoided.

I will be careful and restrict myself to only a few of the reliable contributions. I have in mind Hardy & Littlewood, with their famous conjectures in *Partitio Numerorum*, Dickson, Bouniakowsky, Schinzel, and Sierpiński, with their intriguing hypotheses.

I   Prime Values of Linear Polynomials

Once again, the starting point is Dirichlet's theorem on primes in arithmetic progressions. It states that if $f(X) = bX + a$, with integers a, b such that $a \neq 0$, $b \geq 1$, $\gcd(a, b) = 1$, then there exist infinitely many integers $m \geq 0$ such that $f(m)$ is a prime.

In 1904, Dickson stated the following conjecture, concerning the simultaneous values of several linear polynomials:

(D) *Let $s \geq 1$, $f_i(X) = b_i X + a_i$ with a_i, b_i integers, $b_i \geq 1$ (for $i = 1, \ldots, s$). Assume that the following condition is satisfied:*

(*) *There does not exist any integer $n > 1$ dividing all the products $f_1(k)f_2(k) \cdots f_s(k)$, for every integer k.*

Then there exist infinitely many natural numbers m such that all numbers $f_1(m)$, $f_2(m), \ldots, f_s(m)$ are primes.

The following statement looks weaker than (D):

(D$_0$) *Under the same assumptions for $f_1(X), \ldots, f_s(X)$, there exists a natural number m such that the numbers $f_1(m), \ldots, f_s(m)$ are primes.*

Whereas, at first sight, one might doubt the validity of (D), the statement (D$_0$) is apparently demanding so much less, that it may be more readily accepted. But, the fact is that (D) and (D$_0$) are equivalent.

Indeed, if (D$_0$) is true, there exists $m_1 \geq 0$ such that $f_1(m_1), \ldots$, $f_s(m_1)$ are primes. Let $g_i(X) = f_i(X + 1 + m_1)$ for $i = 1, \ldots, s$. Then (*) is satisfied by $g_1(X), \ldots, g_s(X)$, hence by (D$_0$) there exists $k_1 \geq 0$ such that $g_1(k_1), \ldots, g_s(k_1)$ are primes; let $m_2 = k_1 + 1 + m_1 > m_1$, so $f_1(m_2), \ldots, f_s(m_2)$ are primes. The argument may be repeated, and shows that (D_0) implies (D).

Dickson did not explore the consequences of his conjecture. This was the object of a paper by Schinzel & Sierpiński (1958), which I find so interesting that I will devote a rather large portion of Chapter 6 to it.

As a matter of fact, Schinzel had proposed a more embracing conjecture (the hypothesis (H)) dealing with polynomials not necessarily linear. But, before I discuss the hypothesis (H) and its consequences, I shall indicate the many interesting results which Schinzel & Sierpiński proved, under the assumption that the conjecture (D) is valid.

The impressive consequences of the hypothesis (D), listed below, should convince anyone that the proof of (D) is remote, if ever possible.

(D_1) *Let $s \geq 1$, let $a_1 < a_2 < \cdots < a_s$ be nonzero integers and assume that $f_1(X) = X + a_1, \ldots, f_s(X) = X + a_s$ satisfy condition (*) in (D). Then there exist infinitely many integers $m \geq 1$ such that $m + a_1$, $m + a_2, \ldots, m + a_s$ are consecutive primes.*

The conjecture of Polignac (1849), discussed in Chapter 4, Sections II and III, is a consequence of (D_1):

(D_2) *For every even integer $2k \geq 2$ there exist infinitely many pairs of consecutive primes with difference equal to $2k$. In particular, there exist infinitely many pairs of twin primes.*

Here is an interesting consequence concerning the abundance of twin primes:

(D_3) *For every integer $m \geq 1$, there exist $2m$ consecutive primes which are m couples of twin primes.*

Another quite unexpected consequence concerns primes in arithmetic progressions. In Chapter 4, Section IV, I showed that if a, $a+d, \ldots, a+(n-1)d$ are primes and $1 < n < a$, then d is a multiple of $\prod_{p \leq n} p$.

From (D_1), it follows:

(D$_4$) *Let $1 < n$, let d be a multiple of $\prod_{p \leq n} p$. Then there exist infinitely many arithmetic progressions, with difference d, each consisting of n consecutive primes.*

The reader should compare this very strong statement with what was stated independently of any conjecture in Chapter 4, Section IV. Concerning Sophie Germain primes, it can be deduced from (D):

(D$_5$) *For every $m \geq 1$ there exist infinitely many arithmetic progressions consisting of m Sophie Germain primes.*

In particular, (D) implies the existence of infinitely many Sophie Germain primes, a fact which has never been proved without appealing to a conjecture. I shall return soon to a quantitative statement about the distribution of Sophie Germain primes.

The conjecture (D) is so powerful that it implies also:

(D$_6$) *There exist infinitely many composite Mersenne numbers.*

I recall (see Chapter 2, Section IV) that no one has as yet succeeded in proving (without assuming the conjecture (D)) that there exist infinitely many composite Mersenne numbers. However, it is easy to prove that other sequences, similar to that of Mersenne numbers, contain infinitely many composite numbers. The following result was proposed by Powell as a problem in 1982 (a solution by Israel was published in 1983):

If m, n are integers such that $m > 1$, $mn > 2$ (this excludes $m = 2$, $n = 1$), then there exist infinitely many composite numbers of the form $m^p - n$, where p is a prime number.

Proof. Let q be a prime dividing $mn - 1$, so $q \nmid m$. If p is a prime such that $p \equiv q - 2 \pmod{q - 1}$, then $m(m^p - n) \equiv m(m^{q-2} - n) \equiv 1 - mn \equiv 0 \pmod{q}$, hence q divides $m^p - n$. By Dirichlet's theorem on primes in arithmetic progressions, there exist infinitely many primes p, such that $p \equiv q - 2 \pmod{q - 1}$, hence there exist infinitely many composite integers $m^p - n$, where p is a prime number. $\square$

Another witness of the strength of conjecture (D) is the following:

(D$_7$) *There exist infinitely many Carmichael numbers, each being the product of three distinct primes.*

I recall that Alford, Granville & Pomerance (1994) showed that there exist infinitely many Carmichael numbers, without recurring to conjecture (D). However, it has not been shown that there exist infinitely many Carmichael numbers, each one of which has exactly three prime factors.

Another, even more remarkable consequence of (D) is the famous conjecture of Artin:

(A) *If a is a nonzero integer, which is neither a square nor equal to −1, then there exist infinitely many primes p such that a is a primitive root modulo p.*

Although Artin's conjecture has not yet been proved, substantial progress was achieved in that direction. Firstly, there was the path-breaking paper by Gupta & Ram Murty (1984), and then the crowning result of Heath-Brown (1986), who proved: There are at most two primes and at most three positive square-free integers for which Artin's conjecture fails.

II   Prime Values of Polynomials of Arbitrary Degree

Now I turn to polynomials which may be nonlinear. Historically, the first conjecture was by Bouniakowsky, in 1857, and it concerns one polynomial of degreee at least two:

(B) *Let $f(X)$ be an irreducible polynomial, with integral coefficients, positive leading coefficient and degree at least 2. Assume that the following condition is satisfied:*

(*) *there does not exist any integer $n > 1$ dividing all the values $f(k)$, for every integer k.*

Then there exist infinitely many natural numbers m such that $f(m)$ is a prime.

The reader should know that just as for the conjectures (D) and (D_0), which are equivalent, (B) is equivalent to a conjecture (B_0) that may be stated similarly.

Before I discuss conjecture (B), let it be made clear that there are, in fact, very few results about prime values of polynomials. For instance, it has never been found even one polynomial $f(X)$ of degree greater than 1, such that $|f(n)|$ is a prime for infinitely many natural numbers n.

On the other hand, Sierpiński showed in 1964, that for every $k \geq 1$ there exists an integer b such that $n^2 + b$ is a prime for at least k natural numbers n.

If $f(X)$ has degree $d \geq 2$ and integral coefficients, for every $x \geq 1$ let

$$\pi_{f(X)}(x) = \#\{n \geq 1 \mid |f(n)| \leq x \text{ and } |f(n)| \text{ is a prime}\}.$$

In 1922, Nagell showed that $\lim_{x \to \infty} \pi_{f(X)}(x)/x = 0$, so there are few prime values. In 1931, Heilbronn proved the more precise statement:

There exists a positive constant C (depending on $f(X)$), such that

$$\pi_{f(X)}(x) \leq C \frac{x^{1/d}}{\log x}, \qquad \text{for every } x \geq 1.$$

Not much more seems to be known for abitrary polynomials. However, for special types of polynomials, there are interesting conjectures and extensive computations. I will discuss this in more detail later.

The question of existence of infinitely many primes p such that $f(p)$ is a prime, is even more difficult. In particular, as I have already said, it is not known that there exist infinitely many primes for which the polynomials $f(X) = X + 2$, respectively $f(X) = 2X + 1$, assume prime values (existence of infinitely many twin primes, or infinitely many Sophie Germain primes). However, if happiness comes from almost-primes, then there are good reasons to be contented, once more with sieve methods and the unavoidable book of Halberstam & Richert.

Let me recall the definition of an "almost-prime". Given $k \geq 1$, a natural number $n = p_1 p_2 \cdots p_r$, written as the product of its prime factors (not necessarily distinct) is called a k-almost prime when $r \leq k$. The set of all k-almost-primes is denoted by P_k. Richert (1969) in fact proved:

Let $f(X)$ be a polynomial with integral coefficients, positive leading coefficient, degree $d \geq 1$ (and different from X). Assume that for every prime p, the number $\rho(p)$ of solutions of $f(X) \equiv 0 \pmod{p}$ is less than p; moreover if $p \leq d + 1$ and p does not divide $f(0)$ assume also that $\rho(p) < p - 1$. Then, there exist infinitely many primes p such that $f(p)$ is a $(2d + 1)$-almost-prime.

The following particular case was proved by Rieger (1969): There exist infinitely many primes p such that $p^2 - 2 \in P_5$.

Now, back to Bouniakowsky's hypothesis! Bouniakowsky drew no consequences from his conjecture. Once more, this was the task of Schinzel and Sierpiński, who studied a more general hypothesis, stated one century later, but independently.

The following proposition, which has never been proved, follows easily if the conjecture (B) is assumed to be true:

(B$_1$) *Let a, b, c be relatively prime integers, such that $a \geq 1$ and $a + b$ and c are not simultaneously even. If $b^2 - 4ac$ is not a square, then there exist infinitely many natural numbers m such that $am^2 + bm + c$ is a prime number.*

Proposition (B$_1$) in turn implies:

(B$_2$) *If k is an integer such that $-k$ is not a square, there exist infinitely many natural numbers m such that $m^2 + k$ is a prime number.*

In particular, (B) implies that there exist infinitely many primes of the form $m^2 + 1$.

The "game of primes of the form $m^2 + 1$" is not just as innocent as one would inadvertently think at first. It is deeply related with the class number of real quadratic fields.

The polynomial $X^2 + 1$ is the simplest quadratic polynomial with negative discriminant. A proof that this polynomial assumes infinitely many prime values would represent a gigantic advance. But the problem seems—and definitely is—too difficult. Just compare with the truly remarkable recent theorem of Friedlander & Iwaniec (1998):

There exist infinitely many primes which are the sum of a square and a fourth power.

The proof required deep sieve methods and more. How far is this result from "$m^2 + 1$ infinitely often prime"? The latter would include not only the theorem of Friedlander and Iwaniec, but also that for every $k \geq 1$, there exist infinitely many primes which are the sum of a square and a 2^k-th power.

The following statement was also conjectured by Hardy & Little-wood, in 1923; it may be proved assuming the truth of (B).

(B₃) *Let d be an odd integer, $d > 1$; let k be an integer which is not an e-th power of an integer, for any factor $e > 1$ of d. Then there exist infinitely many natural numbers m such that $m^d + k$ is a prime.*

In his joint paper with Sierpiński, Schinzel proposed the following conjectures:

(H) *Let $s \geq 1$, let $f_1(X), \ldots, f_s(X)$ be irreducible polynomials, with integral coefficients and positive leading coefficient. Assume that the following condition holds:*

   (*) *there does not exist any integer $n > 1$ dividing all the products $f_1(k)f_2(k) \cdots f_s(k)$, for every integer k.*

   Then there exist infinitely many natural numbers m such that all numbers $f_1(m), f_2(m), \ldots, f_s(m)$ are primes.

(H₀) *Under the same assumptions for $f_1(X), \ldots, f_s(X)$, there exists a natural number m such that the numbers $f_1(m), \ldots, f_s(m)$ are primes.*

Once again, (H) and (H₀) are equivalent. If all the polynomials $f_1(X), \ldots, f_s(X)$ have degree 1, these conjectures are Dickson's (D), (D₀). If $s = 1$, they are Bouniakowsky's conjectures (B), (B₀).

I shall not enumerate here all the consequences of this hypothesis, which were proved by the authors, but I like to mention a result by Schinzel in connection with Carmichael's conjecture concerning the valence of Euler's function. Recall the following notation from Chapter 2, Section II, F: for every $m \geq 1$, let

$$V_\varphi(m) = \#\{n \geq 1 \mid \varphi(n) = m\}.$$

Schinzel proved in 1961 that conjecture (H) implies

(H₁) *For every $s > 1$ there exist infinitely many integers $m > 1$ such that $V_\varphi(m) = s$.*

Note that $s \neq 1$, so proposition (H₁) does not contain Carmichael's conjecture.

For my readers who like Pythagorean triangles: You observed that there are many Pythagorean triples (a, b, c) with b even and a, c primes, for example $(3, 4, 5)$, $(5, 12, 13)$, and many more. I am sure that you wondered if there are infinitely many such triples. Maybe you even tried to prove it and quit in frustration. No reason for that. The fact is that no one knows how to prove it—unless you assume that the hypothesis (H) is true.

This is in the paper of Schinzel & Sierpiński, and for your convenience I reproduce the proof. It is first necessary to establish:

(H$_2$) *Let a, b, c, d be integers such that $a > 0$, $d > 0$ and $b^2 - 4ac$ is not a square. Assume also that there exist integers x_0, y_0 such that $\gcd(x_0 y_0, 6ad) = 1$ and $ax_0^2 + bx_0 + c = dy_0$. Then there exist infinitely many primes p, q such that $ap^2 + bp + c = dq$.*

Proof that (H) **implies** (H$_2$). Let $f_1(X) = dX + x_0$, $f_2(X) = adX^2 + (2ax_0 + b)X + y_0$. Since $(2ax_0 + b)^2 - 4ady_0 = (2ax_0 + b)^2 - 4a(ax_0^2 + bx_0 + c) = b^2 - 4ac$ is not a square, the polynomial $f_2(X)$ is irreducible, just like $f_1(X)$.

I verify condition (*). Let $g(X) = f_1(X)f_2(X)$; this polynomial is of degree 3, with leading coefficient ad^2. If there exists a prime p dividing $g(m)$ for every integer m, then p divides $g(m) - g(m-1) = \Delta g(m)$, $g(m-1) - g(m-2) = \Delta g(m-1)$, $g(m-2) - g(m-3) = \Delta g(m-2)$, which are the first differences of $g(X)$, evaluated at m, $m - 1$, $m - 2$. By the same token, p divides $\Delta^2 g(m) = \Delta g(m) - \Delta g(m-1)$, $\Delta^2 g(m-1) = \Delta g(m-1) - \Delta g(m-2)$, and again p divides $\Delta^3 g(m) = \Delta^2 g(m) - \Delta^2 g(m-1)$. But $\Delta^3 g(X) = 6ad^2$. If p also divides $g(0) = x_0 y_0$, then p divides $\gcd(6ad, x_0 y_0) = 1$, which is an absurdity. This shows that condition (*) is satisfied. By (H) there exist infinitely many natural numbers m such that $f_1(X) = p$, $f_2(X) = q$ are primes. Since $af_1(X)^2 + bf_1(X) + c = df_2(X)$, then $ap^2 + bp + c = dq$. □

The above statement (H$_2$) implies:

(H$_3$) *Every rational number $r > 1$ may be written in infinitely many ways in the form $r = (p^2 - 1)/(q - 1)$, where p, q are prime numbers.*

Proof that (H$_2$) **implies** (H$_3$). Let $r = d/a$ with $d > a > 0$. Take $b = 0$, $c = d - a$ in the statement of (H$_2$). Since $b^2 - 4ac =$

$-4a(d - a) < 0$, it is not a square. Let $x_0 = y_0 = 1$, then the hypotheses of (H$_2$) are satified; hence there exist infinitely many primes p, q such that $ap^2 + d - a = dq$, therefore

$$\frac{d}{a} = \frac{d}{a}q - p^2 + 1, \quad \text{hence} \quad r = \frac{p^2 - 1}{q - 1}. \qquad \square$$

And now comes the announced statement

(H$_4$) *There exist infinitely many triples (a, b, c) of positive integers, such that $a^2 + b^2 = c^2$ and a and c are prime numbers.*

Proof that (H$_3$) **implies** (H$_4$). Let $r = 2$, so there exist infinitely many primes p, q such that $2 = (p^2 - 1)/(q - 1)$, hence $p^2 = 2q - 1$ and therefore $p^2 + (q - 1)^2 = q^2$. $\qquad \square$

The triangles $(3, 4, 5)$ and $(5, 12, 13)$ are 2-prime Pythagorean triangles, in the above sense, who are linked by the prime 5. The following question was studied by Dubner, who communicated his results to me in 1999.

I introduce the concept of a *Dubner chain* of Pythagorean triangles. It is a finite (or infinite) sequence $T_1, T_2, T_3, \ldots$ of 2-prime Pythagorean triangles which are linked in this way: the hypotenuse of each triangle is a leg of the next one. As far as I know, it has not been proved that the hypothesis (H) implies the existence of arbitrarily large Dubner chains of triangles.

If $a^2 + b^2 = c^2$, with a prime, then the triple (a, b, c) must be primitive, so that $a = u^2 - v^2$, $b = 2uv$, $c = u^2 + v^2$. For a to be prime it is necessary that $u - v = 1$. Hence $b = 2v^2 + 2v$, $c = 2v^2 + 2v + 1$, thus $c = b + 1$, as in the proof leading to (H$_4$). Therefore the triangles in a Dubner chain become thinner and thinner as a increases. Also, $a^2 = (u + v)^2 = c + b = 2c - 1$, so it is necessary to find primes a such that $c = (a^2 + 1)/2$ is also a prime.

For $k = 2, 3, 4, 5$ and 6, Dubner determined the smallest prime a giving a chain with k triangles, as follows.

k	Smallest a for k triangles
2	3
3	271
4	169219
5	356498179
6	2500282512131

It is far from easy to create long chains of triangles for which it is possible to assert that two sides, measured by large integers, are actually primes. Dubner & Forbes (2001) constructed a Dubner chain with 7 triangles, where $a = 2185103796349763249$ and the last hypotenuse is measured by a prime c with 2310 digits.

The following consequence of (H$_3$) was communicated to me by P.T. Mielke. It concerns triangles with sides measured in integers, but not necessarily with a right-angle.

(H$_5$) *There exist infinitely many triangles with integer sides a, p, q, where p, q are primes and the angle between the sides with measures a, p is $\pi/3$ (respectively $2\pi/3$) radians.*

Proof that (H$_3$) **implies** (H$_5$). By (H$_3$) there exist infinitely many pairs of primes (p, q) such that $4 = (p^2 - 1)/(q - 1)$. For each such pair, note that $p > 2$ and $q = (p^2 + 3)/4$. Let $a = ((p-1)(p+3))/4$, so a is an integer. Then $a - p = ((p+1)(p-3))/4$, and and $p^2 + a^2 - ap = p^2 + a(a - p)$

$$= p^2 + \frac{(p^2 - 1)(p^2 - q)}{16} = \left(\frac{p^2 + 3}{4}\right)^2 = q^2.$$

By the cosine law, this means that the angle between the sides with measures a and p is equal $\pi/3$. The proof is similar when the angle is $2\pi/3$, taking $a = ((p+1)(p-3))/4$.   □

Still in the same paper of 1958, there is the following conjecture by Sierpiński:

(S) *For every integer $n > 1$, let the n^2 integers $1, 2, \ldots, n^2$ be written in an array with n rows, each with n integers, like an $n \times n$ matrix:*

$$
\begin{array}{cccc}
1 & 2 & \cdots & n \\
n+1 & n+2 & \cdots & 2n \\
2n+1 & 2n+2 & \cdots & 3n \\
\vdots & \vdots & \vdots & \vdots \\
(n-1)n+1 & (n-1)n+2 & \cdots & n^2.
\end{array}
$$

Then, there exists a prime number in each row.

Of course, 2 is in the first row. By the theorem of Bertrand and Tschebycheff, there is a prime number in the second row.

More can be said about the first few rows using a strengthening of Bertrand and Tschebycheff's theorem. For example, in 1932, Breusch showed that if $n \geq 48$, then there exists a prime between n and $(9/8)n$. Thus, if $0 < k \leq 7$ and $n \geq 9$, then there exists a prime p such that $kn+1 \leq p \leq (9/8)(kn+1) \leq (k+1)n$. So, there is a prime in each of the first 8 rows.

By the prime number theorem, for every $h \geq 1$ there exists $n_0 = n_0(h) > h$ such that, if $n \geq n_0$, then there exists a prime p such that $n < p < (1 + 1/h)n$. And from this, it follows that, if $n \geq n_0$, then each of the first h rows of the array contains a prime.

Just like the preceding conjectures, (S) has also several interesting consequences.

(S_1) *For every $n \geq 1$, there exist at least two primes p, p' such that $n^2 < p < p' < (n+1)^2$.*

(S_2) *For every $n \geq 1$, there exist at least four primes p, p', p'', p''' such that $n^3 < p < p' < p'' < p''' < (n+1)^3$.*

It should be noted that both statements (S_1), (S_2) have not yet been proved without appealing to the conjecture (S). However, it is easy to show that for every sufficiently large n, there exists a prime p between n^3 and $(n+1)^3$; this is done using Ingham's result that $d_n = p_{n+1} - p_n = O\big(p_n^{(5/8)+\varepsilon}\big)$, for every $\varepsilon > 0$.

Schinzel stated the following "transposed" form of Sierpiński's conjecture:

(S′) *For every integer $n > 1$, let the n^2 integers $1, 2, \dots, n^2$ be written in an array with n rows, each with n integers (just like in (S)). If $1 \leq k \leq n$ and $\gcd(k, n) = 1$, then the kth column contains at least one prime number.*

This time, Schinzel & Sierpiński drew no consequence from the conjecture (S′). I suppose it was Sunday evening and they were tired. However, in 1963, this conjecture was spelled out once more by Kanold.

I conclude with the translation of the following comment in Schinzel & Sierpiński's paper:

We do not know what will be the fate of our hypotheses, however we think that, even if they are refuted, this will not be without profit for number theory.

III Polynomials with Many Successive Composite Values

I now want to report on some results of McCurley, which I find very interesting. According to Bouniakowsky's conjecture, if $f(X)$ is an irreducible polynomial with integral coefficients and satisfying condition (*), then there exists a smallest integer $m \geq 1$ such that $f(m)$ is a prime. Denote it by $p(f)$.

If $f(X) = dX + a$, with $d \geq 2$, $1 \leq a \leq d - 1$, $\gcd(a, d) = 1$, then of course, $p(f)$ exists. With the notation of Chapter 4, Section IV,

$$p(dX + a) = \frac{p(d, a) - a}{d}.$$

Recall that Prachar and Schinzel gave lower bounds for $p(d, a)$.

The work of McCurley provides an extension of the above results for polynomials $f(X)$ of arbitrary degree. An essential tool for McCurley was the following result of Odlyzko (included in the paper by Adleman, Pomerance & Rumely (1983), which was quoted in Chapter 2):

There exists an absolute constant $C > 0$ and infinitely many integers $d > 1$ having at least $e^{C \log d / (\log \log d)}$ factors of the form $p - 1$, where p is an odd prime.

For each d as above, let $p_1, p_2, \ldots, p_r$ be the odd primes such that $p_i - 1$ divides d. Let $k = p_1 p_2 \cdots p_r - 1$ or $k = 3 p_1 p_2 \cdots p_r - 1$, so that $k \equiv 1 \pmod 4$. Let $f(X) = X^d + k$, then $f(X)$ is irreducible and satisfies condition (*). Then, McCurley proved in 1984:

There exists a constant $C' > 0$ such that, if m is an integer satisfying $|m| < e^{C' \log d / (\log \log d)}$, then $f(m)$ is composite.

It should be noted that the proof gives no explicit polynomials. However, in a computer search, the following polynomials were discovered (the first example is due to Shanks, 1971):

Table 27. Polynomials with many initial composite values

$f(X)$	$f(m)$ is composite for all m up to
$X^6 + 1091$	3905
$X^6 + 82991$	7979
$X^{12} + 4094$	170624
$X^{12} + 488669$	616979

The smallest prime value of the last polynomial has no less than 70 digits.

With another method, McCurley showed in 1986:

For every $d \geq 1$ there exists an irreducible polynomial $f(X)$, of degree d, satisfying the condition (*) *and such that if*

$$|m| < e^{C\sqrt{L(f)/(\log L(f))}},$$

then $|f(m)|$ *is composite.*

In the above statement, $L(f)$ is the length of $f(X) = \sum_{k=0}^{d} a_k X^k$, which is defined by $L(f) = \sum_{k=0}^{d} \|a_k\|$ and $\|a_k\|$ is the number of digits in the binary expansion of $|a_k|$, with $\|0\| = 1$. Note that this last result is applicable to polynomials of any degree, and also the proof yields explicit polynomials with the desired property.

McCurley determined $p(X^d + k)$ for several polynomials. From his tables, I note:

Table 28.
Polynomials $X^d + k$ with many initial composite values

d	m	$\max\{p(X^d + k) \mid k \leq m\}$
2	10^6	$p(X^2 + 576239) = 402$
3	10^6	$p(X^3 + 382108) = 297$
4	150000	$p(X^4 + 72254) = 2505$
5	10^5	$p(X^5 + 89750) = 339$

S.M. Williams considered quadratic polynomials with leading coefficient different from 1, and communicated to me the results of his calculations in 1992:

$$p(8X^2 + X + 564135) = 482$$
$$p(4X^2 + X + 530985) = 472$$
$$p(2X^2 + X + 931650) = 443$$
$$p(73X^2 + 7613) = 420.$$

IV   Partitio Numerorum

It is instructive to browse through the paper *Partitio Numerorum, III: On the expression of a number as a sum of primes*, by Hardy & Littlewood (1923). There one finds a conscious systematic attempt—for the first time on such a scale—to derive heuristic formulas for the distribution of primes satisfying various additional conditions.

I shall present here a selection of the probabilistic conjectures, extracted from Hardy & Littlewood's paper (I will keep their labeling, since it is classical). The first conjecture concerns Goldbach's problem:

Conjecture A. *Every sufficiently large even number $2n$ is the sum of two primes. The asymptotic formula for the number of representations is*

$$r_2(2n) \sim C_2 \frac{2n}{(\log 2n)^2} \prod_{\substack{p>2 \\ p|n}} \frac{p-1}{p-2},$$

with

$$C_2 = \prod_{p>2} \left(1 - \frac{1}{(p-1)^2}\right) = 0.66016\ldots.$$

Note that C_2 is the same as the twin primes constant (see Chapter 4, Section III).

The following conjecture deals with (not necessarily consecutive) primes with given difference $2k$, in particular, with twin primes:

Conjecture B. *For every even integer $2k \geq 2$ there exist infinitely many primes p such that $p + 2k$ is also a prime. For $x \geq 1$, let*

$$\pi_{X, X+2k}(x) = \#\{p \text{ prime} \mid p + 2k \text{ is prime and } p + 2k \leq x\}.$$

Then

$$\pi_{X, X+2k}(x) \sim 2C_2 \frac{x}{(\log x)^2} \prod_{\substack{p>2 \\ p|k}} \frac{p-1}{p-2},$$

where C_2 is the twin primes constant.

In particular, if $2k = 2$ this gives the asymptotic estimate for the twin primes counting function, already indicated in Chapter 4, Section III.

The conjecture E is about primes of the form $m^2 + 1$:

Conjecture E. *There exist infinitely many primes of the form $m^2 + 1$. For $x > 1$ let*

$$\pi_{X^2+1}(x) = \#\{p \text{ prime} \mid p \leq x \text{ and } p \text{ is of the form } p = m^2 + 1\}.$$

Then

$$\pi_{X^2+1}(x) \sim C \frac{\sqrt{x}}{\log x}$$

where

$$C = \prod_{p \geq 3} \left(1 - \frac{(-1 \mid p)}{p-1}\right) = 1.37281346\ldots$$

and $(-1 \mid p) = (-1)^{(p-1)/2}$ is the Legendre symbol.

The values predicted by the conjecture are in significant agreement with the actual values of $\pi_{X^2+1}(x)$, as indicated in the table below (the last two were computed by Wunderlich in 1973):

Table 29. Prime numbers of the form $m^2 + 1$

x	$\pi_{X^2+1}(x)$	$C\sqrt{x}/\log x$	Ratio
10^6	112	99	0.8839
10^8	841	745	0.8858
10^{10}	6656	5962	0.8957
10^{12}	54110	49684	0.9182
10^{14}	456362	425861	0.9332

Here is the place to mention the result of Iwaniec (1978): There exist infinitely many integers $m^2 + 1$ which are 2-almost-primes.

And still another conjecture:

Conjecture F. *Let $a > 0$, b, c be integers such that $\gcd(a, b, c) = 1$, $b^2 - 4ac$ is not a square and $a + b$, c are not both even. Then there are infinitely many primes of the form $am^2 + bm + c$ (this was statement* $(\mathrm{B_1})$) *and the number $\pi_{aX^2+bX+c}(x)$ of primes $am^2 + bm + c$ which are less than x is given asymptotically by*

$$\pi_{aX^2+bX+c}(x) \sim \frac{\varepsilon C}{\sqrt{a}} \frac{\sqrt{x}}{\log x} \prod_{\substack{p>2 \\ p\mid\gcd(a,b)}} \frac{p}{p-1}$$

where

$$\varepsilon = \begin{cases} 1 & \text{if } a + b \text{ is odd,} \\ 2 & \text{if } a + b \text{ is even,} \end{cases}$$

$$C = \prod_{\substack{p>2 \\ p\nmid a}} \left(1 - \frac{\left(\dfrac{b^2 - 4ac}{p}\right)}{p-1}\right)$$

and $(b^2 - 4ac \mid p)$ *denotes the Legendre symbol.*

In particular, the conjecture is applicable to primes of the form $m^2 + k$, where $-k$ is not a square.

For the special case of polynomials $f_A(X) = X^2 + X + A$ (with $A \geq 1$ an integer), Conjecture F states that

$$\pi_{X^2+X+A}(x) \sim C(A) \frac{2\sqrt{x}}{\log x},$$

where

$$C(A) = \prod_{p>2} \left(1 - \frac{\left(\dfrac{1 - 4A}{p}\right)}{p-1}\right).$$

Note that for a polynomial $f_A(X)$, the number $\pi^*_{f_A(X)}(N) = \pi^*_{X^2+X+A}(N)$ (as defined in Chapter 3) of values $n \leq N$ such that $f_A(n)$ is prime, is closely related to the value of $\pi_{f_A(X)}(x)$ for $x = N^2$. In this case, asymptotically we have

$$\pi^*_{X^2+X+A}(N) \sim \pi_{X^2+X+A}(N^2) \sim C(A) \frac{N}{\log N}.$$

Therefore a higher value of $C(A)$ suggests that a higher number $\pi^*_{X^2+X+A}(N)$ of primes might be likely to occur.

Shanks (1975) had calculated that if $A = 41$ (which gives Euler's famous polynomial), then $C(41) = 3.3197732$. In 1990, Fung & Williams calculated $C(A)$ and $\pi^*_{X^2+X+A}(10^6)$ for many values of A. Jacobson (1995) extended these computations.

RECORD

For all calculated values up to now, the maximum $C(A) = 5.5338891$ is observed for a certain integer A with 70 digits and was published by Jacobson & Williams in 2003. In his thesis of 1995, Jacobson obtained

$$C(517165153168577) = 5.0976398,$$

which gives

$$\pi^*_{X^2+X+517165153168577}(10^6) = 300923.$$

The previous record of Fung & Williams was

$$C(132874279528931) = 5.0870883,$$

with

$$\pi^*_{X^2+X+132874279528931}(10^6) = 312975.$$

The maximum $\pi^*_{X^2+X+A}(10^6)$ observed so far is

$$\pi^*_{X^2+X+21425625701}(10^6) = 361841$$

with

$$C(21425625701) = 4.7073044.$$

In comparison, we have

$$\pi^*_{X^2+X+41}(10^6) = 261081 \quad \text{with} \quad C(41) = 3.3197732,$$

and for a polynomial considered by N.G.W.H. Beeger in 1939,

$$\pi^*_{X^2+X+27941}(10^6) = 286129 \quad \text{with} \quad C(27941) = 3.6319998.$$

Partitio Numerorum contains many other conjectures, but I want only to mention:

Conjecture N. *There exist infinitely many primes p of the form $p = k^3 + l^3 + m^3$, where k, l, m are positive integers.*

The conjecture includes also an asymptotic statement for the number of triples (k, l, m) such that $p \leq x$.

In a recent paper (2001), Heath-Brown proved the theorem:

There exist infinitely many primes p of the form $p = k^3 + 2l^3$, where k, l are positive integers.

In particular, the existential assertion in Conjecture N is true; however, the method in the proof did not lead to the asymptotic estimate expressed in the conjecture. Dealing with cubes is incomparably harder than with squares, so the theorem of Heath-Brown should be hailed as a susbtantial advance in the problem of representing primes by binary forms.

The conjectures of Hardy and Littlewood gave rise to a considerable amount of computation, intended to determine accurately the constants involved in the formulas, and to verify that the predictions fitted well with the observations. As the constants were often given by slowly convergent infinite products, it was imperative to modify these expressions to make them more easily computable.

Appendix 1

Prime Numbers and Marriage

By M. Ram Murty

There is a famous theorem in combinatorics that is called the *marriage theorem*, due to Philip Hall and proved in 1935. It says that if X is a bipartite graph with partite sets U an V, then we can find a pairing, or matching, of the elements of U with the elements of V if and only if for any subset S of U the size of the "neighbours" of S (that is, those elements of V adjacent to some element of S) has at least the size of S. In other words, we can find a collection of disjoint edges so that every vertex belongs to some edge of the collection.

At first glance, this theorem does not seem all that profound. It is often stated and even proved in matrimonial terms (hence its name) where the partite sets U and V represent men and women respectively and the problem is to pair them up in "friendly" pairs. Despite its simplicity and informal appearance, the theorem has deep applications in combinatorial theory. Our purpose here is to link it to a famous conjecture in number theory.

In 1969, C.A. Grimm formulated a conjecture about consecutive composite numbers. He predicts that if $n+1, \ldots, n+k$ are consecutive composite numbers, then we can find distinct prime numbers p_i so that $p_i \mid n+i$. This can be formulated in terms of matching. Consider the bipartite graph X with partite sets U consisting of the composite numbers $n + 1, \ldots, n + k$, and V consisting of prime divisors of the product $(n + 1) \cdots (n + k)$. We join $n + i$ to a prime p in V if and only if $p \mid n + i$. Then, the question is asking for a matching of U. This can happen if and only if Hall's condition is satisfied.

Erdös & Selfridge (1971) noticed that this conjecture is quite difficult to prove since it has surprising consequences. One of them is that there is always a prime number between two consecutive squares, a conjecture which is out of bounds for even the Riemann hypothesis. The proof of this is not difficult and we give it below.

One consequence of Grimm's conjecture is that if there is no prime in the interval $[n+1, n+k]$, then the number of prime factors dividing the product

$$P(n, k) = (n + 1) \cdots (n + k)$$

is at least k. Even this weaker assertion is unknown and implies that between any two consecutive squares, there is a prime number.

Theorem. *There is a constant $c > 0$ such that if $k \geq c\sqrt{n/\log n}$, then $\omega(P(n, k)) < k$, where $\omega(m)$ denotes the number of distinct prime divisors of m.*

Proof. Let r be the number of prime factors of $P(n, k)$ and denote by p_r the r-th prime. Since $P(n, k) \equiv 0 \pmod{k!}$, we have the obvious inequalities:

$$(n + k)^k \geq P(n, k) \geq k! \prod_{\substack{p > k \\ p \mid P(n,k)}} p \geq k! \prod_{k < p \leq p_r} p.$$

Now, for some constant $c_1 > 0$,

$$\sum_{k < p \leq p_r} \log p = p_r - k + O\left(p_r e^{-c_1 \sqrt{\log p_r}}\right)$$

by the prime number theorem. If $r \geq k$, then this quantity is at least $k \log k$ (again using the prime number theorem), and that says $p_r \sim r \log r + r \log \log r$ as r goes to infinity. Thus, we get

$$(n + k)^k \geq k! k^k (\log k)^k.$$

Using the elementary inequality

$$e^k > k^k / k!$$

we find
$$n + k > (k^2/e) \log k,$$

from which the theorem immediately follows. □

Corollary. *Grimm's conjecture implies that there is a prime between two consecutive square numbers.*

Proof. If Grimm's conjecture is true and say there is no prime between $n = m^2$ and $n+k = (m+1)^2$, then with $k = 2m+1$ we would have $\omega(P(n,k)) \geq k$. By the theorem, this would mean that

$$k = O(\sqrt{n/\log n}),$$

which is not the case. □

It is clear that in the above statements, one can make everything explicit in the sense that all the constants are effective.

This result motivates the study of the function $g(n)$ which is the the largest integer k such that Grimm's conjecture holds for the interval $[n+1, n+k]$. Then clearly $g(n) < 2n$ since the interval will contain two powers of 2. The above theorem says that

$$g(n) = O(\sqrt{n/\log n}).$$

Thus, an interesting problem of research has been to obtain upper and lower bounds for $g(n)$. Using the marriage theorem, Erdős & Selfridge (1971) proved that

$$g(n) \geq (1 + o(1)) \log n.$$

Later, using Baker's method, Ramachandra, Shorey & Tijdeman (1975) proved that

$$\frac{(\log n)^2}{(\log_2 n)^5 (\log_3 n)^2} = O(g(n)).$$

We will now indicate the argument used by Erdős and Selfridge. But first, we show how the marriage theorem implies the following lemma.

Lemma. *For any $n \geq 1$, there are $s + 1$ distinct numbers in the interval $[n+1, n+g(n)+1]$ so that their product T (say) has s distinct prime factors and the largest prime divisor of T is less than $g(n)$.*

Proof. We create a bipartite graph with partite sets given by the numbers in the interval and the primes dividing the product of them. We join an element to a prime if p divides it. By the definition of $g(n)$, Hall's condition must fail for this graph. Thus, there must be $s+1$ distinct numbers in the interval which are adjacent to at most s prime numbers. If we choose s minimal with respect to this property, then any product of s numbers has at least s prime divisors. Thus, T has exactly s prime divisors. Suppose now that T has a prime divisor $p > g(n)$. Then $p \mid n_j$ for some $n_j \in [n+1, n+g(n)+1]$. Since the interval has length $g(n)$, no other integer can be divisible by p. Now, T/n_j is a product of s integers and has at most $s-1$ prime factors. By the minimality of s, T/n_j has at least s prime factors. This is a contradiction. $\qquad\square$

Theorem (Erdös and Selfridge). $g(n) \geq (1 + o(1)) \log n$.

Proof. Now choose the integers $n_1, \ldots, n_{s+1}$, whose product T has exactly s prime divisors as in the previous lemma. For each $p \mid T$, let a_p be the largest power of p dividing some element in the interval and choose m_p from the n_i's such that $p^{a_p} \parallel m_p$. As T has s prime divisors, there is (by the pigeonhole principle) some n_k which is not one of the m_p's. Let

$$n_k = \prod_{p \mid T} p^{e_p}$$

be its unique factorization. Then, by our choice, $e_p \leq a_p$. Thus, p^{e_p} divides both n_k and m_p. Hence, it divides their difference, which is at most $g(n)$. Thus,

$$p^{e_p} \leq g(n),$$

so that $n \leq n_k \leq g(n)^s$ which implies $s \geq (\log n)/(\log g(n))$. But all the prime divisors of T are $\leq g(n)$. Hence, $s \leq \pi(g(n))$ so that by the prime number theorem,

$$s \leq (1 + o(1)) \frac{g(n)}{\log g(n)}.$$

Putting everything together, we get

$$g(n) \geq (1 + o(1)) \log n.$$

This completes the proof. $\qquad\square$

Ramachandra, Shorey & Tijdeman have proved that $(\log n)^{3-\varepsilon} = O(g(n))$ for any $\varepsilon > 0$. A celebrated conjecture of Cramér states that the gap $d_n = p_{n+1} - p_n$ between consecutive prime numbers is $O((\log p_n)^2)$. Thus, Grimm's conjecture would follow from Cramér's conjecture.

Appendix 2

**Citations for Some Possible Prizes for Work
on the Prime Number Theorem**

Suggested by Paul T. Bateman

1. To Pavnuty L. Tschebycheff for his two papers "Sur la fonction qui détermine la totalité des nombres premiers inférieurs à une limite donnée", *Journal de Math.* 17 (1852), 341–365, and "Mémoire sur les nombres premiers", *Journal de Math.* 17 (1852), 366–390.

2. To Bernhard Riemann for his paper "Über die Anzahl der Primzahlen unter einer gegebenen Grösse", *Monatsberichte der Königlichen Preussischen Akademie der Wissenschaften zu Berlin aus dem Jahre 1859* (1860), 671–680.

3. To Jacques Hadamard for his two papers "Étude sur les propriétés des fonctions entières et en particulier d'une fonction considérée par Riemann", *Journal de Math.* 9 (1893), 171–215, and "Sur la distribution des zéros de la fonction $\zeta(s)$ et ses conséquences arithmétiques", *Bulletin de la Soc. Math. de France* 24 (1896), 199–220.

4. To Charles-Jean de la Vallée Poussin for his two papers "Recherches analytiques sur la théorie des nombres premiers; Première partie: La fonction $\zeta(s)$ de Riemann et les nombres premiers en général", *Annales de la Soc. Scientifique de Bruxelles* 20 (1896), 183–256, and "Sur la fonction $\zeta(s)$ de Riemann et le nombre des nombres premiers inférieurs à une limite donnée", *Mémoires couronnés et autres mémoires publiés par l'Acad. Royale des Sciences, des Lettres et des Beaux-Arts de Belgique* 59, No. 1, 1899–1900, 74 pages.

5. To Edmund Landau for his paper "Neuer Beweis des Primzahlsatzes und Beweis des Primidealsatzes", *Math. Annalen* 56 (1903), 645–670, his book *Handbuch der Lehre von der Verteilung der Primzahlen*, Teubner, Leipzig, 1909, and his paper "Über die Wurzeln der Zetafunktion", *Math. Zeitschrift* 20 (1924), 98–104.

6. To John E. Littlewood for his many contributions to analysis and the theory of the distribution of primes, in particular for his two papers "Quelques conséquences de l'hypothèse que la fonction $\zeta(s)$ de Riemann n'a pas de zéros dans le demi-plan $\operatorname{Re}(s) > \frac{1}{2}$", *Comptes Rendus de l'Acad. des Sciences*, Paris, 154 (1912), 263–266, and "Sur la distribution des nombres premiers", *Comptes Rendus de l'Acad. des Sciences*, Paris, 158 (1914), 1869–1872, for his paper with G.H. Hardy, "Contributions to the theory of the Riemann zeta function and the theory of the distribuiton of primes", *Acta Math.* 41 (1917), 119–196, and for his work on the subject treated in E. Landau's paper ""Uber die ζ-Funktion und die L-Funktionen", *Math. Zeitschrift* 20 (1924), 105–125.

7. (1933 Bôcher Prize). To Norbert Wiener for his memoir "Tauberian theorems", *Annals of Math.* (2) 33 (1932), 1–100.

8. To Ivan M. Vinogradov for his papers on exponential sums, particularly the three papers: "On Weyl's sums", *Mat. Sbornik* 42 (1935), 521–529; "A new method of resolving certain general questions of the theory of numbers", *Mat. Sbornik* 43 (1936), 9–19; "A new method of estimation of trigonometrical sums", *Mat. Sbornik* 43 (1936), 175–188.

9. To Arne Beurling for his paper "Analyse de la loi asymptotique de la distribution des nombres premiers généralisés", *Acta Math.* 68 (1937), 255–291.

10. To Atle Selberg for his many contributions to the theory of the zeta function and the distribution of primes, and in particular for his paper "An elementary proof of the prime number theorem", *Annals of Math.* (2) 50 (1949), 305–313.

11. (1951 Cole Prize). To Paul Erdös for his many papers in the theory of numbers, and in particular for his paper "On a new method in elementary number theory which leads to an elementary proof of the prime number theorem", *Proceedings of the National Acad. of Sciences of the U.S.A.* 35 (1949), 374–385.

12. To Donald J. Newman for his contributions to analysis and number theory, in particular his paper "Simple analytic proof of the prime number theorem", *The American Math. Monthly* 87 (1980), 603–696.

Conclusion

Dear Reader, or better, Dear Friend of Numbers:

You, who have come faithfully to this point, may now be thinking about what you read and learned, perhaps already trying some of the numerous problems and putting to work your personal computer, as well as your personal brain.

I hope that this presentation of various topics and facets of the theory of prime numbers has been pleasant and conveyed an approximate picture of the problems being investigated. I hope you perceived that records often express efforts to solve open problems and the calculations involved bring a new light to the questions. But, I hope mostly that you became convinced of the value of the fine theorems produced by so many brilliant minds. Personally, I look with awe to the monumental achievements and with grand bewilderment to all the deep questions which remain unsolved—and for how long?

In writing this book I wanted to produce a work of synthesis, to develop the theory of prime numbers as a discipline where the natural questions are systematically studied.

In the Introduction, I believe, I made clear the reasons for dividing the book into its various parts, which are devoted to the answer of what I consider the imperative questions. This organization is intended to prevent any young students from having the same impression as I had in my early days (it was, alas, long ago ...) that number theory dealt with multiple unrelated problems.

This justifies the general plan but not the choice of details, especially the ones which are missing (we always cry for the dear absent ones ...). It is evident that a choice had to be made. I wanted this book to be small and lightweight, so the hand could hold it—not a bulky brick, which cannot be carried in a pocket, nor read while waiting for or riding in a train (in Canada, we both wait for and ride in trains).

I admit readily that the proofs of the most important theorems are absent. They are usually technically involved and long. Their inclusion would shift the attention from the general structure of the theory towards the particular details. The unhappy reader may find satisfaction by consulting the excellent papers and books noted in the ample bibliography.

Bibliography

General References

The books listed below are highly recommended for the quality of contents and presentation. This list is, of course, not exhaustive.

1909 Landau, E. *Handbuch der Lehre von der Verteilung der Primzahlen.* Teubner, Leipzig, 1909. Reprinted by Chelsea, Bronx, NY, 1974.

1927 Landau, E. *Vorlesungen über Zahlentheorie* (in 3 volumes). S. Hirzel, Leipzig, 1927. Reprinted by Chelsea, Bronx, NY, 1969.

1938 Hardy, G.H. & Wright, E.M. *An Introduction to the Theory of Numbers.* Clarendon Press, Oxford, 1938 (5th edition, 1979).

1952 Davenport, H. *The Higher Arithmetic.* Hutchinson, London, 1952 (7th edition, Cambridge Univ. Press, Cambridge, 1999).

1953 Trost, E. *Primzahlen.* Birkhäuser, Basel, 1953 (2nd edition, 1968).

1957 Prachar, K. *Primzahlverteilung.* Springer-Verlag, Berlin, 1957 (2nd edition, 1978).

1962 Shanks, D. *Solved and Unsolved Problems in Number Theory.* Spartan, Washington, 1962 (3rd edition by Chelsea, Bronx, NY, 1985).

1963 Ayoub, R.G. *An Introduction to the Analytic Theory of Numbers.* Amer. Math. Soc., Providence, RI, 1963.

1964 Sierpiński, W. *Elementary Theory of Numbers.* Hafner, New York, 1964 (2nd edition, North-Holland, Amsterdam, 1988).

1974 Halberstam, H. & Richert, H.E. *Sieve Methods.* Academic Press, New York, 1974.

1975 Ellison, W.J. & Mendès-France, M. *Les Nombres Premiers.* Hermann, Paris, 1975.

1976 Adams, W.W. & Goldstein, L.J. *Introduction to Number Theory.* Prentice-Hall, Englewood Cliffs, NJ, 1976.

1981 Guy, R.K. *Unsolved Problems in Number Theory.* Springer-Verlag, New York, 1981 (2nd edition, 1994).

1982 Hua, L.K. *Introduction to Number Theory.* Springer-Verlag, New York, 1982.

Besides the above references, I like expressly to mention the following books for their fresh and distinct "regard" to the theory of numbers, in particular prime numbers.

1984 Schroeder, M.R. *Number Theory in Science and Communication.* Springer-Verlag, Berlin, 1984 (3rd edition, 1997).

1994 Crandall, R.E. *Projects in Scientific Computation.* Springer-Verlag, New York, 1994.

1996 Bach, E. & Shallit, J. *Algorithmic Number Theory*, Vol. 1: *Efficient Algorithms.* MIT Press, Cambridge, MA, 1996.

2000 Narkiewicz, W. *The Development of Prime Number Theory.* Springer-Verlag, Berlin, 2000.

2001 Crandall, R. & Pomerance, C. *Prime Numbers. A Computational Perspective.* Springer-Verlag, New York, 2001.

Chapter 1

1878 Kummer, E.E. Neuer elementarer Beweis des Satzes, dass die Anzahl aller Primzahlen eine unendliche ist. *Monatsber. Akad. d. Wiss.*, Berlin, 1878/9, 777–778.

1890 Stieltjes, T.J. Sur la théorie des nombres. Étude bibliographique. *Ann. Fac. Sci. Toulouse* 4 (1890), 1–103.

1891 Hurwitz, A. *Übungen zur Zahlentheorie*, 1891–1918 (edited by H. Funk and B. Glaus). E.T.H., Zürich, 1993.

1897 Thue, A. Mindre meddelelser II. Et bevis for at primtalleness antal er unendeligt. *Arch. f. Math. og Naturv.*, Kristiania, 19, No. 4, 1897, 3–5. Reprinted in *Selected Mathematical Papers* (edited by T. Nagell, A. Selberg and S. Selberg), 31–32. Universitetsforlaget, Oslo, 1977.

1924 Pólya, G. & Szegö, G. *Aufgaben und Lehrsätze aus der Analysis*, 2 vols. Springer-Verlag, Berlin, 1924 (4th edition, 1970).

1947 Bellman, R. A note on relatively prime sequences. *Bull. Amer. Math. Soc.* 53 (1947), 778–779.

1955 Furstenberg, H. On the infinitude of primes. *Amer. Math. Monthly* 62 (1955), p. 353.

1959 Golomb, S.W. A connected topology for the integers. *Amer. Math. Monthly* 66 (1959), 663–665.

1963 Mullin, A.A. Recursive function theory (A modern look on an Euclidean idea). *Bull. Amer. Math. Soc.* 69 (1963), p. 737.

1964 Edwards, A.W.F. Infinite coprime sequences. *Math. Gazette* 48 (1964), 416–422.

1967 Samuel, P. *Théorie Algébrique des Nombres*. Hermann, Paris, 1967. English translation published by Houghton-Mifflin, Boston, 1970.

1968 Cox, C.D. & van der Poorten, A.J. On a sequence of prime numbers. *J. Austr. Math. Soc.* 8 (1968), 571–574.

1972 Borning, A. Some results for $k! \pm 1$ and $2 \cdot 3 \cdot 5 \cdots p \pm 1$. *Math. Comp.* 26 (1972), 567–570.

1975 Guy, R.K. & Nowakowski, R. Discovering primes with Euclid. *Delta* 5 (1975), 49–63.

1980 Templer, M. On the primality of $k!+1$ and $2*3*5*\cdots*p+1$. *Math. Comp.* 34 (1980), 303–304.

1980 Washington, L.C. The infinitude of primes via commutative algebras. Unpublished manuscript.

1982 Buhler, J.P., Crandall, R.E. & Penk, M.A. Primes of the form $n! \pm 1$ and $2 \cdot 3 \cdot 5 \cdots p \pm 1$. *Math. Comp.* 38 (1982), 639–643.

1984 Naur, T. Mullin's sequence of primes is not monotonic. *Proc. Amer. Math. Soc.* 90 (1984), 43–44.

1985 Odoni, R.W.K. On the prime divisors of the sequence $w_{n+1} = 1 + w_1 w_2 \cdots w_n$. *J. London Math. Soc.* (2) 32 (1985), 1–11.

1987 Dubner, H. Factorial and primorial primes. *J. Recr. Math.* 19 (1987), 197–203.

1991 Shanks, D. Euclid's primes. *Bull. Inst. Comb. and Appl.* 1 (1991), 33–36.

1993 Caldwell, C. & Dubner, H. Primorial, factorial, and multifactorial primes. *Math. Spectrum* 26 (1993/4), 1–7.

1993 Wagstaff, Jr., S.S. Computing Euclid's primes. *Bull. Inst. Comb. and Appl.* 8 (1993), 23–32.

1995 Caldwell, C. On the primality of $n! \pm 1$ and $2 \times 3 \times 5 \times \cdots \times p \pm 1$. *Math. Comp.* 64 (1995), 889–890.

2000 Narkiewicz, W. *The Development of Prime Number Theory.* Springer-Verlag, Berlin, 2000.

2002 Caldwell, C. & Gallot, Y. On the primality of $n! \pm 1$ and $2 \times 3 \times 5 \times \cdots \times p \pm 1$. *Math. Comp.* 71 (2002), 441-448.

Chapter 2

1801 Gauss, C.F. *Disquisitiones Arithmeticae.* G. Fleischer, Leipzig, 1801. English translation by A.A. Clarke. Yale Univ. Press, New Haven, 1966. Revised English translation by W.C. Waterhouse, Springer-Verlag, New York, 1986.

1844 Eisenstein, F.G. Aufgaben. *J. Reine Angew. Math.* 27 (1844), p. 87. Reprinted in *Mathematische Werke*, Vol. I, p. 112. Chelsea, Bronx, NY, 1975.

1852 Kummer, E.E. Über die Erganzungssätze zu den allgemeinen Reciprocitätsgesetzen. *J. Reine Angew. Math.* 44 (1852), 93–146. Reprinted in *Collected Papers* (edited by A. Weil), Vol. I, 485–538. Springer-Verlag, New York, 1975.

1876 Lucas, E. Sur la recherche des grands nombres premiers. *Assoc. Française p. l'Avanc. des Sciences* 5 (1876), 61–68.

1877 Pepin, T. Sur la formule $2^{2^n} + 1$. *C.R. Acad. Sci. Paris* 85 (1877), 329–331.

1878 Lucas, E. Théorie des fonctions numériques simplement périodiques. *Amer. J. Math.* 1 (1878), 184–240 and 289–321.

1878 Proth, F. Théorèmes sur les nombres premiers. *C.R. Acad. Sci. Paris* 85 (1877), 329–331.

1886 Bang, A.S. Taltheoretiske Undersøgelser. *Tidskrift f. Math.* (5) 4 (1886), 70–80 and 130–137.

1891 Lucas, E. *Théorie des Nombres.* Gauthier-Villars, Paris, 1891. Reprinted by A. Blanchard, Paris, 1961.

1892 Zsigmondy, K. Zur Theorie der Potenzreste. *Monatsh. f. Math. Phys.* 3 (1892), 265–284.

1899 Korselt, A. Problème chinois. *L'Interm. des Math.* 6 (1899), 142–143.

1903 Malo, E. Nombres qui, sans être premiers, vérifient exceptionnellement une congruence de Fermat. *L'Interm. des Math.* 10 (1903), p. 88.

1904 Birkhoff, G.D. & Vandiver, H.S. On the integral divisors of $a^n - b^n$. *Annals of Math.* (2) 5 (1904), 173–180.

1904 Cipolla, M. Sui numeri composti P, che verificano la congruenza di Fermat $a^{P-1} \equiv 1 \pmod{P}$. *Annali di Matematica* (3) 9 (1904), 139–160.

1912 Carmichael, R.D. On composite numbers P which satisfy the Fermat congruence $a^{P-1} \equiv 1 \pmod{P}$. *Amer. Math. Monthly* 19 (1912), 22–27.

1913 Carmichael, R.D. On the numerical factors of the arithmetic forms $\alpha^n \pm \beta^n$. *Annals of Math.* (2) 15 (1913), 30–70.

1913 Dickson, L.E. Finiteness of odd perfect and primitive abundant numbers with n distinct prime factors. *Amer. J. Math.* 35 (1913), 413–422. Reprinted in *The Collected Mathematical Papers* (edited by A.A. Albert), Vol. I, 349–358. Chelsea, Bronx, NY, 1975.

1914 Pocklington, H.C. The determination of the prime or composite nature of large numbers by Fermat's theorem. *Proc. Cambridge Phil. Soc.* 18 (1914/6), 29–30.

1921 Pirandello, L. *Il Fu Mattia Pascal.* Bemporad & Figlio, Firenze, 1921.

1922 Carmichael, R.D. Note on Euler's φ-function. *Bull. Amer. Math. Soc.* 28 (1922), 109–110.

1925 Cunningham, A.J.C. & Woodall, H.J. *Factorization of $y^n \pm 1$, $y = 2, 3, 5, 6, 7, 10, 11, 12$ Up to High Powers (n).* Hodgson, London, 1925.

1929 Pillai, S.S. On some functions connected with $\varphi(n)$. *Bull. Amer. Math. Soc.* 35 (1929), 832–836.

1930 Lehmer, D.H. An extended theory of Lucas' functions. *Annals of Math.* 31 (1930), 419–448. Reprinted in *Selected Papers* (edited by D. McCarthy), Vol. I, 11–48. Ch. Babbage Res. Centre, St. Pierre, Manitoba, Canada, 1981.

1932 Lehmer, D.H. On Euler's totient function. *Bull. Amer. Math. Soc.* 38 (1932), 745–751. Reprinted in *Selected Papers* (edited by D. McCarthy), Vol. I, 319–325. Ch. Babbage Res. Centre, St. Pierre, Manitoba, Canada, 1981.

1932 Western, A.E. On Lucas' and Pepin's tests for the primeness of Mersenne's numbers. *J. London Math. Soc.* 7 (1932), 130–137.

1935 Archibald, R.C. Mersenne's numbers. *Scripta Math.* 3 (1935), 112–119.

1935 Lehmer, D.H. On Lucas' test for the primality of Mersenne numbers. *J. London Math. Soc.* 10 (1935), 162–165. Reprinted in *Selected Papers* (edited by D. McCarthy), Vol. I, 86–89. Ch. Babbage Res. Centre, St. Pierre, Manitoba, Canada, 1981.

1936 Lehmer, D.H. On the converse of Fermat's theorem. *Amer. Math. Monthly* 43 (1936), 347–354. Reprinted in *Selected Papers* (edited by D. McCarthy), Vol. I, 90–95. Ch. Babbage Res. Centre, St. Pierre, Manitoba, Canada, 1981.

1939 Chernick, J. On Fermat's simple theorem. *Bull. Amer. Math. Soc.* 45 (1939), 269–274.

1944 Pillai, S.S. On the smallest primitive root of a prime. *J. Indian Math. Soc.* (N.S.) 8 (1944), 14–17.

1944 Schuh, F. Can $n-1$ be divisible by $\varphi(n)$ when n is composite? (in Dutch). *Mathematica,* Zutphen (B) 12 (1944), 102–107.

1945 Kaplansky, I. Lucas' tests for Mersenne numbers. *Amer. Math. Monthly* 52 (1945), 188–190.

1946 Erdös, P. Problem 4221 (To solve the equation $\varphi(n) = k!$ for every $k \geq 1$). *Amer. Math. Monthly* 53 (1946), p. 537.

1947 Klee, V.L. On a conjecture of Carmichael. *Bull. Amer. Math. Soc.* 53 (1947), 1183–1186.

1947 Lehmer, D.H. On the factors of $2^n \pm 1$. *Bull. Amer. Math. Soc.* 53 (1947), 164–167. Reprinted in *Selected Papers* (edited by D. McCarthy), Vol. III, 1081–1084. Ch. Babbage Res. Centre, St. Pierre, Manitoba, Canada, 1981.

1948 Lambek, J. Solution of problem 4221 (proposed by P. Erdös). *Amer. Math. Monthly* 55 (1948), p. 103.

1948 Ore, O. On the averages of the divisors of a number. *Amer. Math. Monthly* 55 (1948), 615–619.

1948 Steuerwald, R. Über die Kongruenz $a^{n-1} \equiv 1 \pmod{n}$. Sitzungsber. math.-naturw. Kl. Bayer. Akad. Wiss. München, 1948, 69–70.

1949 Erdös, P. On the converse of Fermat's theorem. *Amer. Math. Monthly* 56 (1949), 623–624.

1949 Fridlender, V.R. On the least nth power non-residue (in Russian). *Doklady Akad. Nauk SSSR* (N.S.) 66 (1949), 351–352.

1949 Shapiro, H.N. Note on a theorem of Dickson. *Bull. Amer. Math. Soc.* 55 (1949), 450–452.

1950 Beeger, N.G.W.H. On composite numbers n for which $a^{n-1} \equiv 1 \pmod{n}$, for every a, prime to n. *Scripta Math.* 16 (1950), 133–135.

1950 Giuga, G. Su una presumibile proprietà caratteristica dei numeri primi. *Ist. Lombardo Sci. Lett. Rend. Cl. Sci. Mat. Nat.* (3) 14 (83), (1950), 511–528.

1950 Gupta, H. On a problem of Erdös. *Amer. Math. Monthly* 57 (1950), 326–329.

1950 Kanold, H.-J. Sätze über Kreisteilungspolynome und ihre Anwendungen auf einige zahlentheoretische Probleme, II. *J. Reine Angew. Math.* 187 (1950), 355–366.

1950 Salié, H. Über den kleinsten positiven quadratischen Nichtrest einer Primzahl. *Math. Nachr.* 3 (1949), 7–8.

1950 Somayajulu, B.S.K.R. On Euler's totient function $\varphi(n)$. *Math. Student* 18 (1950), 31–32.

1951 Beeger, N.G.W.H. On even numbers m dividing $2^m - 2$. *Amer. Math. Monthly* 58 (1951), 553–555.

1951 Morrow, D.C. Some properties of D numbers. *Amer. Math. Monthly* 58 (1951), 329–330.

1952 Duparc, H.J.A. On Carmichael numbers. *Simon Stevin* 29 (1952), 21–24.

1952 Grün, O. Über ungerade vollkommene Zahlen. *Math. Zeits.* 55 (1952), 353–354.

1953 Knödel, W. Carmichaelsche Zahlen. *Math. Nachr.* 9 (1953), 343–350.

1953 Selfridge, J.L. Factors of Fermat numbers. *Math. Comp.* 7 (1953), 274–275.

1953 Touchard, J. On prime numbers and perfect numbers. *Scripta Math.* 19 (1953), 35–39.

1954 Kanold, H.-J. Über die Dichten der Mengen der vollkommenen und der befreundeten Zahlen. *Math. Zeits.* 61 (1954), 180–185.

1954 Robinson, R.M. Mersenne and Fermat numbers. *Proc. Amer. Math. Soc.* 5 (1954), 842–846.

1954 Schinzel, A. Quelques théorèmes sur les fonctions $\varphi(n)$ et $\sigma(n)$. *Bull. Acad. Polon. Sci.*, Cl. III, 2 (1954), 467–469.

1954 Schinzel, A. Generalization of a theorem of B.S.K.R. Somayajulu on the Euler's function $\varphi(n)$. *Ganita* 5 (1954), 123–128.

1954 Schinzel, A. & Sierpiński, W. Sur quelques propriétés des fonctions $\varphi(n)$ et $\sigma(n)$. *Bull. Acad. Polon. Sci.*, Cl. III, 2 (1954), 463–466.

1955 Artin, E. The orders of the linear groups. *Comm. Pure and Appl. Math.* 8 (1955), 355–365. Reprinted in *Collected Papers* (edited by S. Lang and J.T. Tate), 387–397. Addison-Wesley, Reading, MA, 1965.

1955 Hornfeck, B. Zur Dichte der Menge der vollkommenen Zahlen. *Arch. Math.* 6 (1955), 442–443.

1955 Laborde, P. A note on the even perfect numbers. *Amer. Math. Monthly* 62 (1955), 348–349.

1956 Hornfeck, B. Bemerkung zu meiner Note über vollkommene Zahlen. *Arch. Math.* 7 (1956), p. 273.

1956 Kanold, H.-J. Über einen Satz von L.E. Dickson, II. *Math. Ann.* 131 (1956), 246–255.

1956 Schinzel, A. Sur l'équation $\varphi(x) = m$. *Elem. Math.* 11 (1956), 75–78.

1956 Schinzel, A. Sur un problème concernant la fonction $\varphi(n)$. *Czechoslovak Math. J.* 6 (81), (1956), 164–165.

1957 Hornfeck, B. & Wirsing, E. Über die Häufigkeit vollkommener Zahlen. *Math. Ann.* 133 (1957), 431–438.

1957 Kanold, H.-J. Über die Verteilung der vollkommenen Zahlen und allgemeinerer Zahlenmengen. *Math. Ann.* 132 (1957), 442–450.

1958 Erdös, P. Some remarks on Euler's φ-function. *Acta Arith.* 4 (1958), 10–19.

1958 Jarden, D. *Recurring Sequences.* Riveon Lematematika, Jerusalem, 1958 (3rd edition, Fibonacci Assoc., San Jose, CA, 1973).

1958 Perisastri, M. A note on odd perfect numbers. *Math. Student* 26 (1958), 179–181.

1958 Schinzel, A. Sur les nombres composés n qui divisent $a^n - a$. *Rend. Circ. Mat. Palermo* (2) 7 (1958), 37–41.

1958 Sierpiński, W. Sur les nombres premiers de la forme $n^n + 1$. *L'Enseign. Math.* (2) 4 (1958), 211–212.

1959 Rotkiewicz, A. Sur les nombres pairs n pour lesquels les nombres $a^n b - a b^n$, respectivement $a^{n-1} - b^{n-1}$ sont divisibles par n. *Rend. Circ. Mat. Palermo* (2) 8 (1959), 341–342.

1959 Satyanarayana, M. Odd perfect numbers. *Math. Student* 27 (1959), 17–18.

1959 Schinzel, A. Sur les nombres composés n qui divisent $a^n - a$. *Rend. Circ. Mat. Palermo* (2) 7 (1958), 1–5.

1959 Wirsing, E. Bemerkung zu der Arbeit über vollkommene Zahlen. *Math. Ann.* 137 (1959), 316–318.

1960 Inkeri, K. Tests for primality. *Annales Acad. Sci. Fennicae*, Ser. A, I, 279, Helsinki, 1960, 19 pages. Reprinted in *Collected Papers of Kustaa Inkeri* (edited by T. Metsänkylä and P. Ribenboim), Queen's Papers in Pure and Appl. Math. 91, 1992, Queen's Univ., Kingston, Ontario, Canada.

1961 Ward, M. The prime divisors of Fibonacci numbers. *Pacific J. Math.* 11 (1961), 379–389.

1962 Burgess, D.A. On character sums and L-series. *Proc. London Math. Soc.* (3) 12 (1962), 193–206.

1962 Crocker, R. A theorem on pseudo-primes. *Amer. Math. Monthly* 69 (1962), p. 540.

1962 Mąkowski, A. Generalization of Morrow's D numbers. *Simon Stevin* 36 (1962), p. 71.

1962 Schinzel, A. The intrinsic divisors of Lehmer numbers in the case of negative discriminant. *Arkiv för Mat.* 4 (1962), 413–416.

1962 Schinzel, A. On primitive prime factors of $a^n - b^n$. *Proc. Cambridge Phil. Soc.* 58 (1962), 555–562.

1962 Shanks, D. *Solved and Unsolved Problems in Number Theory*. Spartan, Washington, 1962 (3rd edition by Chelsea, Bronx, NY, 1985).

1963 Schinzel, A. On primitive prime factors of Lehmer numbers, I. *Acta Arith.* 8 (1963), 211–223.

1963 Schinzel, A. On primitive prime factors of Lehmer numbers, II. *Acta Arith.* 8 (1963), 251–257.

1963 Suryanarayana, D. On odd perfect numbers, II. *Proc. Amer. Math. Soc.* 14 (1963), 896–904.

1964 Biermann, K.-R. Thomas Clausen, Mathematiker und Astronom. *J. Reine Angew. Math.* 216 (1964), 159–198.

1964 Lehmer, E. On the infinitude of Fibonacci pseudo-primes. *Fibonacci Quart.* 2 (1964), 229–230.

1965 Erdös, P. Some recent advances and current problems in number theory. In *Lectures in Modern Mathematics*, Vol. III (edited by T.L. Saaty), 196–244. Wiley, New York, 1965.

1965 Rotkiewicz, A. Sur les nombres de Mersenne dépourvus de facteurs carrés et sur les nombres naturels n tels que $n^2 \mid 2^n - 2$. *Matem. Vesnik* (Beograd) 2 (17), (1965), 78–80.

1966 Grosswald, E. *Topics from the Theory of Numbers.* Macmillan, New York, 1966 (2nd edition, Birkhäuser, Boston, 1984).

1966 Muskat, J.B. On divisors of odd perfect numbers. *Math. Comp.* 20 (1966), 141–144.

1967 Brillhart, J. & Selfridge, J.L. Some factorizations of $2^n \pm 1$ and related results. *Math. Comp.* 21 (1967), 87–96 and p. 751.

1967 Mozzochi, C.J. A simple proof of the Chinese remainder theorem. *Amer. Math. Monthly* 74 (1967), p. 998.

1970 Lieuwens, E. Do there exist composite numbers M for which $k\varphi(M) = M - 1$ holds? *Nieuw Arch. Wisk.* (3) 18 (1970), 165–169.

1970 Parberry, E.A. On primes and pseudo-primes related to the Fibonacci sequence. *Fibonacci Quart.* 8 (1970), 49–60.

1970 Suryanarayana, D. & Hagis, Jr., P. A theorem concerning odd perfect numbers. *Fibonacci Quart.* 8 (1970), 337–346.

1971 Lieuwens, E. *Fermat Pseudo-Primes.* Ph.D. Thesis, Delft, 1971.

1971 Morrison, M.A. & Brillhart, J. The factorization of F_7. *Bull. Amer. Math. Soc.* 77 (1971), p. 264.

1971 Schönhage, A. & Strassen, V. Schnelle Multiplikation grosser Zahlen. *Computing* 7 (1971), 281–292.

1972 Hagis, Jr., P. & McDaniel, W.L. A new result concerning the structure of odd perfect numbers. *Proc. Amer. Math. Soc.* 32 (1972), 13–15.

1972 Mills, W.H. On a conjecture of Ore. *Proc. 1972 Number Th. Conf. in Boulder*, 142–146.

1972 Ribenboim, P. *Algebraic Numbers.* Wiley-Interscience, New York, 1972 (enlarged new edition, Springer-Verlag, 2001).

1972 Rotkiewicz, A. *Pseudoprime Numbers and their Generalizations.* Stud. Assoc. Fac. Sci. Univ. Novi Sad, 1972.

1973 Grosswald, E. Contributions to the theory of Euler's function $\varphi(x)$. *Bull. Amer. Math. Soc.* 79 (1973), 327–341.

1973 Hagis, Jr., P. A lower bound for the set of odd perfect numbers. *Math. Comp.* 27 (1973), 951–953.

1973 Rotkiewicz, A. On pseudoprimes with respect to the Lucas sequences. *Bull. Acad. Polon. Sci.* 21 (1973), 793–797.

1974 Ligh, S. & Neal, L. A note on Mersenne numbers. *Math. Mag.* 47 (1974), 231–233.

1974 Pomerance, C. On Carmichael's conjecture. *Proc. Amer. Math. Soc.* 43 (1974), 297–298.

1974 Schinzel, A. Primitive divisors of the expression $A^n - B^n$ in algebraic number fields. *J. Reine Angew. Math.* 268/269 (1974), 27–33.

1974 Sinha, T.N. Note on perfect numbers. *Math. Student* 42 (1974), p. 336.

1975 Brillhart, J., Lehmer, D.H. & Selfridge, J.L. New primality criteria and factorizations of $2^m \pm 1$. *Math. Comp.* 29 (1975), 620–647.

1975 Guy, R.K. How to factor a number. *Proc. Fifth Manitoba Conf. Numerical Math.*, 1975, 49–89 (Congressus Numerantium, XVI, Winnipeg, Manitoba, 1975).

1975 Hagis, Jr., P. & McDaniel, W.L. On the largest prime divisor of an odd perfect number. *Math. Comp.* 29 (1975), 922–924.

1975 Morrison, M.A. A note on primality testing using Lucas sequences. *Math. Comp.* 29 (1975), 181–182.

1975 Pomerance, C. The second largest prime factor of an odd perfect number. *Math. Comp.* 29 (1975), 914–921.

1975 Pratt, V.R. Every prime has a succinct certificate. *SIAM J. Comput.* 4 (1975), 214–220.

1975 Stewart, C.L. The greatest prime factor of $a^n - b^n$. *Acta Arith.* 26 (1975), 427–433.

1976 Buxton, M. & Elmore, S. An extension of lower bounds for odd perfect numbers. *Notices Amer. Math. Soc.* 23 (1976), p. A55.

1976 Diffie, W. & Hellman, M.E. New directions in cryptography. *IEEE Trans. on Inf. Th.* IT-22 (1976), 644–654.

1976 Erdös, P. & Shorey, T.N. On the greatest prime factor of $2^p - 1$ for a prime p, and other expressions. *Acta Arith.* 30 (1976), 257–265.

1976 Lehmer, D.H. Strong Carmichael numbers. *J. Austral. Math. Soc.* (A) 21 (1976), 508–510. Reprinted in *Selected Papers* (edited by D. McCarthy), Vol. I, 140–142. Ch. Babbage Res. Centre, St. Pierre, Manitoba, Canada, 1981.

1976 Mendelsohn, N.S. The equation $\varphi(x) = k$. *Math. Mag.* 49 (1976), 37–39.

1976 Miller, G.L. Riemann's hypothesis and tests for primality. *J. Comp. Syst. Sci.* 13 (1976), 300–317.

1976 Rabin, M.O. Probabilistic algorithms. In *Algorithms and Complexity* (edited by J.F. Traub), 21–39. Academic Press, New York, 1976.

1976 Yorinaga, M. On a congruential property of Fibonacci numbers. Numerical experiments. Considerations and remarks. *Math. J. Okayama Univ.* 19 (1976), 5–10 and 11–17.

1977 Kishore, M. Odd perfect numbers not divisible by 3 are divisible by at least ten distinct primes. *Math. Comp.* 31 (1977), 274–279.

1977 Kishore, M. *The number of distinct prime factors for which* $\sigma(N) = 2N$, $\sigma(N) = 2N \pm 1$ *and* $\varphi(N) \mid N - 1$. Ph.D. Thesis, Univ. of Toledo, Ohio, 1977, 39 pages.

1977 Malm, D.E.G. On Monte-Carlo primality tests. *Notices Amer. Math. Soc.* 24 (1977), A-529, abstract 77T-A22.

1977 Pomerance, C. On composite n for which $\varphi(n) \mid n - 1$, II. *Pacific J. Math.* 69 (1977), 177–186.

1977 Pomerance, C. Multiply perfect numbers, Mersenne primes and effective computability. *Math. Ann.* 226 (1977), 195–206.

1977 Solovay, R. & Strassen, V. A fast Monte-Carlo test for primality. *SIAM J. Comput.* 6 (1977), 84–85.

1977 Stewart, C.L. On divisors of Fermat, Fibonacci, Lucas, and Lehmer numbers. *Proc. London Math. Soc.* (3) 35 (1977), 425–447.

1977 Stewart, C.L. Primitive divisors of Lucas and Lehmer numbers. In *Transcendence Theory: Advances and Applications* (edited by A. Baker and D.W. Masser), 79–92. Academic Press, London, 1977.

1977 Williams, H.C. On numbers analogous to the Carmichael numbers. *Can. Math. Bull.* 20 (1977), 133–143.

1978 Cohen, G.L. On odd perfect numbers. *Fibonacci Quart.* 16 (1978), 523–527.

1978 Kiss, P. & Phong, B.M. On a function concerning second order recurrences. *Ann. Univ. Sci. Budapest* 21 (1978), 119–122.

1978 Rivest, R.L. Remarks on a proposed cryptanalytic attack on the M.I.T. public-key cryptosystem. *Cryptologia* 2 (1978), 62–65.

1978 Rivest, R.L., Shamir, A. & Adleman, L.M. A method for obtaining digital signatures and public-key cryptosystems. *Comm. ACM* 21 (1978), 120–126.

1978 Williams, H.C. Primality testing on a computer. *Ars Comb.* 5 (1978), 127–185.

1978 Yorinaga, M. Numerical computation of Carmichael numbers. *Math. J. Okayama Univ.* 20 (1978), 151–163.

1979 Chein, E.Z. Non-existence of odd perfect numbers of the form $q_1^{a_1} q_2^{a_2} \cdots q_6^{a_6}$ and $5^{a_1} q_2^{a_2} \cdots q_9^{a_9}$. Ph.D. Thesis, Pennsylvania State Univ., 1979.

1979 Lenstra, Jr., H.W. Miller's primality test. *Inf. Process. Letters* 8 (1979), 86–88.

1980 Baillie, R. & Wagstaff, Jr., S.S. Lucas pseudoprimes. *Math. Comp.* 35 (1980), 1391–1417.

1980 Cohen, G.L. & Hagis, Jr., P. On the number of prime factors of n if $\varphi(n) \mid (n-1)$. *Nieuw Arch. Wisk.* (3) 28 (1980), 177–185.

1980 Hagis, Jr., P. Outline of a proof that every odd perfect number has at least eight prime factors. *Math. Comp.* 35 (1980), 1027–1032.

1980 Monier, L. Evaluation and comparison of two efficient probabilistic primality testing algorithms. *Theoret. Comput. Sci.* 12 (1980), 97–108.

1980 Pomerance, C., Selfridge, J.L. & Wagstaff, Jr., S.S. The pseudoprimes to $25 \cdot 10^9$. *Math. Comp.* 35 (1980), 1003–1026.

1980 Rabin, M.O. Probabilistic algorithm for testing primality. *J. Number Theory* 12 (1980), 128–138.

1980 Wagstaff, Jr., S.S. Large Carmichael numbers. *Math. J. Okayama Univ.* 22 (1980), 33–41.

1980 Wall, D.W. Conditions for $\varphi(N)$ to properly divide $N-1$. In *A Collection of Manuscripts Related to the Fibonacci Sequence* (edited by V.E. Hoggatt and M. Bicknell-Johnson), 205–208. 18th Anniv. Vol., Fibonacci Assoc., San Jose, 1980.

1980 Yorinaga, M. Carmichael numbers with many prime factors. *Math. J. Okayama Univ.* 22 (1980), 169–184.

1981 Brent, R.P. & Pollard, J.M. Factorization of the eighth Fermat number. *Math. Comp.* 36 (1981), 627–630.

1981 Grosswald, E. On Burgess' bound for primitive roots modulo primes and an application to $\Gamma(p)$. *Amer. J. Math.* 103 (1981), 1171–1183.

1981 Hagis, Jr., P. On the second largest prime divisor of an odd perfect number. In *Analytic Number Theory* (edited by M.I. Knopp). Lecture Notes in Math. #899, 254–263. Springer-Verlag, New York, 1981.

1981 Lenstra, Jr., H.W. Primality testing algorithms (after Adleman, Rumely and Williams). In *Séminaire Bourbaki*, exposé No. 576. Lecture Notes in Math. #901, 243–257. Springer-Verlag, Berlin, 1981.

1981 Lüneburg, H. Ein einfacher Beweis für den Satz von Zsigmondy über primitive Primteiler von $A^N - 1$. In *Geometries and Groups* (edited by M. Aigner and D. Jungnickel). Lecture Notes in Math. #893, 219–222. Springer-Verlag, New York, 1981.

1982 Brent, R.P. Succinct proofs of primality for the factors of some Fermat numbers. *Math. Comp.* 38 (1982), 253–255.

1982 Couvreur, C. & Quisquater, J.J. An introduction to fast generation of large prime numbers. *Philips J. Res.* 37 (1982), 231–264; Errata, 38 (1983), p. 77.

1982 Hoogendoorn, P.J. On a secure public-key cryptosystem. In *Computational Methods in Number Theory* (edited by H.W. Lenstra, Jr. and R. Tijdeman), Part I, 159–168. Math. Centre Tracts #154, Amsterdam, 1982.

1982 Lenstra, Jr., H.W. Primality testing. In *Computational Methods in Number Theory* (edited by H.W. Lenstra, Jr. and R. Tijdeman), Part I, 55–77. Math. Centre Tracts #154, Amsterdam, 1982.

1982 Masai, P. & Valette, A. A lower bound for a counterexample in Carmichael's conjecture. *Boll. Un. Mat. Ital.* (6) 1-A (1982), 313–316.

1982 Naur, T. *Integer Factorization.* DAIMI PB-144, Aarhus University, 1982, 129 pages.

1982 Woods, D. & Huenemann, J. Larger Carmichael numbers. *Comput. Math. and Appl.* 8 (1982), 215–216.

1983 Adleman, L.M., Pomerance, C. & Rumely, R.S. On distinguishing prime numbers from composite numbers. *Annals of Math.* (2) 117 (1983), 173–206.

1983 Brillhart, J., Lehmer, D.H., Selfridge, J.L., Tucker-man, B. & Wagstaff, Jr., S.S. *Factorizations of $b^n \pm 1$, $b = 2$, 3, 5, 6, 7, 10, 11, 12 Up to High Powers.* Contemporary Math., Vol. 22, Amer. Math. Soc., Providence, RI, 1983 (2nd edition, 1988, 3rd edition electronical only, 2002).

1983 Hagis, Jr., P. Sketch of a proof that an odd perfect number relatively prime to 3 has at least eleven prime factors. *Math. Comp.* 40 (1983), 399–404.

1983 Kishore, M. Odd perfect numbers not divisible by 3, II. *Math. Comp.* 40 (1983), 405–411.

1983 Naur, T. New integer factorizations. *Math. Comp.* 41 (1983), 687–695.

1983 Pomerance, C. & Wagstaff, Jr., S.S. Implementation of the continued fraction integer factoring algorithm. *Congressus Numerantium* 37 (1983), 99–118.

1983 Powell, B. Problem 6420 (On primitive roots). *Amer. Math. Monthly* 90 (1983), p. 60.

1983 Singmaster, D. Some Lucas pseudoprimes. *Abstracts Amer. Math. Soc.* 4 (1983), p. 197, abstract 83T-10-146.

1983 Yates, S. Titantic primes. *J. Recr. Math.* 16 (1983/4), 250–260.

1984 Cohen, H. & Lenstra, Jr., H.W. Primality testing and Jacobi sums. *Math. Comp.* 42 (1984), 297–330.

1984 Dixon, J.D. Factorization and primality tests. *Amer. Math. Monthly* 91 (1984), 333–352.

1984 Kearnes, K. Solution of problem 6420. *Amer. Math. Monthly* 91 (1984), p. 521.

1984 Nicolas, J.L. Tests de primalité. *Expo. Math.* 2 (1984), 223–234.

1984 Pomerance, C. *Lecture Notes on Primality Testing and Factoring* (Notes by G.M. Gagola, Jr.). Math. Assoc. America, Notes No. 4, 1984, 34 pages.

1984 Williams, H.C. An overview of factoring. In *Advances in Cryptology* (edited by D. Chaum), 71–80. Plenum, New York, 1984.

1984 Yates, S. Sinkers of the Titanics. *J. Recr. Math.* 17 (1984/5), 268–274.

1985 Bedocchi, E. Note on a conjecture on prime numbers. *Rev. Mat. Univ. Parma* (4) 11 (1985), 229–236.

1985 Fouvry, E. Théorème de Brun-Titchmarsh; application au théorème de Fermat. *Invent. Math.* 79 (1985), 383–407.

1985 Riesel, H. *Prime Numbers and Computer Methods for Factorization.* Birkhäuser, Boston, 1985 (2nd edition, 1994).

1985 Wagon, S. Perfect numbers. *Math. Intelligencer* 7, No. 2 (1985), 66–68.

1986 Kiss, P., Phong, B.M. & Lieuwens, E. On Lucas pseudoprimes which are products of *s* primes. In *Fibonacci Numbers and their Applications* (edited by A.N. Philippou, G.E. Bergum and A.F. Horadam), 131–139. Reidel, Dordrecht, 1986.

1986 Wagon, S. Carmichael's "Empirical Theorem". *Math. Intelligencer* 8, No. 2 (1986), 61–62.

1986 Wagon, S. Primality testing. *Math. Intelligencer* 8, No. 3 (1986), 58–61.

1987 Cohen, G.L. On the largest component of an odd perfect number. *J. Austral. Math. Soc.* (A) 42 (1987), 280–286.

1987 Cohen, H. & Lenstra, A.K. Implementation of a new primality test. *Math. Comp.* 48 (1987), 103–121 and S1–S4.

1987 Koblitz, N. *A Course in Number Theory and Cryptography.* Springer-Verlag, New York, 1987.

1987 Lenstra, Jr., H.W. Factoring integers with elliptic curves. *Annals of Math.* 126 (1987), 649–673.

1987 Li Yan & Du Shiran *Chinese Mathematics. A Concise History* (English translation by J.N. Crossley and A.W.C. Lun). Clarendon Press, Oxford, 1987.

1987 Pomerance, C. Very short primality proofs. *Math. Comp.* 48 (1987), 315–322.

1988 Brillhart, J., Montgomery, P.L. & Silverman, R.D. Tables of Fibonacci and Lucas factorizations, and Supplement. *Math. Comp.* 50 (1988), 251–260 and S1–S15.

1988 Young, J. & Buell, D.A. The twentieth Fermat number is composite. *Math. Comp.* 50 (1988), 261–263.

1989 Bateman, P.T., Selfridge, J.L. & Wagstaff, Jr., S.S. The new Mersenne conjecture. *Amer. Math. Monthly* 96 (1989), 125–128.

1989 Brent, R.P. & Cohen, G.L. A new lower bound for odd perfect numbers. *Math. Comp.* 53 (1989), 431–437 and S7–S24.

1989 Bressoud, D.M. *Factorization and Primality Testing.* Springer-Verlag, New York, 1989.

1989 Dubner, H. A new method for producing large Carmichael numbers. *Math. Comp.* 53 (1989), 411–414.

1989 Lemos, M. *Criptografia, Números Primos e Algoritmos.* 17° Colóquio Brasileiro de Matemática, Inst. Mat. Pura e Aplic., Rio de Janeiro, 1989, 72 pages.

1990 Lenstra, A.K. & Lenstra, Jr., H.W. Algorithms in number theory. In *Handbook of Theoretical Computer Science* (edited by J. van Leeuwen, A. Meyer, M. Nivat, M. Paterson and D. Perrin). North-Holland, Amsterdam, 1990.

1991 Brent, R.P., Cohen, G.L. & te Riele, H.J.J. Improved techniques for lower bounds for odd perfect numbers. *Math. Comp.* 57 (1991), 857–868.

1991 Frasnay, C. Extension à l'anneau $\mathbb{Z}_p$ du théorème de Lucas, sur les coefficients binomiaux. *Singularité* 2 (1991), 13–15.

1992 Yates, S. Collecting gigantic and titanic primes. *J. Recr. Math.* 24 (1992), 187–195.

1993 Atkin, A.O.L. & Morain, F. Elliptic curves and primality proving. *Math. Comp.* 61 (1993), 29–68.

1993 Cohen, H. *A Course in Computational Number Theory.* Springer-Verlag, New York, 1993.

1993 Jaeschke, G. On strong pseudoprimes to several bases. *Math. Comp.* 61 (1993), 915–926.

1993 Lenstra, A.K., Lenstra, Jr., H.W., Manasse, M.S. & Pollard, J.M. The number field sieve. In *The Development of the Number Field Sieve* (edited by A.K. Lenstra and H.W. Lenstra, Jr.). Lecture Notes in Math. #1554, 11–42. Springer-Verlag, New York, 1993.

1993 Pomerance, C. Carmichael numbers. *Nieuw Arch. Wisk.* (4) 11 (1993), 199–209.

1993 Williams, H.C. How was F_6 factored? *Math. Comp.* 61 (1993), 463–474.

1994 Alford, W.R., Granville, A. & Pomerance, C. There are infinitely many Carmichael numbers. *Annals of Math.* (2) 140 (1994), 703–722.

1994 Heath-Brown, D.R. Odd perfect numbers. *Math. Proc. Cambridge Phil. Soc.* 115 (1994), 191–196.

1994 Schlafly, A. & Wagon, S. Carmichael's conjecture on the Euler function is valid below $10^{10000000}$. *Math. Comp.* 63 (1994), 415–419.

1994 Williams, H.C. & Shallit, J.O. Factoring integers before computers. In *Mathematics of Computation, 1943-1993: A Half-Century of Computational Mathematics* (edited by W. Gautschi). Proc. Symp. Appl. Math., Vol. 48, 481–531. Amer. Math. Soc., Providence, RI, 1994.

1995 Crandall, R.E., Doenias, J., Norrie, C. & Young, J. The twenty-second Fermat number is composite. *Math. Comp.* 64 (1995), 863–868.

1995 Gostin, G.B. New factors of Fermat numbers. *Math. Comp.* 64 (1995), 393–395.

1995 McIntosh, R.J. On the converse of Wolstenholme's theorem. *Acta Arith.* 71 (1995), 381–389.

1995 Trevisan, V. & Carvalho, J.B. The composite character of the twenty-second Fermat number. *J. Supercomp.* 9 (1995), 179–182.

1996 Borwein, D., Borwein, J.M., Borwein, P.B. & Girgensohn, R. Giuga's conjecture on primality. *Amer. Math. Monthly* 103 (1996), 40–50.

1996 Löh, G. & Niebuhr W. A new algorithm for constructing large Carmichael numbers. *Math. Comp.* 65 (1996), 823–836.

1998 Ford, K. The distribution of totients. *Hardy-Ramanujan J.* 2 (1998), 67–151.

1998 Hagis, Jr., P. & Cohen, G.L. Every odd perfect number has a prime factor which exceeds 10^6. *Math. Comp.* 67 (1998), 1323–1330.

1998 Pinch, R.G.E. The Carmichael numbers up to 10^{16}. Unpublished manuscript.

1998 Williams, H.C. *Édouard Lucas and Primality Testing.* J. Wiley and Sons, New York, 1998.

1998 Young, J. Large primes and Fermat factors. *Math. Comp.* 67 (1998), 1735–1738.

1999 Brent, R.P. Factorization of the tenth Fermat number. *Math. Comp.* 68 (1999), 429–451.

1999 Cook, R.J. Bounds for odd perfect numbers. In *Number Theory* (edited by R. Gupta). CRM Proc. Lecture Notes # 19, 67–71. Amer. Math. Soc., Providence, RI, 1999.

1999 Coutinho, S.C. *The Mathematics of Ciphers: Number Theory and RSA Cryptography.* A.K. Peters, Natick, MA, 1999.

1999 Ford, K. The number of solutions of $\varphi(x) = m$. *Annals of Math.* (2) 150 (1999), 283–311.

1999 Grytczuk, A. & Wojtowicz, M. There are no small odd perfect numbers, and Erratum. *C.R. Acad. Sci. Paris* 328 (1999), 1101–1105, and 330 (2000), p. 533.

1999 Iannucci, D.E. The second largest prime divisor of an odd perfect number exceeds ten thousand. *Math. Comp.* 68 (1999), 1749–1760.

1999 Woltman, G. On the discovery of the 38th known Mersenne prime. *Fibonacci Quart.* 37 (1999), 367–370.

2000 Brent, R.P., Crandall, R.E., Dilcher, K. & van Halewyn, C. Three new factors of Fermat numbers. *Math. Comp.* 69 (2000), 1297–1304.

2000 Caldwell, C. & Dubner, H. Primes in π. Preprint, 2000.

2000 Iannucci, D.E. The third largest prime divisor of an odd perfect number exceeds one hundred. *Math. Comp.* 69 (2000), 867–879.

2001 Crandall, R.E. & Pomerance, C. *Prime Numbers. A Computational Perspective.* Springer-Verlag, New York, 2001.

2001 Křížek, M., Luca, F. & Somer, L. *17 Lectures on Fermat Numbers. From Number Theory to Geometry.* Springer-Verlag, New York, 2001.

2001 Ribenboim, P. *Classical Theory of Algebraic Numbers.* Springer-Verlag, New York, 2001.

2001 Zhang, Z. Finding strong pseudoprimes to several bases. *Math. Comp.* 70 (2001), 863–872.

2002 Agrawal, M., Kayal, N. & Saxena, N. PRIMES is in P. Preprint, 2002.

2002 Morain, F. Primalité théorique et primalité pratique ou AKS vs. ECCP. Preprint, 2002.

2003 Bailey, D.H. Some background on Kanada's recent pi calculation. Preprint, 2003.

2003 Crandall, R.E., Mayer, E.W. & Papadopuolos, J.S. The twenty-fourth Fermat number is composite. *Math. Comp.* 72 (2003), 1555–1572.

2003 Wagstaff, Jr., S.S. *Cryptanalysis of Number Theoretic Ciphers.* Chapman & Hall/CRC, Boca Raton, FL, 2003.

Chapter 3

1912 Frobenius, F.G. Über quadratische Formen, die viele Primzahlen darstellen. *Sitzungsber. d. Königl. Akad. d. Wiss. zu Berlin*, 1912, 966–980. Reprinted in *Gesammelte Abhandlungen*, Vol. III, 573–587. Springer-Verlag, Berlin, 1968.

1912 Rabinowitsch, G. Eindeutigkeit der Zerlegung in Primzahlfaktoren in quadratischen Zahlkörpern. *Proc. Fifth Intern. Congress Math.*, Cambridge, Vol. 1, 1912, 418–421.

1933 Lehmer, D.H. On imaginary quadratic fields whose class number is unity. *Bull. Amer. Math. Soc.* 39 (1933), p. 360.

1934 Heilbronn, H. & Linfoot, E.H. On the imaginary quadratic corpora of class-number one. *Quart. J. Pure and Appl. Math.*, Oxford, (2) 5 (1934), 293–301.

1936 Lehmer, D.H. On the function $x^2 + x + A$. *Sphinx* 6 (1936), 212–214.

1938 Skolem, T. *Diophantische Gleichungen.* Springer-Verlag, Berlin, 1938.

1947 Mills, W.H. A prime-representing function. *Bull. Amer. Math. Soc.* 53 (1947), p. 604.

1951 Wright, E.M. A prime-representing function. *Amer. Math. Monthly* 58 (1951), 616–618.

1952 Heegner, K. Diophantische Analysis und Modulfunktionen. *Math. Zeits.* 56 (1952), 227–253.

1960 Putnam, H. An unsolvable problem in number theory. *J. Symb. Logic* 1960 (220–232).

1962 Cohn, H. *Advanced Number Theory.* J. Wiley and Sons, New York, 1962. Reprinted by Dover, New York, 1980.

1964 Willans, C.P. On formulae for the nth prime. *Math. Gazette* 48 (1964), 413–415.

1966 Baker, A. Linear forms in the logarithms of algebraic numbers. *Mathematika* 13 (1966), 204–216.

1967 Stark, H.M. A complete determination of the complex quadratic fields of class-number one. *Michigan Math. J.* 14 (1967), 1–27.

1968 Deuring, M. Imaginäre quadratische Zahlkörper mit der Klassenzahl Eins. *Invent. Math.* 5 (1968), 169–179.

1969 Dudley, U. History of a formula for primes. *Amer. Math. Monthly* 76 (1969), 23–28.

1969 Stark, H.M. A historical note on complex quadratic fields with class-number one. *Proc. Amer. Math. Soc.* 21 (1969), 254–255.

1971 Baker, A. Imaginary quadratic fields with class number 2. *Annals of Math.* (2) 94 (1971), 139–152.

1971 Baker, A. On the class number of imaginary quadratic fields. *Bull. Amer. Math. Soc.* 77 (1971), 678–684.

1971 Gandhi, J.M. Formulae for the nth prime. *Proc. Washington State Univ. Conf. on Number Theory*, 96–106. Pullman, WA, 1971.

1971 Matijasevič, Yu.V. Diophantine representation of the set of prime numbers (in Russian). *Dokl. Akad. Nauk SSSR* 196 (1971), 770–773. English translation by R.N. Goss, *Soviet Math. Dokl.* 12 (1971), 354–358.

1971 Stark, H.M. A transcendence theorem for class-number problems. *Annals of Math.* (2) 94 (1971), 153–173.

1972 Vanden Eynden, C. A proof of Gandhi's formula for the nth prime. *Amer. Math. Monthly* 79 (1972), p. 625.

1973 Davis, M. Hilbert's tenth problem is unsolvable. *Amer. Math. Monthly* 80 (1973), 233–269.

1973 Karst, E. New quadratic forms with high density of primes. *Elem. Math.* 28 (1973), 116–118.

1973 Weinberger, P.J. Exponents of the class groups of complex quadratic fields. *Acta Arith.* 22 (1973), 117–124.

1974 Golomb, S.W. A direct interpretation of Gandhi's formula. *Amer. Math. Monthly* 81 (1974), 752–754.

1974 Hendy, M.D. Prime quadratics associated with complex quadratic fields of class number two. *Proc. Amer. Math. Soc.* 43 (1974), 253–260.

1974 Szekeres, G. On the number of divisors of $x^2 + x + A$. *J. Number Theory* 6 (1974), 434–442.

1975 Ernvall, R. A formula for the least prime greater than a given integer. *Elem. Math.* 30 (1975), 13–14.

1975 Jones, J.P. Diophantine representation of the Fibonacci numbers. *Fibonacci Quart.* 13 (1975), 84–88.

1975 Matijasevič, Yu. & Robinson, J. Reduction of an arbitrary diophantine equation to one in 13 unknowns. *Acta Arith.* 27 (1975), 521–553.

1976 Jones, J.P., Sato, D., Wada, H. & Wiens, D. Diophantine representation of the set of prime numbers. *Amer. Math. Monthly* 83 (1976), 449–464.

1977 Goldfeld, D.M. The conjectures of Birch and Swinnerton-Dyer and the class numbers of quadratic fields. *Astérisque* 41/42 (1977), 219–227.

1977 Matijasevič, Yu.V. Primes are nonnegative values of a polynomial in 10 variables. *Zapiski Sem. Leningrad Mat. Inst. Steklov* 68 (1977), 62–82. English translation by L. Guy and J.P. Jones, *J. Soviet Math.* 15 (1981), 33–44.

1979 Jones, J.P. Diophantine representation of Mersenne and Fermat primes, *Acta Arith.* 35 (1979), 209–221.

1981 Ayoub, R.G. & Chowla, S. On Euler's polynomial. *J. Number Theory* 13 (1981), 443–445.

1983 Gross, B. & Zagier, D. Points de Heegner et dérivées de fonctions *L. C.R. Acad. Sci. Paris* 297 (1983), 85–87.

1986 Gross, B.H. & Zagier, D.B. Heegner points and derivatives of *L*-series. *Invent. Math.* 84 (1986), 225–320.

1986 Sasaki, R. On a lower bound for the class number of an imaginary quadratic field. *Proc. Japan Acad.* A (Math. Sci.) 62 (1986), 37–39.

1986 Sasaki, R. A characterization of certain real quadratic fields. *Proc. Japan Acad.* A (Math. Sci.) 62 (1986), 97–100.

1988 Ribenboim, P. Euler's famous prime generating polynomial and the class number of imaginary quadratic fields. *L'Enseign. Math.* 34 (1988), 23–42.

1989 Flath, D.E. *Introduction to Number Theory.* Wiley-Interscience, New York, 1989.

1989 Goetgheluck, P. On cubic polynomials giving many primes. *Elem. Math.* 44 (1989), 70–73.

1990 Louboutin, S. Prime producing quadratic polynomials and class-numbers of real quadratic fields, and Addendum. *Can. J. Math.* 42 (1990), 315–341 and p. 1131.

1991 Louboutin, S. Extensions du théorème de Frobenius-Rabinovitsch. *C.R. Acad. Sci. Paris* 312 (1991), 711–714.

1995 Boston, N. & Greenwood, M.L. Quadratics representing primes. *Amer. Math. Monthly* 102 (1995), 595–599.

1995 Lukes, R.F., Patterson, C.D. & Williams, H.C. Numerical sieving devices: Their history and applications. *Nieuw Arch. Wisk.* (4) 13 (1995), 113–139.

1996 Mollin, R.A. *Quadratics.* CRC Press, Boca Raton, FL, 1996.

1996 Mollin, R.A. An elementary proof of the Rabinowitsch-Mollin-Williams criterion for real quadratic fields. *J. Math. Sci.* 7 (1996), 17–27.

1997 Mollin, R.A. Prime-producing quadratics. *Amer. Math. Monthly* 104 (1997), 529–544.

2000 Ribenboim, P. *My Numbers, My Friends.* Springer-Verlag, New York, 2000.

2003 Dress, F. & Landreau, B. Polynômes prenant beaucoup de valeurs premières. Preprint, 2003.

2003 Jacobson, Jr., M.J. & Williams, H.C. New quadratic polynomials with high densities of prime values. *Math. Comp.* 72 (2003), 499–519.

Chapter 4

1849 de Polignac, A. Recherches nouvelles sur les nombres premiers. *C.R. Acad. Sci. Paris* 29 (1849), 397–401; Rectification: 738–739.

1885 Meissel, E.D.F. Berechnung der Menge von Primzahlen, welche innerhalb der ersten Milliarde natürlicher Zahlen vorkommen. *Math. Ann.* 25 (1885), 251–257.

1892 Sylvester, J.J. On arithmetical series. *Messenger of Math.* 21 (1892), 1–19 and 87–120. Reprinted in *Gesammelte Abhandlungen*, Vol. III, 573–587. Springer-Verlag, New York, 1968.

1901 von Koch, H. Sur la distribution des nombres premiers. *Acta Math.* 24 (1901), 159–182.

1901 Torelli, G. Sulla totalità dei numeri primi fino ad un limite assegnato. *Atti Reale Accad. Sci. Fis. Mat. Napoli* (2) 11 (1901), 1–222.

1901 Wolfskehl, P. Über eine Aufgabe der elementaren Arithmetik. *Math. Ann.* 54 (1901), 503–504.

1903 Gram, J.-P. Note sur les zéros de la fonction $\zeta(s)$ de Riemann. *Acta Math.* 27 (1903), 289–304.

1909 Landau, E. *Handbuch der Lehre von der Verteilung der Primzahlen.* Teubner, Leipzig, 1909. Reprinted by Chelsea, Bronx, NY, 1974.

1909 Lehmer, D.N. *Factor Table for the First Ten Millions.* Carnegie Inst., Publication #105, Washington, 1909. Reprinted by Hafner, New York, 1956.

1914 Lehmer, D.N. *List of Prime Numbers from 1 to 10,006,721.* Carnegie Inst., Publication #165, Washington, 1914. Reprinted by Hafner, New York, 1956.

1914 Littlewood, J.E. Sur la distribution des nombres premiers. *C.R. Acad. Sci. Paris* 158 (1914), 1869–1872.

1919 Brun, V. Le crible d'Eratosthène et le théorème de Goldbach. *C.R. Acad. Sci. Paris* 168 (1919), 544–546.

1919 Brun, V. La série $\frac{1}{5} + \frac{1}{7} + \frac{1}{11} + \frac{1}{13} + \frac{1}{17} + \frac{1}{19} + \frac{1}{29} + \frac{1}{31} + \frac{1}{41} + \frac{1}{43} + \frac{1}{59} + \frac{1}{61} + \cdots$ où les dénominateurs sont "nombres premiers jumeaux" est convergente ou finie. *Bull. Sci. Math.* (2) 43 (1919), 100–104 and 124–128.

1919 Ramanujan, S. A proof of Bertrand's postulate. *J. Indian Math. Soc.* 11 (1919), 181–182. Reprinted in *Collected Papers of Srinivasa Ramanujan* (edited by G.H. Hardy and P.V. Seshu Aiyar), 208–209. Cambridge Univ. Press, Cambridge, 1927. Reprinted by Chelsea, Bronx, NY, 1962.

1920 Brun, V. Le crible d'Eratosthène et le théorème de Goldbach. *Videnskapsselskapets Skrifter Kristiania, Mat.-nat. Kl.*, 1920, No. 3, 36 pages.

1923 Hardy, G.H. & Littlewood, J.E. Some problems of "Partitio Numerorum", III: On the expression of a number as a sum of primes. *Acta Math.* 44 (1923), 1–70. Reprinted in *Collected Papers of G.H. Hardy*, Vol. I, 561–630. Clarendon Press, Oxford, 1966.

1930 Hoheisel, G. Primzahlprobleme in der Analysis. *Sitzungsberichte Berliner Akad. d. Wiss.*, 1930, 580–588.

1930 Schnirelmann, L. Über additive Eigenschaften von Zahlen. *Ann. Inst. Polytechn. Novočerkask*, 14, 1930, 3–28 and *Math. Ann.* 107 (1933), 646–690.

1931 Westzynthius, E. Über die Verteilung der Zahlen, die zu den n ersten Primzahlen teilerfremd sind. *Comm. Phys. Math. Helsingfors* (5) 25 (1931), 1–37.

1932 Erdös, P. Beweis eines Satzes von Tschebyscheff. *Acta Sci. Math. Szeged* 5 (1932), 194–198.

1933 Skewes, S. On the difference $\pi(x) - \mathrm{li}(x)$. *J. London Math. Soc.* 8 (1933), 277–283.

1934 Ishikawa, H. Über die Verteilung der Primzahlen. *Sci. Rep. Tokyo Bunrika Daigaku* (A) 2 (1934), 27–40.

1934 Romanoff, N.P. Über einige Sätze der additiven Zahlentheorie. *Math. Ann.* 109 (1934), 668–678.

1935 Erdös, P. On the difference of consecutive primes. *Quart. J. Pure and Appl. Math.*, Oxford, (2) 6 (1935), 124–128.

1936 Tschudakoff, N.G. On the zeros of Dirichlet's L-functions (in Russian). *Mat. Sbornik* 1 (1936), 591–602.

1937 Cramér, H. On the order of magnitude of the difference between consecutive prime numbers. *Acta Arith.* 2 (1937), 23–46.

1937 Ingham, A.E. On the difference between consecutive primes. *Quart. J. Pure and Appl. Math.*, Oxford, (2) 8 (1937), 255–266.

1937 Landau, E. *Über einige neuere Fortschritte der additiven Zahlentheorie.* Cambridge Univ. Press, Cambridge, 1937. Reprinted by Stechert-Hafner, New York, 1964.

1937 van der Corput, J.G. Sur l'hypothèse de Goldbach pour presque tous les nombres pairs. *Acta Arith.* 2 (1937), 266–290.

1937 Vinogradov, I.M. Representation of an odd number as the sum of three primes (in Russian). *Dokl. Akad. Nauk SSSR* 15 (1937), 169–172.

1938 Estermann, T. Proof that almost all even positive integers are sums of two primes. *Proc. London Math. Soc.* 44 (1938), 307–314.

1938 Poulet, P. Table des nombres composés vérifiant le théorème de Fermat pour le module 2, jusqu' à 100.000.000. *Sphinx* 8 (1938), 42–52. Corrections: *Math. Comp.* 25 (1971), 944–945, and 26 (1972), p. 814.

1938 Rankin, R.A. The difference between consecutive prime numbers. *J. London Math. Soc.* 13 (1938), 242–247.

1938 Rosser, J.B. The nth prime is greater than $n \log n$. *Proc. London Math. Soc.* 45 (1938), 21–44.

1938 Tschudakoff, N.G. On the density of the set of even integers which are not representable as a sum of two odd primes (in Russian). *Izv. Akad. Nauk SSSR*, Ser. Mat., 1 (1938), 25–40.

1939 van der Corput, J.G. Über Summen von Primzahlen und Primzahlquadraten. *Math. Ann.* 116 (1939), 1–50.

1940 Erdös, P. The difference of consecutive primes. *Duke Math. J.* 6 (1940), 438–441.

1944 Chowla, S. There exists an infinity of 3-combinations of primes in A. P. *Proc. Lahore Phil. Soc.* 6 (1944), 15–16.

1944 Linnik, Yu.V. On the least prime in an arithmetic progression I. The basic theorem (in Russian). *Mat. Sbornik* 15 (57), (1944), 139–178.

1946 Brauer, A. On the exact number of primes below a given limit. *Amer. Math. Monthly* 9 (1946), 521–523.

1947 Khinchin, A.Ya. *Three Pearls of Number Theory.* Original Russian edition in OGIZ, Moscow, 1947. Translation into English published by Graylock Press, Baltimore, 1952.

1947 Rényi, A. On the representation of even numbers as the sum of a prime and an almost prime. *Dokl. Akad. Nauk SSSR* 56 (1947), 455–458.

1949 Clement, P.A. Congruences for sets of primes. *Amer. Math. Monthly* 56 (1949), 23–25.

1949 Erdös, P. On a new method in elementary number theory which leads to an elementary proof of the prime number theorem. *Proc. Nat. Acad. Sci. USA* 35 (1949), 374–384.

1949 Erdös, P. On the converse of Fermat's theorem. *Amer. Math. Monthly* 56 (1949), 623–624.

1949 Moser, L. A theorem on the distribution of primes. *Amer. Math. Monthly* 56 (1949), 624–625.

1949 Richert, H.E. Über Zerfällungen in ungleiche Primzahlen. *Math. Zeits.* 52 (1949), 342–343.

1949 Selberg, A. An elementary proof of Dirichlet's theorem about primes in an arithmetic progression. *Annals of Math.* 50 (1949), 297–304.

1949 Selberg, A. An elementary proof of the prime number theorem. *Annals of Math.* 50 (1949), 305–313.

1950 Erdös, P. On almost primes. *Amer. Math. Monthly* 57 (1950), 404–407.

1950 Erdös, P. On integers of the form $2^k + p$ and some related problems, *Summa Bras. Math.* 2 (1950), 113–123.

1950 Hasse, H. *Vorlesungen über Zahlentheorie.* Springer-Verlag, Berlin, 1950.

1950 Selberg, A. An elementary proof of the prime number theorem for arithmetic progressions. *Can. J. Math.* 2 (1950), 66–78.

1951 Titchmarsh, E.C. *The Theory of the Riemann Zeta Function.* Clarendon Press, Oxford, 1951.

1956 Borodzkin, K.G. On the problem of I.M. Vinogradov's constant (in Russian). *Proc. Third Math. Congress*, Moscow, 1 (1956), p. 3.

1956 Erdös, P. On pseudo-primes and Carmichael numbers. *Publ. Math. Debrecen* 4 (1956), 201–206.

1957 Leech, J. Note on the distribution of prime numbers. *J. London Math. Soc.* 32 (1957), 56–58.

1957 Pan, C.D. On the least prime in an arithmetic progression. *Sci. Record* (N.S.) 1 (1957), 311–313.

1958 Pan, C.D. On the least prime in an arithmetic progression. *Acta Sci. Natur. Univ. Pekinensis* 4 (1958), 1–34.

1958 Schinzel, A. & Sierpiński, W. Sur certaines hypothèses concernant les nombres premiers. *Acta Arith.* 4 (1958), 185–208; Erratum, 5 (1959), p. 259.

1959 Killgrove, R.B. & Ralston, K.E. On a conjecture concerning the primes. *Math. Comp.* 13 (1959), 121–122.

1959 Lehmer, D.H. On the exact number of primes less than a given limit. *Illinois J. Math.* 3 (1959), 381–388. Reprinted in *Selected Papers* (edited by D. McCarthy), Vol. III, 1104–1111. Ch. Babbage Res. Centre, St. Pierre, Manitoba, Canada, 1981.

1959 Schinzel, A. Démonstration d'une conséquence de l'hypothèse de Goldbach. *Compositio Math.* 14 (1959), 74–76.

1960 Golomb, S.W. The twin prime constant. *Amer. Math. Monthly* 67 (1960), 767–769.

1961 Prachar, K. Über die kleinste Primzahl einer arithmetischen Reihe. *J. Reine Angew. Math.* 206 (1961), 3–4.

1961 Schinzel, A. Remarks on the paper "Sur certaines hypothèses concernant les nombres premiers". *Acta Arith.* 7 (1961), 1–8.

1961 Wrench, Jr., J.W. Evaluation of Artin's constant and the twin-prime constant. *Math. Comp.* 15 (1961), 396–398.

1962 Rosser, J.B. & Schoenfeld, L. Approximate formulas for some functions of prime numbers. *Illinois J. Math.* 6 (1962), 64–94.

1962 Schinzel, A. Remark on a paper of K. Prachar "Über die kleinste Primzahl einer arithmetischen Reihe". *J. Reine Angew. Math.* 210 (1962), 121–122.

1963 Ayoub, R.G. *An Introduction to the Theory of Numbers.* Amer. Math. Soc., Providence, RI, 1963.

1963 Kanold, H.-J. Elementare Betrachtungen zur Primzahltheorie. *Arch. Math.* 14 (1963), 147–151.

1963 Rankin, R.A. The difference between consecutive prime numbers, V. *Proc. Edinburgh Math. Soc.* (2) 13 (1963), 331–332.

1963 Rotkiewicz, A. Sur les nombres pseudo-premiers de la forme $ax + b$. *C.R. Acad. Sci. Paris* 257 (1963), 2601–2604.

1963 Walfisz, A.Z. *Weylsche Exponentialsummen in der neueren Zahlentheorie.* VEB Deutscher Verlag d. Wiss., Berlin, 1963.

1964 Grosswald, E. A proof of the prime number theorem. *Amer. Math. Monthly* 71 (1964), 736–743.

1964 Shanks, D. On maximal gaps between successive primes. *Math. Comp.* 18 (1964), 646–651.

1965 Chen, J.R. On the least prime in an arithmetical progression. *Sci. Sinica* 14 (1965), 1868–1871.

1965 Gelfond, A.O. & Linnik, Yu.V. *Elementary Methods in Analytic Number Theory.* Translated by A. Feinstein, revised and edited by L.J. Mordell. Rand McNally, Chicago, 1965.

1965 Rotkiewicz, A. Les intervalles contenant les nombres pseudo-premiers. *Rend. Circ. Mat. Palermo* (2) 14 (1965), 278–280.

1965 Stein, M.L. & Stein, P.R. New experimental results on the Goldbach conjecture. *Math. Mag.* 38 (1965), 72–80.

1965 Stein, M.L. & Stein, P.R. Experimental results on additive 2-bases. *Math. Comp.* 19 (1965), 427–434.

1966 Bombieri, E. & Davenport, H. Small differences between prime numbers. *Proc. Roy. Soc.* (A) 293 (1966), 1–18.

1967 Lander, L.J. & Parkin, T.R. On first appearance of prime differences. *Math. Comp.* 21 (1967), 483–488.

1967 Lander, L.J. & Parkin, T.R. Consecutive primes in arithmetic progression. *Math. Comp.* 21 (1967), p. 489.

1967 Rotkiewicz, A. On the pseudo-primes of the form $ax + b$. *Proc. Cambridge Phil. Soc.* 63 (1967), 389–392.

1967 Szymiczek, K. On pseudoprimes which are products of distinct primes. *Amer. Math. Monthly* 74 (1967), 35–37.

1969 Montgomery, H.L. Zeros of L-functions. *Invent. Math.* 8 (1969), 346–354.

1969 Rosser, J.B., Yohe, J.M. & Schoenfeld, L. Rigorous computation of the zeros of the Riemann zeta-function (with discussion). *Inform. Processing* 68 (Proc. IFIP Congress, Edinburgh, 1968), Vol. I, 70–76. North-Holland, Amsterdam, 1969.

1970 Diamond, H.G. & Steinig, J. An elementary proof of the prime number theorem with a remainder term. *Invent. Math.* 11 (1970), 199–258.

1972 Huxley, M.N. On the difference between consecutive primes. *Invent. Math.* 15 (1972), 164–170.

1972 Huxley, M.N. *The Distribution of Prime Numbers.* Oxford Univ. Press, Oxford, 1972.

1972 Rotkiewicz, A. On a problem of W. Sierpiński. *Elem. Math.* 27 (1972), 83–85.

1973 Brent, R.P. The first occurrence of large gaps between successive primes. *Math. Comp.* 27 (1973), 959–963.

1973/1978 Chen, J.R. On the representation of a large even integer as the sum of a prime and the product of at most two primes, I and II. *Sci. Sinica* 16 (1973), 157–176, and 21 (1978), 421–430.

1973 Montgomery, H.L. The pair correlation of zeros of the zeta function. *Analytic Number Theory* (Proc. Symp. Pure Math., Vol. XXIV, St. Louis, 1972), 181–193. Amer. Math. Soc., Providence, RI, 1973.

1973 Montgomery, H.L. & Vaughan, R.C. The large sieve. *Mathematika* 20 (1973), 119–134.

1974 Ayoub, R.G. Euler and the zeta-function. *Amer. Math. Monthly* 81 (1974), 1067–1086.

1974 Edwards, H.M. *Riemann's Zeta Function.* Academic Press, New York, 1974.

1974 Halberstam, H. & Richert, H.E. *Sieve Methods.* Academic Press, New York, 1974.

1974 Hensley, D. & Richards, I. Primes in intervals. *Acta Arith.* 25 (1974), 375–391.

1974 Levinson, N. More than one third of zeros of Riemann's zeta function are on $\sigma = 1/2$. *Adv. in Math.* 13 (1984), 383–436.

1974 Mąkowski, A. On a problem of Rotkiewicz on pseudoprimes. *Elem. Math.* 29 (1974), p. 13.

1974 Richards, I. On the incompatibility of two conjectures concerning primes; a discussion of the use of computers in attacking a theoretical problem. *Bull. Amer. Math. Soc.* 80 (1974), 419–438.

1974 Shanks, D. & Wrench, J.W. Brun's constant. *Math. Comp.* 28 (1974), 293–299.

1975 Brent, R.P. Irregularities in the distribution of primes and twin primes. *Math. Comp.* 29 (1975), 43–56.

1975 Montgomery, H.L. & Vaughan, R.C. The exceptional set in Goldbach's problem. *Acta Arith.* 27 (1975), 353–370.

1975 Ross, P.M. On Chen's theorem that each large even number has the form $p_1 + p_2$ or $p_1 + p_2 p_3$. *J. London Math. Soc.* (2) 10 (1975), 500–506.

1975 Rosser, J.B. & Schoenfeld, L. Sharper bounds for the Chebyshev functions $\theta(x)$ and $\psi(x)$. *Math. Comp.* 29 (1975), 243–269.

1975 Swift, J.D. Table of Carmichael numbers to 10^9. *Math. Comp.* 29 (1975), 338–339.

1975 Udrescu, V.S. Some remarks concerning the conjecture $\pi(x + y) \leq \pi(x) + \pi(y)$. *Rev. Roumaine Math. Pures Appl.* 20 (1975), 1201–1209.

1976 Apostol, T.M. *Introduction to Analytic Number Theory.* Springer-Verlag, New York, 1976.

1976 Brent, R.P. Tables concerning irregularities in the distribution of primes and twin primes to 10^{11}. *Math. Comp.* 30 (1976), p. 379.

1977 Hudson, R.H. A formula for the exact number of primes below a given bound in any arithmetic progression. *Bull. Austral. Math. Soc.* 16 (1977), 67–73.

1977 Hudson, R.H. & Brauer, A. On the exact number of primes in the arithmetic progressions $4n \pm 1$ and $6n \pm 1$. *J. Reine Angew. Math.* 291 (1977), 23–29.

1977 Huxley, M.N. Small differences between consecutive primes, II. *Mathematika* 24 (1977), 142–152.

1977 Jutila, M. On Linnik's constant. *Math. Scand.* 41 (1977), 45–62.

1977 Langevin, M. Méthodes élémentaires en vue du théorème de Sylvester. *Sém. Delange-Pisot-Poitou*, 17e année, 1975/76, fasc. 1, exp. No. G12, 9 pages, Paris, 1977.

1977 Weintraub, S. Seventeen primes in arithmetic progression. *Math. Comp.* 31 (1977), p. 1030.

1978 Bays, C. & Hudson, R.H. On the fluctuations of Littlewood for primes of the form $4n \pm 1$. *Math. Comp.* 32 (1978), 281–286.

1978 Heath-Brown, D.R. Almost-primes in arithmetic progressions and short intervals. *Math. Proc. Cambridge Phil. Soc.* 83 (1978), 357–375.

1979 Heath-Brown, D.R. & Iwaniec, H. On the difference between consecutive powers. *Bull. Amer. Math. Soc.* (N.S.) 1 (1979), 758–760.

1979 Iwaniec, H. & Jutila, M. Primes in short intervals. *Arkiv för Mat.* 17 (1979), 167–176.

1979 Pomerance, C. The prime number graph. *Math. Comp.* 33 (1979), 399–408.

1979 Ribenboim, P. *13 Lectures on Fermat's Last Theorem.* Springer-Verlag, New York, 1979.

1979 Wagstaff, Jr., S.S. Greatest of the least primes in arithmetic progressions having a given modulus. *Math. Comp.* 33 (1979), 1073–1080.

1979 Yorinaga, M. Numerical computation of Carmichael numbers, II. *Math. J. Okayama Univ.* 21 (1979), 183–205.

1980 Brent, R.P. The first occurrence of certain large prime gaps. *Math. Comp.* 35 (1980), 1435–1436.

1980 Chen, J.R. & Pan, C.D. The exceptional set of Goldbach numbers, I. *Sci. Sinica* 23 (1980), 416–430.

1980 Light, W.A., Forrest, J., Hammond, N. & Roe, S. A note on Goldbach's conjecture. *BIT* 20 (1980), p. 525.

1980 Newman, D.J. Simple analytic proof of the prime number theorem. *Amer. Math. Monthly* 87 (1980), 693–696.

1980 Pintz, J. On Legendre's prime number formula. *Amer. Math. Monthly* 87 (1980), 733–735.

1980 Pomerance, A. A note on the least prime in an arithmetic progression. *J. Number Theory* 12 (1980), 218–223.

1980 Pomerance, C., Selfridge, J.L. & Wagstaff, Jr., S.S. The pseudoprimes to $25 \cdot 10^9$. *Math. Comp.* 35 (1980), 1003–1026.

1980 van der Poorten, A.J. & Rotkiewicz, A. On strong pseudoprimes in arithmetic progressions. *J. Austral. Math. Soc.* (A) 29 (1980), 316–321.

1981 Graham, S. On Linnik's constant. *Acta Arith.* 39 (1981), 163–179.

1981 Heath-Brown, D.R. Three primes and an almost prime in arithmetic progression. *J. London Math. Soc.* (2) 23 (1981), 396–414.

1981 Pomerance, C. On the distribution of pseudoprimes. *Math. Comp.* 37 (1981), 587–593.

1982 Diamond, H.G. Elementary methods in the study of the distribution of prime numbers. *Bull. Amer. Math. Soc.* (N.S.) 7 (1982), 553–589.

1982 Pomerance, C. A new lower bound for the pseudoprimes counting function. *Illinois J. Math.* 26 (1982), 4–9.

1982 Pritchard, P.A. 18 primes in arithmetic progression. *J. Recr. Math.* 15 (1982/3), p. 288.

1982 Weintraub, S. A prime gap of 682 and a prime arithmetic sequence. *BIT* 22 (1982), p. 538.

1983 Chen, J.R. The exceptional value of Goldbach numbers, II. *Sci. Sinica* (A) 26 (1983), 714–731.

1983 Powell, B. Problem 6429 (Difference between consecutive primes). *Amer. Math. Monthly* 90 (1983), p. 338.

1983 Riesel, H. & Vaughan, R.C. On sums of primes. *Arkiv för Mat.* 21 (1983), 45–74.

1983 Robin, G. Estimation de la fonction de Tschebychef θ sur le $k^{\text{ième}}$ nombre premier et grandes valeurs de la fonction $\omega(n)$, nombre de diviseurs premiers de n. *Acta Arith.* 42 (1983), 367–389.

1984 Daboussi, H. Sur le théorème des nombres premiers. *C.R. Acad. Sci. Paris* 298 (1984), 161–164.

1984 Davies, R.O. Solution of problem 6429. *Amer. Math. Monthly* 91 (1984), p. 64.

1984 Iwaniec, H. & Pintz, J. Primes in short intervals. *Monatsh. Math.* 98 (1984), 115–143.

1984 Schroeder, M.R. *Number Theory in Science and Communication.* Springer-Verlag, New York, 1984.

1984 Wang, Y. *Goldbach Conjecture.* World Scientific Publ., Singapore, 1984.

1985 Heath-Brown, D.R. The ternary Goldbach problem. *Rev. Mat. Iberoamer.* 1 (1985), 45–58.

1985 Ivić, A. *The Riemann Zeta-Function.* J. Wiley and Sons, New York, 1985.

1985 Lagarias, J.C., Miller, V.S. & Odlyzko, A.M. Computing $\pi(x)$: The Meissel-Lehmer method. *Math. Comp.* 44 (1985), 537–560.

1985 Lou, S. & Yao, Q. The upper bound of the difference between consecutive primes. *Kexue Tongbao* 8 (1985), 128–129.

1985 Maier, H. Primes in short intervals. *Michigan Math. J.* 32 (1985), 221–225.

1985 Odlyzko, A.M. & te Riele, H.J.J. Disproof of the Mertens conjecture. *J. Reine Angew. Math.* 357 (1985), 138–160.

1985 Powell, B. Problem 1207 (A generalized weakened Goldbach theorem). *Math. Mag.* 58 (1985), p. 46.

1985 Pritchard, P.A. Long arithmetic progressions of primes: some old, some new. *Math. Comp.* 45 (1985), 263–267.

1985 te Riele, H.J.J. Some historical and other notes about the Mertens conjecture and its recent disproof. *Nieuw Arch. Wisk.* (4) 3 (1985), 237–243.

1986 Bombieri, E., Friedlander, J.B. & Iwaniec, H. Primes in arithmetic progression to large moduli, I. *Acta Math.* 156 (1986), 203–251.

1986 Finn, M.V. & Frohliger, J.A. Solution of problem 1207. *Math. Mag.* 59 (1986), 48–49.

1986 Mozzochi, C.J. On the difference between consecutive primes. *J. Number Theory* 24 (1986), 181–187.

1986 van de Lune, J., te Riele, H.J.J. & Winter, D.T. On the zeros of the Riemann zeta function in the critical strip, IV. *Math. Comp.* 46 (1986), 667–681.

1986 Wagon, S. Where are the zeros of zeta of s? *Math. Intelligencer* 8, No. 4 (1986), 57–62.

1987 Pintz, J. An effective disproof of the Mertens conjecture. *Astérisque* 147/148 (1987), 325–333.

1987 te Riele, H.J.J. On the sign of the difference $\pi(x) - \mathrm{li}(x)$. *Math. Comp.* 48 (1987), 323–328.

1988 Erdős, P., Kiss, P. & Sárközy, A. A lower bound for the counting function of Lucas pseudoprimes. *Math. Comp.* 51 (1988), 315–323.

1988 Odlyzko, A.M. & Schönhage, A. Fast algorithms for multiple evaluations of the Riemann zeta function. *Trans. Amer. Math. Soc.* 309 (1988), 797–809.

1988 Patterson, S.J. *Introduction to the Theory of the Riemann Zeta-Function.* Cambridge Univ. Press, Cambridge, 1988.

1989 Chen, J.R. & Liu, J.M. On the least prime in an arithmetical progression, III and IV. *Sci. China* (A) 32 (1989), 654–673 and 792–807.

1989 Chen, J.R. & Wang, T.Z. On the odd Goldbach problem. *Acta Math. Sinica* 32 (1989), 702–718.

1989 Conrey, J.B. At least two fifths of the zeros of the Riemann zeta function are on the critical line. *Bull. Amer. Math. Soc.* 20 (1989), 79–81.

1989 Granville, A., van de Lune, J. & te Riele, H.J.J. Checking the Goldbach conjecture on a vector computer. In *Number Theory and Applications* (edited by R.A. Mollin). Kluwer, Dordrecht, 1989, 423–432.

1989 Young, J. & Potler, A. First occurrence of prime gaps. *Math. Comp.* 52 (1989), 221–224.

1990 Granville, A. & Pomerance, C. On the least prime in certain arithmetic progressions. *J. London Math. Soc.* (2) 41 (1990), 193–200.

1990 Jaeschke, G. The Carmichael numbers to 10^{12}. *Math. Comp.* 55 (1990), 383–389.

1990 Parady, B.K., Smith, J.F. & Zarantonello, S. Largest known twin primes. *Math. Comp.* 55 (1990), 381–382.

1991 Weintraub, S. A prime gap of 784. *J. Recr. Math.* 23, No. 1 (1991), 6–7.

1992 Golomb, S. Problem 10208. *Amer. Math. Monthly* 99 (1992), p. 266.

1992 Granville, A. Primality testing and Carmichael numbers. *Notices Amer. Math. Soc.* 39 (1992), 696–700.

1992 Heath-Brown, D.R. Zero-free regions for Dirichlet L-functions and the least prime in an arithmetic progression. *Proc. London Math. Soc.* (3) 64 (1992), 265–338.

1992 Pinch, R.G.E. The pseudoprimes up to 10^{12}. Unpublished manuscript.

1993 Deshouillers, J.-M., Granville, A., Narkiewicz, W. & Pomerance, C. An upper bound in Goldbach's problem. *Math. Comp.* 61 (1993), 209–213.

1993 Lou, S. & Yao, Q. The number of primes in a short interval. *Hardy-Ramanujan J.* 16 (1993), 21–43.

1993 Odlyzko, A. Iterated absolute values of differences of consecutive primes. *Math. Comp.* 61 (1993), 373–380.

1993 Pinch, R.G.E. The Carmichael numbers up to 10^{15}. *Math. Comp.* 61 (1993), 381–391.

1993 Pomerance, C. Carmichael numbers. *Nieuw Arch. Wisk.* (4) 11 (1993), 199–209.

1993 Sinisalo, M.K. Checking the Goldbach conjecture up to 4×10^{11}. *Math. Comp.* 61 (1993), 931–934.

1994 Alford, W.R., Granville, A. & Pomerance, C. There are infinitely many Carmichael numbers. *Annals of Math.* 140 (1994), 703–722.

1994 Lou, S. & Yao, Q. Estimates of sums of Dirichlet series. *Hardy-Ramanujan J.* 17 (1994), 1–31.

1995 Ford, K. Solution of problem 10208. *Amer. Math. Monthly* 102 (1995), 361–362.

1995 Nicely, T.R. Enumeration to 10^{14} of the twin primes and Brun's constant. *Virginia J. of Sci.* 46 (1995), 195–204.

1995 Pritchard, P.A., Moran, A. & Thyssen, A. Twenty-two primes in arithmetic progression. *Math. Comp.* 64 (1995), 1337–1339.

1995 Ramaré, O. On Snirel'man's constant. *An. Scuola Norm. Sup. Pisa, Cl. Sci.* (4) 22 (1995), 645–706.

1996 Chen, J.R. & Wang, T.Z. The Goldbach problem for odd numbers. *Acta Math. Sinica* 39 (1996), 169–174.

1996 Connes, A. Formule de trace en géométrie non-commutative et hypothèse de Riemann. *C.R. Acad. Sci. Paris* 323 (1996), 1231–1236.

1996 Deléglise, M. & Rivat, J. Computing $\pi(x)$: The Meissel, Lehmer, Lagarias, Miller, Odlyzko method. *Math. Comp.* 65 (1996), 235–245.

1996 Indlekofer, H.-J. & Járai, A. Largest known twin primes. *Math. Comp.* 65 (1996), 427–428.

1996 Massias, J.-P. & Robin, G. Bornes effectives pour certaines fonctions concernant les nombres premiers. *J. Théor. Nombres Bordeaux* 8 (1996), 215–242.

1996 Ramaré, O. & Rumely, R. Primes in arithmetic progressions. *Math. Comp.* 65 (1996), 397–425.

1996 Shiu, D. *Prime Numbers in Arithmetical Progressions.* Ph.D. Thesis, Univ. of Oxford, 1996.

1997 Deshouillers, J.-M., Effinger, G., te Riele, H. & Zinoviev, D. A complete Vinogradov 3-primes theorem under the Riemann Hypothesis. *Electron. Res. Announc. Amer. Math. Soc.* 3 (1997), 99–104.

1997 Dubner, H. & Nelson, H. Seven consecutive primes in arithmetic progression. *Math. Comp.* 66 (1997), 1743–1749.

1997 Forbes, T. A large pair of twin primes. *Math. Comp.* 66 (1997), 451–455.

1997 Mollin, R.A. Prime-producing quadratics. *Amer. Math. Monthly* 104 (1997), 529–544.

1998 Deléglise, M. & Rivat, J. Computing $\psi(x)$. *Math. Comp.* 67 (1998), 1691–1696.

1998 Deshoulliers, J.-M., te Riele, H.J.J. & Saouter, Y. New experimental results concerning the Goldbach conjecture. In *Proc. Third Int. Symp. on Algorithmic Number Th.* (edited by J.P. Buhler). Lecture Notes in Computer Sci. # 1423, 204–215. Springer-Verlag, New York, 1998.

1998 Dusart, P. Autour de la fonction qui compte le nombre de nombres premiers. Ph.D. Thesis, Université de Limoges, 1998, 171 pages.

1998 Pinch, R.G.E. The Carmichael numbers up to 10^{16}. Unpublished manuscript.

1998 Saouter, Y. Checking the odd Goldbach conjecture up to 10^{20}. *Math. Comp.* 67 (1998), 863–866.

1999 Dusart, P. The kth prime is greater than $k(\ln k + \ln \ln k - 1)$ for $k \geq 2$. *Math. Comp.* 68 (1999), 411–415.

1999 Dusart, P. Inégalités explicites pour $\psi(x)$, $\theta(x)$, $\pi(x)$ et les nombres premiers. *C.R. Math. Rep. Acad. Sci. Canada* 21 (1999), 53–59.

1999 Forbes, T. Prime clusters and Cunningham chains. *Math. Comp.* 68 (1999), 1739–1747.

1999 Indlekofer, H.-J. & Járai, A. Largest known twin primes and Sophie Germain primes. *Math. Comp.* 68 (1999), 1317–1324.

1999 Nicely, T.R. New maximal prime gaps and first occurrences. *Math. Comp.* 68 (1999), 1311–1315.

1999 Nicely, T.R. Enumeration to 1.6×10^{15} of the prime quadruplets. Preprint, 1999.

1999 Nicely, T.R. & Nyman, B. First occurrence of a prime gap of 1000 or greater. Preprint, 1999.

2000 Bays, C. & Hudson, R.H. A new bound for the smallest x with $\pi(x) > \text{li}(x)$. *Math. Comp.* 69 (2000), 1285–1296.

2000 Pinch, R.G.E. The pseudoprimes up to 10^{13}. In *Proc. Fourth Int. Symp. on Algorithmic Number Th.* (edited by W. Bosma). Lecture Notes in Computer Sci. # 1838, 459–474. Springer-Verlag, New York, 2000.

2000 Shiu, D.K.L. Strings of congruent primes. *J. London Math. Soc.* (2) 61 (2000), 359–373.

2001 Baker, R.C., Harman, G. & Pintz, J. The difference between consecutive primes, II. *Proc. London Math. Soc.* (3) (83), 2001, 532–562.

2001 Kadiri, H. Une région explicite sans zéros pour la fonction zeta de Riemann. Preprint, 2001.

2001 Nicely, T.R. A new error analysis for Brun's constant. *Virginia J. of Sci.* 52 (2001), 45–55.

2001 Odlyzko, A.M. The 10^{22}-nd zero of the Riemann zeta function. In *Dynamical, Spectral, and Arithmetic Zeta Functions* (edited by M.L. Lapidus and M. van Frankenhuysen), 139–144. Contemporary Math., Vol. 290, Amer. Math. Soc., Providence, RI, 2001

2001 Richstein, J. Verifying the Goldbach conjecture up to $4 \cdot 10^{14}$. *Math. Comp.* 70 (2001), 1745–1749.

2002 Dubner, H., Forbes, T., Lygeros, N., Mizony, M., Nelson, H. & Zimmermann, P. Ten consecutive primes in arithmetic progression. *Math. Comp.* 71 (2002), 1323–1328.

2002 Dusart, P. Sur la conjecture $\pi(x + y) \leq \pi(x) + \pi(y)$. *Acta Arith.*, 102 (2002), 295–308.

2003 Ramaré, O. & Saouter, Y. Short effective intervals containing primes. *J. Number Theory* 98 (2003), 10–33.

Chapter 5

1948 Gunderson, N.G. *Derivation of Criteria for the First Case of Fermat's Last Theorem and the Combination of these Criteria to Produce a New Lower Bound for the Exponent.* Ph.D. Thesis, Cornell University, 1948, 111 pages.

1951 Dénes, P. An extension of Legendre's criterion in connection with the first case of Fermat's last theorem. *Publ. Math. Debrecen* 2 (1951), 115–120.

1953 Goldberg, K. A table of Wilson quotients and the third Wilson prime. *J. London Math. Soc.* 28 (1953), 252–256.

1954 Ward, M. Prime divisors of second order recurring sequences. *Duke Math. J.* 21 (1954), 607–614.

1956 Obláth, R. Une propriété des puissances parfaites. *Mathesis* 65 (1956), 356–364.

1956 Riesel, H. Några stora primtal. *Elementa* 39 (1956), 258–260.

1958 Jarden, D. *Recurring Sequences.* Riveon Lematematika, Jerusalem, 1958 (3rd edition, 1973).

1958 Robinson, R.M. A report on primes of the form $k \cdot 2^n + 1$ and on factors of Fermat numbers. *Proc. Amer. Math. Soc.* 9 (1958), 673–681.

1960 Sierpiński, W. Sur un problème concernant les nombres $k \cdot 2^n + 1$. *Elem. Math.* 15 (1960), 73–74.

1964 Graham, R.L. A Fibonacci-like sequence of composite numbers. *Math. Mag.* 37 (1964), 322–324.

1964 Riesel, H. Note on the congruence $a^{p-1} \equiv 1 \pmod{p^2}$. *Math. Comp.* 18 (1964), 149–150.

1964 Siegel, C.L. Zu zwei Bemerkungen Kummers. *Nachr. Akad. d. Wiss. Göttingen, Math. Phys. Kl.*, II, 1964, 51–62. Reprinted in *Gesammelte Abhandlungen* (edited by K. Chandrasekharan and H. Maas), Vol. III, 436–442. Springer-Verlag, Berlin, 1966.

1965 Kloss, K.E. Some number theoretic calculations. *J. Res. Nat. Bureau of Stand.* B, 69 (1965), 335–336.

1966 Hasse, H. Über die Dichte der Primzahlen p, für die eine vorgegebene ganzrationale Zahl $a \neq 0$ von gerader bzw. ungerader Ordnung mod p ist. *Math. Ann.* 168 (1966), 19–23.

1966 Kruyswijk, D. *On the congruence $u^{p-1} \equiv 1 \pmod{p^2}$* (in Dutch). Math. Centrum Amsterdam, 1966, 7 pages.

1969 Riesel, H. Lucasian criteria for the primality of $N = h \cdot 2^n - 1$. *Math. Comp.* 23 (1969), 869–875.

1971 Brillhart, J., Tonascia, J. & Weinberger, P.J. On the Fermat quotient. In *Computers in Number Theory* (edited by A.L. Atkin and B.J. Birch), 213–222. Academic Press, New York, 1971.

1975 Johnson, W. Irregular primes and cyclotomic invariants. *Math. Comp.* 29 (1975), 113–120.

1976 Hooley, C. *Application of Sieve Methods to the Theory of Numbers.* Cambridge Univ. Press, Cambridge, 1976.

1978 Wagstaff, Jr., S.S. The irregular primes to 125000. *Math. Comp.* 32 (1978), 583–591.

1978 Williams, H.C. Some primes with interesting digit patterns. *Math. Comp.* 32 (1978), 1306–1310.

1979 Erdös, P. & Odlyzko, A.M. On the density of odd integers of the form $(p - 1)2^{-n}$ and related questions. *J. Number Theory* 11 (1979), 257–263.

1979 Ribenboim, P. *13 Lectures on Fermat's Last Theorem.* Springer-Verlag, New York, 1979.

1979 Williams, H.C. & Seah, E. Some primes of the form $(a^n - 1)/(a - 1)$. *Math. Comp.* 33 (1979), 1337–1342.

1980 Newman, M., Shanks, D. & Williams, H.C. Simple groups of square order and an interesting sequence of primes. *Acta Arith.* 38 (1980), 129–140.

1980 Powell, B. Primitive densities of certain sets of primes. *J. Number Theory* 12 (1980), 210–217.

1981 Lehmer, D.H. On Fermat's quotient, base two. *Math. Comp.* 36 (1981), 289–290.

1982 Powell, B. Problem E 2956 (The existence of small prime solutions of $x^{p-1} \not\equiv 1 \pmod{p^2}$). *Amer. Math. Monthly* 89 (1982), p. 498.

1982 Yates, S. *Repunits and Repetends.* Star Publ. Co., Boynton Beach, FL, 1982.

1983 Jaeschke, G. On the smallest k such that $k \cdot 2^N + 1$ are composite. *Math. Comp.* 40 (1983), 381–384; Corrigendum, 45 (1985), p. 637.

1983 Keller, W. Factors of Fermat numbers and large primes of the form $k \cdot 2^n + 1$. *Math. Comp.* 41 (1983), 661–673.

1983 Ribenboim, P. 1093. *Math. Intelligencer* 5, No. 2 (1983), 28–34.

1985 Lagarias, J.C. The set of primes dividing the Lucas numbers has density $\frac{2}{3}$. *Pacific J. Math.* 118 (1985), 19–23.

1986 Tzanakis, N. Solution to problem E 2956. *Amer. Math. Monthly* 93 (1986), p. 569.

1986 Williams, H.C. & Dubner, H. The primality of R1031. *Math. Comp.* 47 (1986), 703–711.

1987 Granville, A. *Diophantine Equations with Variable Exponents with Special Reference to Fermat's Last Theorem.* Ph.D. Thesis, Queen's University, Kingston, 1987, 207 pages.

1987 Rotkiewicz, A. Note on the diophantine equation $1 + x + x^2 + \cdots + x^n = y^m$. *Elem. Math.* 42 (1987), p. 76.

1988 Brillhart, J., Montgomery, P.L. & Silverman, R.D. Tables of Fibonacci and Lucas factorizations, and Supplement. *Math. Comp.* 50 (1988), 251–260 and S1–S15.

1988 Gonter, R.H. & Kundert, E.G. *Wilson's theorem* $(n-1)! \equiv -1 \pmod{p^2}$ *has been computed up to* 10,000,000. Fourth SIAM Conference on Discrete Mathematics, San Francisco, June 1988.

1988 Granville, A. & Monagan, M.B. The first case of Fermat's last theorem is true for all prime exponents up to 714,591,416,091, 389. *Trans. Amer. Math. Soc.* 306 (1988), 329–359.

1989 Dubner, H. Generalized Cullen numbers. *J. Recr. Math.* 21 (1989), 190–194.

1989 Löh, G. Long chains of nearly doubled primes. *Math. Comp.* 53 (1989), 751–759.

1989 Tanner, J.W. & Wagstaff, Jr., S.S. New bound for the first case of Fermat's last theorem. *Math. Comp.* 53 (1989), 743–750.

1990 Brown, J., Noll, L.C., Parady, B.K., Smith, J.F., Smith, G.W. & Zarantonello, S. Letter to the editor. *Amer. Math. Monthly* 97 (1990), p. 214.

1990 Knuth, D. A Fibonacci-like sequence of composite numbers. *Math. Mag.* 63 (1990), 21–25.

1991 Aaltonen, M. & Inkeri, K. Catalan's equation $x^p - y^q = 1$ and related congruences. *Math. Comp.* 56 (1991), 359–370. Reprinted in *Collected Papers of Kustaa Inkeri* (edited by T. Metsänkylä and P. Ribenboim), Queen's Papers in Pure and Appl. Math. 91, 1992, Queen's Univ., Kingston, Ontario, Canada.

1991 Fee, G. & Granville, A. The prime factors of Wendt's binomial circulant determinant. *Math. Comp.* 57 (1991), 839–848.

1991 Keller, W. Woher kommen die größten derzeit bekannten Primzahlen? *Mitt. Math. Ges. Hamburg* 12 (1991), 211–229.

1992 Buhler, J.P., Crandall, R.E. & Sompolski, R.W. Irregular primes to one million. *Math. Comp.* 59 (1992), 717–722.

1993 Buhler, J.P., Crandall, R.E., Ernvall, R. & Metsänkylä, T. Irregular primes and cyclotomic invariants to four million. *Math. Comp.* 61 (1993), 151–153.

1993 Dubner, H. Generalized repunit primes. *Math. Comp.* 61 (1993), 927–930.

1993 Montgomery, P.L. New solutions of $a^{p-1} \equiv 1 \pmod{p^2}$. *Math. Comp.* 61 (1993), 361–363.

1994 Crandall, R. & Fagin, B. Discrete weighted transforms and large-integer arithmetic. *Math. Comp.* 62 (1994), 305–324.

1994 Gonter, R.H. & Kundert, E.G. *All prime numbers up to* 18,876,041 *have been tested without finding a new Wilson prime.* Unpublished manuscript, Amherst, MA, 1994, 10 pages.

1994 Suzuki, J. On the generalized Wieferich criteria. *Proc. Japan Acad. Sci.* A (Math. Sci.), 70 (1994), 230–234.

1995 Keller, W. New Cullen primes. *Math. Comp.* 64 (1995), 1733–1741.

1995 Keller, W. & Niebuhr, W. Supplement to "New Cullen primes". *Math. Comp.* 64 (1995), S39–S46.

1997 Crandall, R., Dilcher, K. & Pomerance, C. A search for Wieferich and Wilson primes. *Math. Comp.* 66 (1997), 433–449.

1997 Ernvall, R. & Metsänkylä, T. On the p-divisibility of Fermat quotients. *Math. Comp.* 66 (1997), 1353–1365.

1999 Dubner, H. & Keller, W. New Fibonacci and Lucas primes. *Math. Comp.* 68 (1999), 417–427 and S1–S12.

1999 Forbes, T. Prime clusters and Cunningham chains. *Math. Comp.* 68 (1999), 1739–1747.

1999 Ribenboim, P. *Fermat's Last Theorem for Amateurs.* Springer-Verlag, New York, 1999.

2000 Pinch, R.G.E. The pseudoprimes up to 10^{13}. In *Proc. Fourth Int. Symp. on Algorithmic Number Th.* (edited by W. Bosma). Lecture Notes in Computer Sci. #1838, 459–474. Springer-Verlag, New York, 2000.

2001 Buhler, J., Crandall, R., Ernvall, R., Metsänkylä, T. & Shokrollahi, M.A. Irregular primes and cyclotomic invariants to 12 million. *J. Symbolic Comp.* 31 (2001), 89–96.

2001 Keller, W. & Richstein, J. Solutions of the congruence $a^{p-1} \equiv 1 \pmod{p^r}$. Preprint, 2001.

2002 Dubner, H. Repunit R49081 is a probable prime. *Math. Comp.* 71 (2002), 833–835.

2002 Dubner, H. & Gallot, Y. Distribution of generalized Fermat prime numbers. *Math. Comp.* 71 (2002), 825–832.

2002 Izotov, A.S. Second-order linear recurrences of composite numbers. *Fibonacci Quart.* 40 (2002), 266–268.

2002 Sellers, J.A. & Williams, H.C. On the infinitude of composite NSW numbers. *Fibonacci Quart.* 40 (2002), 253–254.

2003 Knauer, J. & Richstein, J. The continuing search for Wieferich primes. Preprint, 2003.

Chapter 6

1857 Bouniakowsky, V. Nouveaux théorèmes relatifs à la distribution des nombres premiers et à la décomposition des entiers en facteurs. *Mém. Acad. Sci. St. Petersbourg* (6), *Sci. Math. Phys.*, 6 (1857), 305–329.

1904 Dickson, L.E. A new extension of Dirichlet's theorem on prime numbers. *Messenger of Math.* 33 (1904), 155–161.

1922 Nagell, T. Zur Arithmetik der Polynome. *Abhandl. Math. Sem. Univ. Hamburg* 1 (1922), 179–194.

1923 Hardy, G.H. & Littlewood, J.E. Some problems in "Partitio Numerorum", III: On the expression of a number as a sum of primes. *Acta Math.* 44 (1923), 1–70. Reprinted in *Collected Papers of G.H. Hardy*, Vol. I, 561–630. Clarendon Press, Oxford, 1966.

1931 Heilbronn, H. Über die Verteilung der Primzahlen in Polynomen. *Math. Ann.* 104 (1931), 794–799.

1932 Breusch, R. Zur Verallgemeinerung des Bertrandschen Postulates, dass zwischen x und $2x$ stets Primzahlen liegen. *Math. Zeits.* 34 (1932), 505–526.

1939 Beeger, N.G.W.H. Report on some calculations of prime numbers. *Nieuw Arch. Wisk.* 20 (1939), 48–50.

1958 Schinzel, A. & Sierpiński, W. Sur certaines hypothèses concernant les nombres premiers. Remarque. *Acta Arith.* 4 (1958), 185–208, and 5 (1959), p. 259.

1961 Schinzel, A. Remarks on the paper "Sur certaines hypothèses concernant les nombres premiers". *Acta Arith.* 7 (1961), 1–8.

1964 Sierpiński, W. Les binômes $x^2 + n$ et les nombres premiers. *Bull. Soc. Roy. Sci. Liège* 33 (1964), 259–260.

1969 Rieger, G.J. On polynomials and almost-primes. *Bull. Amer. Math. Soc.* 75 (1969), 100–103.

1971 Shanks, D. A low density of primes. *J. Recr. Math.* 4 (1971/2), 272–275.

1973 Wunderlich, M.C. On the Gaussian primes on the line $\text{Im}(x) = 1$. *Math. Comp.* 27 (1973), 399–400.

1974 Halberstam, H. & Richert, H.E. *Sieve Methods.* Academic Press, New York, 1974.

1975 Shanks, D. Calculation and applications of Epstein zeta functions. *Math. Comp.* 29 (1975), 271–287.

1978 Iwaniec, H. Almost-primes represented by quadratic polynomials. *Invent. Math.* 47 (1978), 171–188.

1982 Powell, B. Problem 6384 (Numbers of the form $m^p - n$). *Amer. Math. Monthly* 89 (1982), p. 278.

1983 Israel, R.B. Solution of problem 6384. *Amer. Math. Monthly* 90 (1983), p. 650.

1984 Gupta, R. & Ram Murty, P.M. A remark on Artin's conjecture. *Invent. Math.* 78 (1984), 127–130.

1984 McCurley, K.S. Prime values of polynomials and irreducibility testing. *Bull. Amer. Math. Soc.* 11 (1984), 155–158.

1986 Heath-Brown, D.R. Artin's conjecture for primitive roots. *Quart. J. Math. Oxford* (2) 37 (1986), 27–38.

1986 McCurley, K.S. The smallest prime value of $x^n + a$. *Can. J. Math.* 38 (1986), 925–936.

1986 McCurley, K.S. Polynomials with no small prime values. *Proc. Amer. Math. Soc.* 97 (1986), 393–395.

1990 Fung, G.W. & Williams, H.C. Quadratic polynomials which have a high density of prime values. *Math. Comp.* 55 (1990), 345–353.

1994 Alford, W.R., Granville, A. & Pomerance, C. There are infinitely many Carmichael numbers. *Annals of Math.* (2) 140 (1994), 703–722.

1995 Jacobson, Jr., M.J. *Computational Techniques in Quadratic Fields.* M.A. Thesis, University of Manitoba, Winnipeg, 1995.

1998 Friedlander, J. & Iwaniec, H. The polynomial $x^2 + y^4$ captures its primes. *Annals of Math.* (2) 148 (1998), 945–1040.

2001 Dubner, H. & Forbes, T. Prime Pythagorean triangles. *J. Integer Seq.* 4 (2001), Art. 01.2.3, 1–11 (electronic).

2001 Heath-Brown, D.R. Primes represented by $x^3 + 2y^3$. *Acta Math.* 186 (2001), 1–84.

2003 Jacobson, Jr., M.J. & Williams, H.C. New quadratic polynomials with high densities of prime values. *Math. Comp.* 72 (2003), 499–519.

Appendix 1

1935 Hall, P. On representatives of subsets. *J. London Math. Soc.* 10 (1935), 26–30.

1969 Grimm, C.A. A conjecture on consecutive composite numbers. *Amer. Math. Monthly* 76 (1969), 1126–1128.

1971 Erdös, P. & Selfridge, J.L. Some problems on the prime factors of consecutive integers, II. *Proc. Washington State Univ. Conf. on Number Theory* (edited by J.H. Jordan and W.A. Webb), 13–21. Pullman, WA, 1971.

1975 Ramachandra, K., Shorey, T.N. & Tijdeman, R. On Grimm's problem relating to factorisation of a block of consecutive integers. *J. Reine Angew. Math.* 273 (1975), 109–124.

Appendix 2

1981 Bateman, P.T. Major figures in the history of the prime number theorem. *Abstracts Amer. Math. Soc.* 2 (1981), 87th Annual Meeting, San Francisco, p. 2.

Web Site Sources

General Sources

Caldwell, C. The Prime Pages. Prime number research, records, and resources. `http://www.utm.edu/research/primes`
Weisstein, E. The World of Mathematics. Number theory. `http://mathworld.wolfram.com/topics/NumberTheory.html`

Chapter 2

Keller, W. Prime factors $k.2^n + 1$ of Fermat numbers F_m and complete factoring status.
`http://www.prothsearch.net/fermat.html`
Caldwell, C. The largest known Mersenne primes.
`http://www.utm.edu/research/primes/largest.html#Mersenne`
Woltman, G. Status of Great Internet Mersenne Prime Search.
`http://www.mersenne.org/status.htm`
Martin, M. Largest primes verified with the ECPP algorithm.
`http://www.ellipsa.net/primes/top20.html`
Kelly, B. Fibonacci and Lucas factorizations.
`http://home.att.net/~blair.kelly/mathematics/fibonacci/`

Chapter 4

Gourdon, X. & Sebah, P. Counting the number of primes.
http://numbers.computation.free.fr/Constants/Primes
/countingPrimes.html

Wedeniwski, S. Verification of the Riemann hypothesis.
http://www.zetagrid.net/zeta/rh.html

Nicely, T.R. First occurrence prime gaps.
http://www.trnicely.net/gaps/gaplist.html

Nicely, T.R. Counts of twin prime pairs.
http://www.trnicely.net/counts.html

Caldwell, C. The largest known twin primes.
http://www.utm.edu/research/primes/largest.html#twin

Forbes, T. Prime k-tuplets.
http://www.ltkz.demon.co.uk/ktuplets.htm

Oliveira e Silva, T. Goldbach conjecture verification.
http://www.ieeta.pt/ tos/goldbach.html

Chapter 5

Caldwell, C. The largest known Sophie Germain primes.
http://www.utm.edu/research/primes/largest.html#Sophie

Keller, W. & Richstein, J. Fermat quotients $q_p(a)$ that are divisible by p. http://www.informatik.uni-giessen.de/staff
/richstein/cnth/FermatQuotient.html

Helm, L. & Norris, D. A distributed attack on the Sierpinski problem. http://www.seventeenorbust.com/

Ballinger, R. & Keller, W. The Riesel problem: Definition and status. http://www.prothsearch.net/rieselprob.html

Caldwell, C. The largest known (non-Mersenne) primes.
http://primes.utm.edu/primes/lists/all.txt

Leyland, P. Factorization of Cullen and Woodall numbers.
http://research.microsoft.com/~pleyland/factorization
/cullen_woodall/cw.htm

Löh, G. Generalized Cullen primes.
http://www.rrz.uni-hamburg.de/RRZ/G.Loeh/gc/status.html

Primes up to 10,000

2	3	5	7	11	13	17	19	23	29
31	37	41	43	47	53	59	61	67	71
73	79	83	89	97	101	103	107	109	113
127	131	137	139	149	151	157	163	167	173
179	181	191	193	197	199	211	223	227	229
233	239	241	251	257	263	269	271	277	281
283	293	307	311	313	317	331	337	347	349
353	359	367	373	379	383	389	397	401	409
419	421	431	433	439	443	449	457	461	463
467	479	487	491	499	503	509	521	523	541
547	557	563	569	571	577	587	593	599	601
607	613	617	619	631	641	643	647	653	659
661	673	677	683	691	701	709	719	727	733
739	743	751	757	761	769	773	787	797	809
811	821	823	827	829	839	853	857	859	863
877	881	883	887	907	911	919	929	937	941
947	953	967	971	977	983	991	997	1009	1013
1019	1021	1031	1033	1039	1049	1051	1061	1063	1069
1087	1091	1093	1097	1103	1109	1117	1123	1129	1151
1153	1163	1171	1181	1187	1193	1201	1213	1217	1223
1229	1231	1237	1249	1259	1277	1279	1283	1289	1291

1297	1301	1303	1307	1319	1321	1327	1361	1367	1373
1381	1399	1409	1423	1427	1429	1433	1439	1447	1451
1453	1459	1471	1481	1483	1487	1489	1493	1499	1511
1523	1531	1543	1549	1553	1559	1567	1571	1579	1583
1597	1601	1607	1609	1613	1619	1621	1627	1637	1657
1663	1667	1669	1693	1697	1699	1709	1721	1723	1733
1741	1747	1753	1759	1777	1783	1787	1789	1801	1811
1823	1831	1847	1861	1867	1871	1873	1877	1879	1889
1901	1907	1913	1931	1933	1949	1951	1973	1979	1987
1993	1997	1999	2003	2011	2017	2027	2029	2039	2053
2063	2069	2081	2083	2087	2089	2099	2111	2113	2129
2131	2137	2141	2143	2153	2161	2179	2203	2207	2213
2221	2237	2239	2243	2251	2267	2269	2273	2281	2287
2293	2297	2309	2311	2333	2339	2341	2347	2351	2357
2371	2377	2381	2383	2389	2393	2399	2411	2417	2423
2437	2441	2447	2459	2467	2473	2477	2503	2521	2531
2539	2543	2549	2551	2557	2579	2591	2593	2609	2617
2621	2633	2647	2657	2659	2663	2671	2677	2683	2687
2689	2693	2699	2707	2711	2713	2719	2729	2731	2741
2749	2753	2767	2777	2789	2791	2797	2801	2803	2819
2833	2837	2843	2851	2857	2861	2879	2887	2897	2903
2909	2917	2927	2939	2953	2957	2963	2969	2971	2999
3001	3011	3019	3023	3037	3041	3049	3061	3067	3079
3083	3089	3109	3119	3121	3137	3163	3167	3169	3181
3187	3191	3203	3209	3217	3221	3229	3251	3253	3257
3259	3271	3299	3301	3307	3313	3319	3323	3329	3331
3343	3347	3359	3361	3371	3373	3389	3391	3407	3413
3433	3449	3457	3461	3463	3467	3469	3491	3499	3511
3517	3527	3529	3533	3539	3541	3547	3557	3559	3571
3581	3583	3593	3607	3613	3617	3623	3631	3637	3643
3659	3671	3673	3677	3691	3697	3701	3709	3719	3727
3733	3739	3761	3767	3769	3779	3793	3797	3803	3821
3823	3833	3847	3851	3853	3863	3877	3881	3889	3907
3911	3917	3919	3923	3929	3931	3943	3947	3967	3989
4001	4003	4007	4013	4019	4021	4027	4049	4051	4057
4073	4079	4091	4093	4099	4111	4127	4129	4133	4139
4153	4157	4159	4177	4201	4211	4217	4219	4229	4231
4241	4243	4253	4259	4261	4271	4273	4283	4289	4297
4327	4337	4339	4349	4357	4363	4373	4391	4397	4409

4421	4423	4441	4447	4451	4457	4463	4481	4483	4493
4507	4513	4517	4519	4523	4547	4549	4561	4567	4583
4591	4597	4603	4621	4637	4639	4643	4649	4651	4657
4663	4673	4679	4691	4703	4721	4723	4729	4733	4751
4759	4783	4787	4789	4793	4799	4801	4813	4817	4831
4861	4871	4877	4889	4903	4909	4919	4931	4933	4937
4943	4951	4957	4967	4969	4973	4987	4993	4999	5003
5009	5011	5021	5023	5039	5051	5059	5077	5081	5087
5099	5101	5107	5113	5119	5147	5153	5167	5171	5179
5189	5197	5209	5227	5231	5233	5237	5261	5273	5279
5281	5297	5303	5309	5323	5333	5347	5351	5381	5387
5393	5399	5407	5413	5417	5419	5431	5437	5441	5443
5449	5471	5477	5479	5483	5501	5503	5507	5519	5521
5527	5531	5557	5563	5569	5573	5581	5591	5623	5639
5641	5647	5651	5653	5657	5659	5669	5683	5689	5693
5701	5711	5717	5737	5741	5743	5749	5779	5783	5791
5801	5807	5813	5821	5827	5839	5843	5849	5851	5857
5861	5867	5869	5879	5881	5897	5903	5923	5927	5939
5953	5981	5987	6007	6011	6029	6037	6043	6047	6053
6067	6073	6079	6089	6091	6101	6113	6121	6131	6133
6143	6151	6163	6173	6197	6199	6203	6211	6217	6221
6229	6247	6257	6263	6269	6271	6277	6287	6299	6301
6311	6317	6323	6329	6337	6343	6353	6359	6361	6367
6373	6379	6389	6397	6421	6427	6449	6451	6469	6473
6481	6491	6521	6529	6547	6551	6553	6563	6569	6571
6577	6581	6599	6607	6619	6637	6653	6659	6661	6673
6679	6689	6691	6701	6703	6709	6719	6733	6737	6761
6763	6779	6781	6791	6793	6803	6823	6827	6829	6833
6841	6857	6863	6869	6871	6883	6899	6907	6911	6917
6947	6949	6959	6961	6967	6971	6977	6983	6991	6997
7001	7013	7019	7027	7039	7043	7057	7069	7079	7103
7109	7121	7127	7129	7151	7159	7177	7187	7193	7207
7211	7213	7219	7229	7237	7243	7247	7253	7283	7297
7307	7309	7321	7331	7333	7349	7351	7369	7393	7411
7417	7433	7451	7457	7459	7477	7481	7487	7489	7499
7507	7517	7523	7529	7537	7541	7547	7549	7559	7561
7573	7577	7583	7589	7591	7603	7607	7621	7639	7643
7649	7669	7673	7681	7687	7691	7699	7703	7717	7723
7727	7741	7753	7757	7759	7789	7793	7817	7823	7829

7841	7853	7867	7873	7877	7879	7883	7901	7907	7919
7927	7933	7937	7949	7951	7963	7993	8009	8011	8017
8039	8053	8059	8069	8081	8087	8089	8093	8101	8111
8117	8123	8147	8161	8167	8171	8179	8191	8209	8219
8221	8231	8233	8237	8243	8263	8269	8273	8287	8291
8293	8297	8311	8317	8329	8353	8363	8369	8377	8387
8389	8419	8423	8429	8431	8443	8447	8461	8467	8501
8513	8521	8527	8537	8539	8543	8563	8573	8581	8597
8599	8609	8623	8627	8629	8641	8647	8663	8669	8677
8681	8689	8693	8699	8707	8713	8719	8731	8737	8741
8747	8753	8761	8779	8783	8803	8807	8819	8821	8831
8837	8839	8849	8861	8863	8867	8887	8893	8923	8929
8933	8941	8951	8963	8969	8971	8999	9001	9007	9011
9013	9029	9041	9043	9049	9059	9067	9091	9103	9109
9127	9133	9137	9151	9157	9161	9173	9181	9187	9199
9203	9209	9221	9227	9239	9241	9257	9277	9281	9283
9293	9311	9319	9323	9337	9341	9343	9349	9371	9377
9391	9397	9403	9413	9419	9421	9431	9433	9437	9439
9461	9463	9467	9473	9479	9491	9497	9511	9521	9533
9539	9547	9551	9587	9601	9613	9619	9623	9629	9631
9643	9649	9661	9677	9679	9689	9697	9719	9721	9733
9739	9743	9749	9767	9769	9781	9787	9791	9803	9811
9817	9829	9833	9839	9851	9857	9859	9871	9883	9887
9901	9907	9923	9929	9931	9941	9949	9967	9973	

Index of Tables

Index of Records

Index of Records

Index of Names

Index of Names

Subject Index